# Student Solutions Manual

# Functions and Change
## A Modeling Approach to College Algebra

**SIXTH EDITION**

**Bruce Crauder**
Oklahoma State University

**Benny Evans**
Oklahoma State University

**Alan Noell**
Oklahoma State University

Prepared by

**Bruce Crauder**
Oklahoma State University

**Benny Evans**
Oklahoma State University

**Alan Noell**
Oklahoma State University

CENGAGE
Learning·

Australia • Brazil • Mexico • Singapore • United Kingdom • United States

ISBN: 978-1-337-11140-9

**Cengage Learning**
20 Channel Center Street
Boston, MA 02210
USA

Cengage Learning is a leading provider of customized learning solutions with office locations around the globe, including Singapore, the United Kingdom, Australia, Mexico, Brazil, and Japan. Locate your local office at: **www.cengage.com/global.**

Cengage Learning products are represented in Canada by Nelson Education, Ltd.

To learn more about Cengage Learning Solutions, visit **www.cengage.com.**

Purchase any of our products at your local college store or at our preferred online store **www.cengagebrain.com.**

Printed in the United States of America
Print Number: 01     Print Year: 2016

# Contents

# Solution Guide for Prologue: Calculator Arithmetic

## CALCULATOR ARITHMETIC

1. **Valentine's Day**: To find the percentage we first calculate

$$\frac{\text{Average female expenditure}}{\text{Average male expenditure}} = \frac{\$96.58}{\$190.53} = 0.5069.$$

Thus the average female expenditure was 50.69% of the average male expenditure.

3. **A billion dollars**: A stack of a billion one-dollar bills would be $0.0043 \times 1{,}000{,}000{,}000 = 4{,}300{,}000$ inches high. In miles this height is

$$4{,}300{,}000 \text{ inches} \times \frac{1 \text{ foot}}{12 \text{ inches}} \times \frac{1 \text{ mile}}{5280 \text{ feet}} = 67.87 \text{ miles}.$$

So the stack would be 67.87 miles high.

5. **10% discount and 10% tax**: The sales price is 10% off of the original price of \$75.00, so the sales price is $75.00 - 0.10 \times 75.00 = 67.50$ dollars. Adding in the sales tax of 10% on this sales price, we'll need to pay $67.50 + 0.10 \times 67.50 = 74.25$ dollars.

7. **A bad investment**: The total value of your investment today is:

$$\text{Original investment } - 7\% \text{ loss } = 720 - 0.07 \times 720 = \$669.60.$$

9. **Pay raise**: The percent pay raise is obtained from

$$\frac{\text{Amount of raise}}{\text{Original hourly pay}}.$$

The raise was $9.50 - 9.25 = 0.25$ dollar while the original hourly pay is \$9.25, so the fraction is $\frac{0.25}{9.25} = 0.0270$. Thus we have received a raise of 2.70%.

11. **Trade discount**:

   (a) The cost price is $9.99 - 40\% \times 9.99 = 5.99$ dollars.

   (b) The difference between the suggested retail price and the cost price is $65.00 - 37.00 = 28.00$ dollars. We want to determine what percentage of \$65 this difference represents. We find the percentage by division: $\frac{28.00}{65.00} = 0.4308$ or 43.08%. This is the trade discount used.

13. **Present value**: We are given that the future value is $5000 and that $r = 0.12$. Thus the present value is

$$\frac{\text{Future value}}{1+r} = \frac{5000}{1+0.12} = 4464.29 \text{ dollars.}$$

15. **The Rule of 72**:

(a) The Rule of 72 says our investment should double in

$$\frac{72}{\% \text{ interest rate}} = \frac{72}{13} = 5.54 \text{ years.}$$

(b) Using Part (a), the future value interest factor is

$$(1 + \text{ interest rate})^{\text{years}} = (1+0.13)^{5.54} = 1.97.$$

This is less than the doubling future value interest factor of 2.

(c) Using our value from Part (b), the future value of a $5000 investment is

$$\text{Original investment} \times \text{ future value interest factor } = 5000 \times 1.97 = \$9850.$$

So our investment did not exactly double using the Rule of 72.

17. **The size of the Earth**:

(a) The equator is a circle with a radius of approximately 4000 miles. The distance around the equator is its circumference, which is

$$2\pi \times \text{ radius } = 2\pi \times 4000 = 25{,}132.74 \text{ miles,}$$

or approximately 25,000 miles.

(b) The volume of the Earth is

$$\frac{4}{3}\pi \times \text{ radius }^3 = \frac{4}{3}\pi \times 4000^3 = 268{,}082{,}573{,}100 \text{ cubic miles.}$$

Note that the calculator gives 2.680825731E11, which is the way the calculator writes numbers in scientific notation. It means $2.680825731 \times 10^{11}$ and should be written as such. That is about 268 billion cubic miles or $2.68 \times 10^{11}$ cubic miles.

(c) The surface area of the Earth is about

$$4\pi \times \text{ radius }^2 = 4\pi \times 4000^2 = 201{,}061{,}929.8 \text{ square miles,}$$

or approximately 201,000,000 square miles.

19. **The length of Earth's orbit**:

    (a) If the orbit is a circle then its circumference is the distance traveled. That circumference is

    $$2\pi \times \text{ radius } = 2\pi \times 93 = 584.34 \text{ million miles},$$

    or about 584 million miles. This can also be calculated as

    $$2\pi \times \text{ radius } = 2\pi \times 93{,}000{,}000 = 584{,}336{,}233.6 \text{ miles}.$$

    (b) Velocity is distance traveled divided by time elapsed. The velocity is given by

    $$\frac{\text{Distance traveled}}{\text{Time elapsed}} = \frac{584.34 \text{ million miles}}{1 \text{ year}} = 584.34 \text{ million miles per year},$$

    or about 584 million miles per year. This can also be calculated as

    $$\frac{584{,}336{,}233.6 \text{ miles}}{1 \text{ year}} = 584{,}336{,}233.6 \text{ miles per year}.$$

    (c) There are 24 hours per day and 365 days per year. So there are $24 \times 365 = 8760$ hours per year.

    (d) The velocity in miles per hour is

    $$\frac{\text{Miles traveled}}{\text{Hours elapsed}} = \frac{584.34}{8760} = 0.0667 \text{ million miles per hour}.$$

    This is approximately 67,000 miles per hour. This can also be calculated as

    $$\frac{\text{Miles traveled}}{\text{Hours elapsed}} = \frac{584{,}336{,}233.6}{8760} = 66{,}705.05 \text{ miles per hour}.$$

21. **Newton's second law of motion**: A man with a mass of 75 kilograms weighs $75 \times 9.8 = 735$ newtons. In pounds this is $735 \times 0.225$, or about 165.38.

23. **Frequency of musical notes**: The frequency of the next higher note than middle C is $261.63 \times 2^{1/12}$, or about 277.19 cycles per second. The D note is one note higher, so its frequency in cycles per second is

    $$(261.63 \times 2^{1/12}) \times 2^{1/12},$$

    or about 293.67.

25. **Lean body weight in females**: The lean body weight of a young adult female who weighs 132 pounds and has wrist diameter of 2 inches, abdominal circumference of 27 inches, hip circumference of 37 inches, and forearm circumference of 7 inches is

    $$19.81 + 0.73 \times 132 + 21.2 \times 2 - 0.88 \times 27 - 1.39 \times 37 + 2.43 \times 7 = 100.39 \text{ pounds}.$$

    It follows that her body fat weighs $132 - 100.39 = 31.61$ pounds. To compute the body fat percent we calculate $\dfrac{31.61}{132}$ and find 23.95%.

27. **Relativistic length**: The apparent length of the rocket ship is given by the formula $200\sqrt{1-r^2}$, where $r$ is the ratio of the ship's velocity to the speed of light. Since the ship is travelling at 99% of the speed of light, this means that $r = 0.99$. Plugging this into the formula yields that the 200 meter spaceship will appear to be only $200\sqrt{1-0.99^2} = 200\sqrt{(1-0.99 \wedge 2)} = 28.21$ meters long.

29. **Advantage Cash card**:

   (a) The Advantage Cash card gives a discount of 5% and you pay no sales tax, so you pay $1.00 less 5%, which is $0.95.

   (b) If you pay cash, you must also pay sales tax of 7.375%, so you pay a total of $1.00 plus 7.375%, which is $1.07375 to five decimal places, or $1.07.

   (c) If you open an Advantage Cash card for $300, you get a bonus of 5%. Now 5% of $300 is $300 \times .05 = 15$ dollars, so your card balance is $315. You also get a discount of 5% off the retail price and pay no sales tax, so you can purchase a total retail value such that 95% of it equals $315, that is

$$\text{Retail value} \times 0.95 = 315,$$

   so the retail value is $315/0.95 = 331.58$ dollars.

   (d) If you have $300 cash, then you can buy a retail value such that when you add to it 7.375% for sales tax, you get $300. So

$$\text{Retail value} \times 1.07375 = 300,$$

   and thus the retail value you can buy is $300/1.07375 = 279.39$ dollars.

   (e) From Part (d) to Part (c), the increase is $331.58 - 279.39 = 52.19$, so the percentage increase is $52.19/279.39 \times 100\% = 18.68\%$. In practical terms, this means that using the Advantage Cash card allows you to buy 18.68% more food than using cash.

## Skill Building Exercises

S-1. **Basic calculations**: In typewriter notation, $\dfrac{2.6 \times 5.9}{6.3}$ is $(2.6 \times 5.9) \div 6.3$, which equals 2.434... and so is rounded to two decimal places as 2.43.

S-3. **Basic calculations**: In typewriter notation, $\dfrac{e}{\sqrt{\pi}}$ is $e \div (\sqrt{(\pi)})$, which equals 1.533... and so is rounded to two decimal places as 1.53.

S-5. **Parentheses and grouping**: When we add parentheses, $\dfrac{7.3 - 6.8}{2.5 + 1.8}$ becomes $\dfrac{(7.3 - 6.8)}{(2.5 + 1.8)}$, which, in typewriter notation, becomes $(7.3 - 6.8) \div (2.5 + 1.8)$. This equals 0.116... and so is rounded to two decimal places as 0.12.

S-7. **Parentheses and grouping**: When we add parentheses, $\dfrac{\sqrt{6+e}+1}{3}$ becomes $\dfrac{(\sqrt{(6+e)}+1)}{3}$, which, in typewriter notation, becomes $(\sqrt{(6+e)}+1) \div 3$. This equals 1.317... and so is rounded to two decimal places as 1.32.

S-9. **Subtraction versus sign**: Noting which are negative signs and which are subtraction signs, we see that $\dfrac{-3}{4-9}$ means $\dfrac{negative\ 3}{4\ subtract\ 9}$. Adding parentheses and putting it into typewriter notation yields *negative* $3 \div (4\ subtract\ 9)$, which equals 0.6.

S-11. **Subtraction versus sign**: Noting which are negative signs and which are subtraction signs, we see that $-\sqrt{8.6-3.9}$ means *negative* $\sqrt{8.6\ subtract\ 3.9}$. In typewriter notation this is

$$negative\ \sqrt{(8.6\ subtract\ 3.9)},$$

which equals $-2.167...$ and so is rounded to two decimal places as $-2.17$.

S-13. **Chain calculations**:

a. To do this as a chain calculation, we first calculate $\dfrac{3}{7.2+5.9}$ and then complete the calculation by adding the second fraction to this first answer. In typewriter notation $\dfrac{3}{7.2+5.9}$ is $3 \div (7.2+5.9)$, which is calculated as 0.2290076336; this is used as *Ans* in the next part of the calculation. Turning to the full expression, we calculate it as $Ans + \dfrac{7}{6.4 \times 2.8}$ which is, in typewriter notation, $Ans + 7 \div (6.4 \times 2.8)$. This is 0.619..., which rounds to 0.62.

b. To do this as a chain calculation, we first calculate the exponent, $1 - \dfrac{1}{36}$, and then the full expression becomes

$$\left(1+\frac{1}{36}\right)^{Ans}.$$

In typewriter notation, the first calculation is $1 - 1 \div 36$, and the second is $(1 + 1 \div 36) \wedge Ans$. This equals 1.026... and so is rounded to two decimal places as 1.03.

S-15. **Evaluate expression**: In typewriter notation, $\dfrac{5.2}{7.3+0.2^{4.5}}$ is $5.2 \div (7.3+0.2 \wedge 4.5)$, which equals 0.712... and so is rounded to two decimal places as 0.71.

S-17. **Arithmetic**: Writing in typewriter notation, we have $(2 \wedge 3.2 - 1) \div (\sqrt{(3)}+4) = 1.43$, rounded to two decimal places.

S-19. **Arithmetic**: Writing in typewriter notation, we have $(2 \wedge negative\ 3 + \sqrt{(7)} + \pi)(e \wedge 2 + 7.6 \div 6.7) = 50.39$, rounded to two decimal places, where *negative* means to use a minus sign.

**S-21. Evaluating formulas:** To evaluate the formula $\dfrac{A-B}{A+B}$ we plug in the values for $A$ and $B$ to yield $\dfrac{4.7-2.3}{4.7+2.3} = (4.7-2.3) \div (4.7+2.3) = 0.34$, rounded to two decimal places.

**S-23. Evaluating formulas:** To evaluate the formula $\sqrt{x^2+y^2}$ we plug in the values for $x$ and $y$ to yield $\sqrt{1.7^2+3.2^2} = \sqrt{(1.7 \wedge 2 + 3.2 \wedge 2)} = 3.62$, rounded to two decimal places.

**S-25. Evaluating formulas:** To evaluate the formula $(1-\sqrt{A})(1+\sqrt{B})$ we plug in the values for $A$ and $B$ to yield $(1-\sqrt{3})(1+\sqrt{5}) = (1-\sqrt{(3)})(1+\sqrt{(5)}) = -2.37$, rounded to two decimal places.

**S-27. Evaluating formulas:** To evaluate the formula $\sqrt{b^2-4ac}$ we plug in the values for $b$, $a$, and $c$ to yield $\sqrt{7^2-4\times2\times0.07} = \sqrt{(7 \wedge 2 - 4 \times 2 \times 0.07)} = 6.96$, rounded to two decimal places.

**S-29. Evaluating formulas:** To evaluate the formula $(x+y)^{-x}$ we plug in the values for $x$ and $y$ to yield $(3+4)^{-3} = (3+4) \wedge (\text{ negative } 3) = 0.0029$, rounded to four decimal places.

**S-31. Lending money:** The interest due is $I = Prt$ where $P = 5000$, $r = 0.05$, which is 5% as a decimal, and $t = 3$, so $I = 5000 \times 0.05 \times 3 = 750$ dollars.

**S-33. Temperature:** Since $F = \dfrac{9}{5}C + 32$, then when $C = 32$, $F = \dfrac{9}{5}32 + 32 = (9 \div 5) \times 32 + 32 = 89.60$ degrees Fahrenheit.

**S-35. Future value:** The future value is given by $F = P(1+r)^t$, where $P = 1000$, $r = 0.06$, and $t = 5$, so plugging these in gives a future value of $F = 1000(1+0.06)^5 = 1000(1+0.06) \wedge 5 = 1338.23$ dollars.

**S-37. Carbon 14:** The amount of carbon 14 is given by $C = 5 \times 0.5^{t/5730}$, so for $t = 5000$, the amount of carbon 14 remaining is $C = 5 \times 0.5^{5000/5730} = 5 \times 0.5 \wedge (5000 \div 5730) = 2.73$ grams.

## Prologue Review Exercises

1. **Parentheses and grouping:** In typewriter notation, $\dfrac{5.7+8.3}{5.2-9.4}$ is $(5.7+8.3) \div (5.2-9.4)$, which equals $-3.333\ldots$ and so is rounded to two decimal places as $-3.33$.

2. **Evaluate expression:** In typewriter notation, $\dfrac{8.4}{3.5+e^{-6.2}}$ is $8.4 \div (3.5+e \wedge (\text{ negative } 6.2))$, which equals $2.398\ldots$ and so is rounded to two decimal places as $2.40$.

3. **Evaluate expression**: In typewriter notation, $\left(7 + \dfrac{1}{e}\right)^{\left(\frac{5}{2+\pi}\right)}$ is $(7 + 1 \div e) \wedge (5 \div (2 + \pi))$, which equals 6.973... and so is rounded to two decimal places as 6.97. This can also be done as a chain calculation.

4. **Gas mileage**: The number of gallons required to travel 27 miles is

$$g = \frac{27}{15} = 1.8 \text{ gallons.}$$

The number of gallons required to travel 250 miles is

$$g = \frac{250}{15} = 16.67 \text{ gallons.}$$

5. **Kepler's third law**: The mean distance from Pluto to the sun is

$$D = 93 \times 249^{2/3} = 3680.86 \text{ million miles,}$$

or about 3681 million miles. For Earth we have $P = 1$ year, and the mean distance is

$$D = 93 \times 1^{2/3} = 93 \text{ million miles.}$$

6. **Traffic signal**: If the approach speed is 80 feet per second then the length of the yellow light should be

$$n = 1 + \frac{80}{30} + \frac{100}{80} = 4.92 \text{ seconds.}$$

# Solution Guide for Chapter 1: Functions

## 1.1 FUNCTIONS GIVEN BY FORMULAS

1. **Movie tickets**:

    (a) Because 2009 is 9 years after the year 2000, the expression $C(9)$ is the average cost, in dollars, of a movie ticket in 2009.

    (b) Because 2012 is 12 years after the year 2000, the average cost of a movie ticket in 2012 expressed in functional notation is $C(12)$.

    (c) The average cost of a movie ticket in 2012 is

    $$C(12) = 5.40 + 0.22 \times 12 = 8.04 \text{ dollars.}$$

    Therefore, the average cost of a movie ticket in 2012 is $8.04.

3. **Speed from skid marks**:

    (a) In functional notation the speed for a 60-foot-long skid mark is $S(60)$. The value is

    $$S(60) = 5.05\sqrt{60} = 39.12 \text{ miles per hour.}$$

    Therefore, the speed at which the skid mark will be 60 feet long is 39.12 miles per hour.

    (b) The expression $S(100)$ represents the speed, in miles per hour, at which an emergency stop will on dry pavement leave a skid mark that is 100 feet long.

5. **Adult weight from puppy weight**:

    (a) In functional notation the adult weight of a puppy that weighs 6 pounds at 14 weeks is $W(14, 6)$.

    (b) The predicted adult weight of a puppy that weighs 6 pounds at 14 weeks is

    $$W(14, 6) = 52 \times \frac{6}{14} = 22.29 \text{ pounds.}$$

    Therefore, the predicted adult weight for this puppy is 22.29 pounds.

7. **Tax owed**:

(a) In functional notation the tax owed on a taxable income of $13,000 is $T(13,000)$. The value is

$$T(13,000) = 0.11 \times 13,000 - 500 = 930 \text{ dollars.}$$

(b) The tax owed on a taxable income of $14,000 is

$$T(14,000) = 0.11 \times 14,000 - 500 = 1040 \text{ dollars.}$$

Using the answer to Part (a), we see that the tax increases by $1040 - 930 = 110$ dollars.

(c) The tax owed on a taxable income of $15,000 is

$$T(15,000) = 0.11 \times 15,000 - 500 = 1150 \text{ dollars.}$$

Thus the tax increases by $1150 - 1040 = 110$ dollars again.

9. **Flying ball**:

(a) In functional notation the velocity 1 second after the ball is thrown is $V(1)$. The value is

$$V(1) = 40 - 32 \times 1 = 8 \text{ feet per second.}$$

Because the upward velocity is positive, the ball is rising.

(b) The velocity 2 seconds after the ball is thrown is

$$V(2) = 40 - 32 \times 2 = -24 \text{ feet per second.}$$

Because the upward velocity is negative, the ball is falling.

(c) The velocity 1.25 seconds after the ball is thrown is

$$V(1.25) = 40 - 32 \times 1.25 = 0 \text{ feet per second.}$$

Because the velocity is 0, we surmise from Parts (a) and (b) that the ball is at the peak of its flight.

(d) Using the answers to Parts (a) and (b), we see that from 1 second to 2 seconds the velocity changes by

$$V(2) - V(1) = -24 - 8 = -32 \text{ feet per second.}$$

Because

$$V(3) = 40 - 32 \times 3 = -56 \text{ feet per second,}$$

from 2 seconds to 3 seconds the velocity changes by

$$V(3) - V(2) = -56 - (-24) = -32 \text{ feet per second.}$$

Because

$$V(4) = 40 - 32 \times 4 = -88 \text{ feet per second,}$$

from 3 seconds to 4 seconds the velocity changes by

$$V(4) - V(3) = -88 - (-56) = -32 \text{ feet per second.}$$

Over each of these 1-second intervals the velocity changes by $-32$ feet per second. In practical terms, this means that the velocity decreases by 32 feet per second for each second that passes. This indicates that the *downward* acceleration of the ball is constant at 32 feet per second per second, which makes sense because the acceleration due to gravity is constant near the surface of Earth.

11. **A population of deer**:

(a) Now $N(0)$ represents the number of deer initially on the reserve and

$$N(0) = \frac{12.36}{0.03 + 0.55^0} = 12 \text{ deer.}$$

So there were 12 deer in the initial herd.

(b) We calculate using

$$N(10) = \frac{12.36}{0.03 + 0.55^{10}} = 379.92 \text{ deer.}$$

This says that after 10 years there should be about 380 deer in the reserve.

(c) The number of deer in the herd after 15 years is represented by $N(15)$, and this value is

$$N(15) = \frac{0.36}{0.03 + 0.55^{15}} = 410.26 \text{ deer.}$$

This says that there should be about 410 deer in the reserve after 15 years.

(d) The difference in the deer population from the tenth to the fifteenth year is given by $N(15) - N(10) = 410.26 - 379.92 = 30.34$. Thus the population increased by about 30 deer from the tenth to the fifteenth year.

13. **Radioactive substances**:

(a) The amount of carbon 14 left after 800 years is expressed in functional notation as $C(800)$. This is calculated as

$$C(800) = 5 \times 0.5^{800/5730} = 4.54 \text{ grams.}$$

(b) There are many ways to do this part of the exercise. The simplest is to note that half the amount is left when the exponent of 0.5 is 1 since then the 5 is multiplied by $0.5 = \dfrac{1}{2}$. The exponent in the formula is 1 when $t = 5730$ years.

Another way to do this part is to experiment with various values for $t$, increasing the value when the answer is less than 2.5 and decreasing it when the answer comes out more than 2.5. Students are in fact discovering and executing a crude version of the bisection method.

**15. What if interest is compounded more often than monthly?**

(a) We would expect our monthly payment to be higher if the interest is compounded daily since additional interest is charged on interest which has been compounded.

(b) Continuous compounding should result in a larger monthly payment since the interest is compounded at an even faster rate than with daily compounding.

(c) We are given that $P = 7800$ and $t = 48$. Because the APR is 8.04% or 0.0804, we compute that

$$r = \frac{\text{APR}}{12} = \frac{0.0804}{12} = 0.0067.$$

Thus the monthly payment is

$$M(7800, 0.0067, 48) = \frac{7800(e^{0.0067} - 1)}{1 - e^{-0.0067 \times 48}} = 190.67 \text{ dollars}.$$

Our monthly payment here is 10 cents higher than if interest is compounded monthly as in Example 1.2 (where the payment was $190.57).

**17. How much can I borrow?**

(a) Since we will be paying $350 per month for 4 years, then we will be making 48 payments, or $t = 48$. Also, $r$ is the monthly interest rate of 0.75%, or 0.0075 as a decimal. The amount of money we can afford to borrow in this case is given in functional notation by $P(350, 0.0075, 48)$. It is calculated as

$$P(350, 0.0075, 48) = 350 \times \frac{1}{0.0075} \times \left(1 - \frac{1}{(1 + 0.0075)^{48}}\right) = \$14{,}064.67.$$

(b) If the monthly interest rate is 0.25% then we can afford to borrow

$$P(350, 0.0025, 48) = 350 \times \frac{1}{0.0025} \times \left(1 - \frac{1}{(1 + 0.0025)^{48}}\right) = \$15{,}812.54.$$

(c) If we make monthly payments over 5 years then we will make 60 payments in all. So now we can afford to borrow

$$P(350, 0.0025, 60) = 350 \times \frac{1}{0.0025} \times \left(1 - \frac{1}{(1 + 0.0025)^{60}}\right) = \$19{,}478.33.$$

19. **Brightness of stars**: Here we have $m_1 = -1.45$ and $m_2 = 2.04$. Thus

$$t = 2.512^{m_2 - m_1} = 2.512^{2.04 - (-1.45)} = 2.512^{3.49} = 24.89.$$

Hence Sirius appears 24.89 times brighter than Polaris.

21. **Parallax**: We are given that $p = 0.751$. Thus the distance from Alpha Centauri to the sun is about

$$d(0.751) = \frac{3.26}{0.751} = 4.34 \text{ light-years.}$$

23. **Mitscherlich's equation**:

(a) We are given that $b = 1$. Thus the percentage (as a decimal) of maximum yield is

$$Y(1) = 1 - 0.5^1 = 0.5.$$

Hence 50% of maximum yield is produced if 1 baule is applied.

(b) In functional notation the percentage (as a decimal) of maximum yield produced by 3 baules is $Y(3)$. The value is

$$Y(3) = 1 - 0.5^3 = 0.875,$$

or about 0.88. This is 88% of maximum yield.

(c) Now 500 pounds of nitrogen per acre corresponds to $\dfrac{500}{223}$ baules, so the percentage (as a decimal) of maximum yield is $1 - 0.5^{500/223}$, or about 0.79. This is 79% of maximum yield.

25. **Thermal conductivity**: We are given that $k = 0.85$ for glass and that $t_1 = 24$, $t_2 = 5$.

(a) Because $d = 0.007$, the heat flow is

$$Q = \frac{0.85(24 - 5)}{0.007} = 2307.14 \text{ watts per square meter.}$$

(b) The total heat loss is

$$\text{Heat flow} \times \text{Area of window} = 2307.14 \times 2.5,$$

or about 5767.85 watts.

27. **Fault rupture length**: Here we have $M = 6.5$, so the expected length is

$$L(6.5) = 0.0000017 \times 10.47^{6.5} = 7.25 \text{ kilometers.}$$

29. **Equity in a home**:

    (a) The monthly interest rate as a decimal is $r = \text{APR}/12 = 0.06/12 = 0.005$.

    (b) The mortgage is for $P = 400{,}000$, the term is 30 years, so in months $t = 30 \times 12 = 360$, $r = 0.005$, and after 20 years of payments, $k = 20 \times 12 = 240$, so in functional notation it is $E(240)$. The value of $E(240)$ is

    $$400{,}000 \times \frac{(1 + 0.005)^{240} - 1}{(1 + 0.005)^{360} - 1},$$

    which equals 183,985.66 dollars.

    (c) To find the equity after $y$ years, use $k = 12y$ months, so the formula for $E$ in terms of $y$ is

    $$E(y) = 400{,}000 \times \frac{(1 + 0.005)^{12y} - 1}{(1 + 0.005)^{360} - 1}.$$

31. **Adjustable rate mortgage—exact payments**:

    (a) To find the equity after 24 months, we use $k = 24$. Here $P = 325{,}000$, $r = \text{APR}/12 = 0.045/12 = 0.00375$, and $t = 30 \times 12 = 360$, so the equity accrued after 24 months is

    $$E = 325{,}000 \times \frac{(1 + 0.00375)^{24} - 1}{(1 + 0.00375)^{360} - 1} = 10{,}726.84 \text{ dollars.}$$

    (b) The new mortgage amount is $P = 325{,}000 - 10{,}726.84 = 314{,}273.16$, the new interest rate is $r = \text{APR}/12 = 0.07/12 = 0.00583$, the new term is $t = 28 \times 12 = 336$, and so the new monthly payment is

    $$M = \frac{Pr(1 + r)^t}{(1 + r)^t - 1} = \frac{314{,}273.16 \times 0.00583 \times (1 + 0.00583)^{336}}{(1 + 0.00583)^{336} - 1} = 2135.00 \text{ dollars.}$$

## Skill Building Exercises

**S-1. Evaluating formulas**: To evaluate $f(x) = \dfrac{\sqrt{x + 1}}{x^2 + 1}$ at $x = 2$, simply substitute 2 for $x$. Thus the value of $f$ at 2 is $\dfrac{\sqrt{2 + 1}}{2^2 + 1}$, which equals 0.346... and so is rounded to 0.35.

**S-3. Evaluating formulas**: To evaluate $g(x, y) = \dfrac{x^3 + y^3}{x^2 + y^2}$ at $x = 4.1$, $y = 2.6$, simply substitute 4.1 for $x$ and 2.6 for $y$. Thus the value of $g$ when $x = 4.1$ and $y = 2.6$ is $\dfrac{4.1^3 + 2.6^3}{4.1^2 + 2.6^2}$, which equals 3.669... and so is rounded to 3.67.

**S-5. Evaluating formulas**: To get the function value $f(6.1)$, substitute 6.1 for $s$ in the formula $f(s) = \dfrac{s^2 + 1}{s^2 - 1}$. Thus $f(6.1) = \dfrac{6.1^2 + 1}{6.1^2 - 1}$, which equals 1.055... and so is rounded to 1.06.

**S-7. Evaluating formulas**: To get the function value $h(3, 2.2, 9.7)$, substitute 3 for $x$, 2.2 for $y$, and 9.7 for $z$ in the formula $h(x, y, z) = \dfrac{x^y}{z}$. Thus $h(3, 2.2, 9.7) = \dfrac{3^{2.2}}{9.7}$, which equals 1.155... and so is rounded to 1.16.

**S-9. Evaluating formulas**: To evaluate $f(x, y) = \dfrac{1.2^x + 1.3^y}{\sqrt{x+y}}$ at $x = 3$ and $y = 4$, simply substitute 3 for $x$ and 4 for $y$. Thus the value of $f$ when $x = 3$ and $y = 4$ is $\dfrac{1.2^3 + 1.3^4}{\sqrt{3+4}}$, which equals 1.732... and so is rounded to 1.73.

**S-11. Evaluating formulas**: To get the function value $W(2.2, 3.3, 4.4)$, substitute 2.2 for $a$, 3.3 for $b$, and 4.4 for $c$ in the formula $W(a, b, c) = \dfrac{a^b - b^a}{c^a - a^c}$. Thus $H(2.2, 3.3, 4.4) = \dfrac{2.2^{3.3} - 3.3^{2.2}}{4.4^{2.2} - 2.2^{4.4}}$, which equals 0.0555... and so is rounded to 0.06.

**S-13. Evaluating formulas**: We simply substitute 3 for $x$: $f(3) = 3^{-3} - \dfrac{3^2}{3+1}$, which equals $-2.212...$ and so is rounded to $-2.21$.

**S-15. Evaluating formulas**: We simply substitute 0 for $t$ to obtain $C(0) = 0.1 + 2.78e^{-0.37 \times 0}$, which equals 2.88; we substitute 10 for $t$ to obtain $C(10) = 0.1 + 2.78e^{-0.37 \times 10}$, which equals 0.17.

**S-17. Evaluating formulas**: To evaluate $C(g, d) = \dfrac{gd}{32}$ at $g = 52.3$ and $d = 13.5$, simply substitute 52.3 for $g$ and 13.5 for $d$. Thus the value of $C(52.3, 13.5)$ is $\dfrac{52.3 \times 13.5}{32}$, which equals 22.064... and so is rounded to 22.06.

**S-19. Evaluating formulas**: We simply substitute 0 for $t$ to obtain $R(0) = 325 - 280e^{-0.005 \times 0}$, which equals 45; we substitute 30 for $t$ to obtain $R(30) = 325 - 280e^{-0.005 \times 30}$, which equals 84.00.

**S-21. Evaluating formulas**: To get the function value $P(500, 0.06, 30)$, substitute 500 for $F$, 0.06 for $r$, and 30 for $t$ in the formula $P(F, r, t) = \dfrac{F}{(1+r)^t}$. Thus $P(500, 0.06, 30) = \dfrac{500}{(1+0.06)^{30}}$, which equals 87.055... and so is rounded to 87.06.

**S-23. Evaluating formulas**: To get the function value $d(3.1, 0.5)$, substitute 3.1 for $m$ and 0.5 for $M$, in the formula $d(m, M) = 3.26 \times 10^{(m-M+5)/5}$. Thus $d(3.1, 0.5) = 3.26 \times 10^{(3.1-0.5+5)/5}$, which equals 107.948... and so is rounded to 107.95.

**S-25. Profit**: To express the expected profit, note that the variable $t$ is measured in years, so 2 years and 6 months corresponds to $t = 2.5$, and so the expected profit is expressed by $p(2.5)$.

**S-27. Speed of fish**: Since $s(L)$ is the top speed of a fish $L$ inches long, $s(13)$ is the top speed of a fish 13 inches long.

S-29. **Doubling time**: In practical terms, $D(5000, 0.06)$ is the time required for an investment of $P = 5000$ dollars at an APR of $r = 0.06$, so 6%, to double in value.

S-31. **A bird population**: In practical terms $N(7)$ is the number of birds in the population 7 years after observation began.

S-33. **Dow Jones Industrial Average**: Because 2 p.m. is 120 minutes after noon, in practical terms $D(120)$ is the Dow Jones Industrial Average at 2 p.m. today.

## 1.2  FUNCTIONS GIVEN BY TABLES

1. **Minimum wage**:

   (a) According to the table, $m(1990)$ is $6.66.

   (b) In functional notation the minimum wage in 1985 is $m(1985)$.

   (c) Because 1985 is halfway between 1980 and 1990, we can estimate $m(1985)$ by averaging:
   $$\frac{m(1980) + m(1990)}{2} = \frac{8.46 + 6.66}{2} = 7.56.$$
   Hence we estimate that the minimum wage in 1985 was about $7.56.

3. **Box office hits**:

   (a) According to the table, $M(2014)$ is *American Sniper*, and $B(2014)$ is 350.13 million dollars.

   (b) In functional notation the amount for the movie with the highest gross in 2013 is $B(2013)$.

5. **Choosing a bat**:

   (a) Here $B(55)$ is the recommended bat length for a man weighing between 161 and 170 pounds if his height is 55 inches. According to the table, that length is 31 inches.

   (b) In functional notation the recommended bat length for a man weighing between 161 and 170 pounds if his height is 63 inches is $B(63)$.

7. **The American food dollar**:

   (a) Here $P(1989) = 30\%$. This means that in 1989 Americans spent 30% of their food dollars eating out.

(b) The expression $P(1999)$ is the percent of the American food dollar spent eating away from home in 1999. Since 1999 falls halfway between 1989 and 2009, our estimate for $P(1999)$ is the average of $P(1989)$ and $P(2009)$, or

$$\frac{P(1989) + P(2009)}{2} = \frac{30 + 34}{2} = 32.$$

Approximately 32% of the American food dollar in 1999 was spent eating out.

(c) The average rate of change per year in percentage of the food dollar spent away from home from 1989 to 2009 is

$$\frac{P(2009) - P(1989)}{2009 - 1989} = \frac{34 - 30}{20} = 0.2,$$

or 0.2 percentage point per year.

(d) The expression $P(2004)$ is the percent of the American food dollar spent eating away from home in 2004. We estimate it as

$$P(2004) = P(1989) + 15 \times \text{ Yearly change } = 30 + 15 \times 0.2 = 33,$$

or 33%.

(e) Assuming the increase in $P$ continues at the same rate of about 0.2 percentage point per year as we calculated in Part (c), then we estimate

$$P(2014) = P(2009) + 5 \times \text{ Yearly change } = 34 + 5 \times 0.2 = 35,$$

or 35%.

9. **Internet access:**

(a) Here $I(2000) = 113$ million. This means that 113 million adult Americans had internet access in 2000.

(b) The average rate of change per year from 2003 to 2009 is

$$\frac{\text{Change in } I}{\text{Time elapsed}} = \frac{I(2009) - I(2003)}{6} = \frac{196 - 166}{6} = 5 \text{ million per year.}$$

(c) Since the average rate of change per year from 2003 to 2009 is 5 million per year, we estimate

$$I(2007) = I(2003) + 4 \times \text{ Yearly change } = 166 + 4 \times 5 = 186 \text{ million adult Americans.}$$

11. **A troublesome snowball:** Here $W(t)$ is the volume of dirty water soaked into the carpet, so its limiting value is the total volume of water frozen in the snowball. The limiting value is reached when the snowball has completely melted.

13. **Carbon 14:**

   (a) The average yearly rate of change for the first 5000 years is

   $$\frac{\text{Amount of change}}{\text{Years elapsed}} = \frac{C(5) - C(0)}{5000} = \frac{2.73 - 5}{5000} = -4.54 \times 10^{-4} \text{ gram per year.}$$

   That is $-0.000454$ gram per year.

   (b) We use the average yearly rate of change from Part (a):

   $$C(1.236) = C(0) + 1236 \times \text{ yearly rate of change } = 5 + 1236 \times -0.000454 = 4.44 \text{ grams.}$$

   (c) The limiting value is zero since all of the carbon 14 will eventually decay.

15. **Effective percentage rate for various compounding periods:**

   (a) We have that $n = 1$ represents compounding yearly, $n = 2$ represents compounding semiannually, $n = 12$ represents compounding monthly, $n = 365$ represents compounding daily, $n = 8760$ represents compounding hourly, and $n = 525,600$ represents compounding every minute.

   (b) We have that $E(12)$ is the EAR when compounding monthly, and $E(12) = 12.683\%$.

   (c) If interest is compounded daily then the EAR is $E(365)$. So the interest accrued in one year is

   $$8000 \times E(365) = 8000 \times 0.12747 = \$1019.76.$$

   (d) If interest were compounded continuously then the EAR would probably be about 12.750%. As the length of the compounding period decreases, the EAR given in the table appears to stabilize at this value.

17. **Growth in height:**

   (a) In functional notation, the height of the man at age 13 is given by $H(13)$.

   From ages 10 to 15, the average yearly growth rate in height is

   $$\frac{\text{Inches increased}}{\text{Years elapsed}} = \frac{67.0 - 55.0}{5} = 2.4 \text{ inches per year.}$$

   Since age 13 is 3 years after age 10, we can estimate $H(13)$ as

   $$H(10) + 3 \times \text{ yearly growth } = 55.0 + 3 \times 2.4 = 62.2 \text{ inches.}$$

   (b)  i. We calculate the average yearly growth rate for each 5-year period just as we calculated 2.4 inches per year as the average yearly growth rate from ages 10 to 15 in Part (a). The average yearly growth rate is measured in inches per year.

| Age change | 0 to 5 | 5 to 10 | 10 to 15 | 15 to 20 | 20 to 25 |
|---|---|---|---|---|---|
| Average yearly growth rate | 4.2 | 2.5 | 2.4 | 1.3 | 0.1 |

ii. The man grew the most from age 0 to age 5.

iii. The trend is that as the man gets older, he grows more slowly.

(c) It is reasonable to guess that 74 or 75 inches is the limiting value for the height of this man. He grew only 0.5 inches from ages 20 to 25, so it is reasonable to expect little or no further growth from age 25 on.

19. **Tax owed**:

(a) The average rate of change over the first interval is

$$\frac{T(16,200) - T(16,000)}{16,200 - 16,000} = \frac{888 - 870}{200} = 0.09 \text{ dollar per dollar.}$$

Continuing in this way, we get the following table, where the rate of change is measured in dollars per dollar.

| Interval | 16,000 to 16,200 | 16,200 to 16,400 | 16,400 to 16,600 |
|---|---|---|---|
| Rate of change | 0.09 | 0.09 | 0.09 |

(b) The average rate of change has a constant value of 0.09 dollar per dollar. This suggests that, at every income level in the table, for every increase of $1 in taxable income the tax owed increases by $0.09, or 9 cents.

(c) Because the average rate of change is a nonzero constant and thus does not tend to 0, we would expect $T$ not to have a limiting value but rather to increase at a constant rate as $I$ increases.

21. **Yellowfin tuna**:

(a) The average rate of change in weight is

$$\frac{W(110) - W(100)}{110 - 100} = \frac{56.8 - 42.5}{10} = 1.43 \text{ pounds per centimeter.}$$

(b) The average rate of change in weight is

$$\frac{W(180) - W(160)}{180 - 160} = \frac{256 - 179}{20} = 3.85 \text{ pounds per centimeter.}$$

(c) Examining the table shows that the rate of change in weight is smaller for small tuna than it is for large tuna. Hence an extra centimeter of length makes more difference in weight for a large tuna.

(d) To estimate the weight of a yellowfin tuna that is 167 centimeters long we use the average rate of change we found in Part (b):

$$W(160) + 7 \times 3.85 = 179 + 7 \times 3.85 = 205.95 \text{ pounds.}$$

Hence the weight of a yellowfin tuna that is 167 centimeters long is 205.95, or about 206.0, pounds.

(e) Here we are thinking of the weight as the variable and the length as a function of the weight. The average rate of change in length is

$$\frac{\text{Length at 256 pounds} - \text{Length at 179 pounds}}{256 - 179} = \frac{180 - 160}{256 - 179} = 0.26,$$

so the average rate of change is 0.26 centimeter per pound. Note that this number is the reciprocal of the answer from Part (b).

(f) To estimate the length of a yellowfin tuna that weighs 225 pounds we use the average rate of change we found in Part (e):

Length at 179 pounds $+ (225 - 179) \times 0.26 = 160 + 46 \times 0.26 = 171.96$ centimeters.

Hence the length of a yellowfin tuna that weighs 225 pounds is 171.96, or about 172, centimeters.

23. **Widget production**:

(a) The average rate of change over the first interval is

$$\frac{W(20) - W(10)}{10} = \frac{37.5 - 25.0}{10} = 1.25 \text{ thousand widgets per worker.}$$

Continuing in this way, we get the following table, where the rate of change is measured in thousands of widgets per worker.

| Interval | 10 to 20 | 20 to 30 | 30 to 40 | 40 to 50 |
|---|---|---|---|---|
| Rate of change | 1.25 | 0.63 | 0.31 | 0.15 |

(b) The average rate of change decreases and approaches 0 as we go across the table. This means that the increase in production gained from adding another worker gets smaller and smaller as the level of workers employed moves higher and higher. Eventually there is very little benefit in employing an extra worker.

(c) To estimate how many widgets will be produced if there are 55 full-time workers we use the entry from the table for the average rate of change over the last interval:

$$W(50) + 5 \times 0.15 = 48.4 + 5 \times 0.15 = 49.15 \text{ thousand widgets.}$$

Hence the number of widgets produced by 55 full-time workers is about 49.2 thousand, or 49,200.

(d) Because the average rate of change is decreasing, the actual increase in production in going from 50 to 55 workers is likely to be less than what the average rate of change from 40 to 50 suggests. Thus our estimate is likely to be too high.

25. **The Margaria-Kalamen test:**

(a) The average rate of change per year in excellence level from 25 years to 35 years old is
$$\frac{168 - 210}{35 - 25} = -4.2 \text{ points per year.}$$

(b) We estimate the power score needed for a 27-year-old man using the score for a 25-year-old man and the average rate of change from Part (a):
$$210 + 2 \times -4.2 = 201.6 \text{ points.}$$

Hence the power score that would merit an excellent rating for a 27-year-old man is 201.6, or about 202, points.

(c) The decrease in power score for excellent rating over these three periods is the greatest in the second period (35 years to 45 years), so we would expect to see the greatest decrease in leg power from 35 to 45 years old.

27. **Home equity:**

(a) $E(10)$ is the equity accrued after 10 years of payments. Its value, according to the table, is \$27,734.

(b) We calculate the average rate of change for each 5-year period below:

| 5-year interval | 0 to 5 | 5 to 10 | 10 to 15 | 15 to 20 | 20 to 25 | 25 to 30 |
|---|---|---|---|---|---|---|
| Average rate of change | 2361.60 | 3185.20 | 4296.60 | 5795.40 | 7817.00 | 10,544.20 |

(c) The equity increases more rapidly late in the life of the mortgage.

(d) To estimate the equity accrued after 17 years, we add 2 years average rate of change (from 15 to 20 years) to the equity accrued after 15 years:
$$E(17) = E(15) + 2 \times 5795.40 = 49{,}217 + 2 \times 5795.40 = 60{,}807.80 \text{ dollars.}$$

(e) No, it wouldn't make sense to estimate past 30 years, since the mortgage has a term of 30 years.

28. **Defense spending:**

(a) The average yearly rate of change in defense spending from 1990 to 1995 is
$$\frac{\text{Change from 1990 to 1995}}{\text{Years elapsed}} = \frac{D(5) - D(0)}{5 - 0} = \frac{310.0 - 328.4}{5} = -3.68 \text{ billion dollars per year.}$$

(b) We estimate $D(3)$ by starting from $D(0)$ and adding three years' average rate of change:

$$D(3) = D(0) + 3 \times -3.68 = 328.4 + 3 \times -3.68 = 317.36 \text{ billion dollars.}$$

This means that federal defense spending was about 317.4 billion dollars in 1993.

(c) The average yearly rate of change in defense spending from 2005 to 2010 is

$$\frac{\text{Change from 2005 to 2010}}{\text{Years elapsed}} = \frac{D(20) - D(15)}{20 - 15} = \frac{843.8 - 565.5}{5} = 55.66 \text{ billion dollars per year.}$$

(d) We estimate $D(22)$ by starting from $D(20)$ and adding two years' average rate of change:

$$D(22) = D(20) + 2 \times 55.66 = 843.8 + 2 \times 55.66 = 955.12 \text{ billion dollars.}$$

29. **A home experiment**: Answers will vary greatly. In general, there will be initially a small percentage of bread surface covered with mold. That percentage will quickly rise as the mold covers much of the bread surface. There are usually a few small patches which the mold covers more slowly. Ultimately all the bread surface is covered with mold. Here is a typical data table:

| Time | 8 am | 4 pm | 12 am | 8 am | 4 pm | 12 am |
|------|------|------|-------|------|------|-------|
| Mold | 10%  | 25%  | 60%   | 98%  | 100% | 100%  |

## Skill Building Exercises

S-1. **Function values**: According to the table, when $t = 10$, then $N = 17.6$ and so $N(10) = 17.6$.

S-3. **Function values**: According to the table, when $t = 30$, then $N = 44.6$ and so $N(30) = 44.6$.

S-5. **Function values**: According to the table, when $t = 50$, then $N = 53.2$ and so $N(50) = 53.2$.

S-7. **Function values**: According to the table, when $t = 70$, then $N = 53.9$ and so $N(70) = 53.9$.

S-9. **Averaging**: We can estimate the value of $N(25)$ by finding the average of $N(20)$ and $N(30)$ since 25 is the average of 20 and 30. The average is $\dfrac{N(20) + N(30)}{2} = \dfrac{23.8 + 44.6}{2}$, which equals 34.2.

**S-11. Averaging**: We can estimate the value of $N(45)$ by finding the average of $N(40)$ and $N(50)$ since 45 is the average of 40 and 50. The average is $\dfrac{N(40) + N(50)}{2} = \dfrac{51.3 + 53.2}{2}$, which equals 52.25. We round to 52.3 since the table has one decimal place of accuracy.

**S-13. Averaging**: We can estimate the value of $N(65)$ by finding the average of $N(60)$ and $N(70)$ since 65 is the average of 60 and 70. The average is $\dfrac{N(60) + N(70)}{2} = \dfrac{53.7 + 53.9}{2}$, which equals 53.8.

**S-15. Average rate of change**: The average rate of change in $N$ from $t = 20$ to $t = 30$ is given by the change in $N$ divided by the change in $t$:

$$\frac{N(30) - N(20)}{30 - 20} = \frac{44.6 - 23.8}{10} = 2.08.$$

Thus the average rate of change in $N$ is 2.08. To estimate the value of $N(27)$, we begin at $N(20)$ and add 7 times the average rate of change:

$$N(27) = N(20) + 7 \times 2.08 = 23.8 + 7 \times 2.08 = 38.36,$$

which we round to 38.4, since the table has one decimal place of accuracy.

**S-17. Average rate of change**: The average rate of change in $N$ from $t = 40$ to $t = 50$ is given by the change in $N$ divided by the change in $t$:

$$\frac{N(50) - N(40)}{50 - 40} = \frac{53.2 - 51.3}{10} = 0.19.$$

Thus the average rate of change in $N$ is 0.19. To estimate the value of $N(42)$, we begin at $N(40)$ and add 2 times the average rate of change:

$$N(42) = N(40) + 2 \times 0.19 = 51.3 + 2 \times 0.19 = 51.68,$$

which we round to 51.7, since the table has one decimal place of accuracy.

**S-19. Average rate of change**: The average rate of change in $N$ from $t = 60$ to $t = 70$ is given by the change in $N$ divided by the change in $t$:

$$\frac{N(70) - N(60)}{70 - 60} = \frac{53.9 - 53.7}{10} = 0.02.$$

Thus the average rate of change in $N$ is 0.02. To estimate the value of $N(64)$, we begin at $N(60)$ and add 4 times the average rate of change:

$$N(64) = N(60) + 4 \times 0.02 = 53.7 + 4 \times 0.02 = 53.78,$$

which we round to 53.8, since the table has one decimal place of accuracy.

**S-21. Function values**: Here $f(0)$ is 5.7 since that is the corresponding value in the table.

S-23. **Function values:** Here $f(10)$ is 1.1 since that is the corresponding value in the table.

S-25. **Function values:** Here $f(20)$ is $-7.9$ since that is the corresponding value in the table.

S-27. **Average rate of change:** The average rate of change in $f$ from $x = 5$ to $x = 10$ is given by the change in $f$ divided by the change in $x$:

$$\frac{f(10) - f(5)}{10 - 5} = \frac{1.1 - 4.3}{5} = -0.64.$$

Thus the average rate of change in $f$ is $-0.64$. To estimate the value of $f(7)$, we calculate $f(7)$ as $f(5)$ plus 2 times the average rate of change:

$$f(7) = f(5) + 2 \times -0.64 = 4.3 + 2 \times -0.64,$$

which equals 3.02, or about 3.0.

S-29. **Average rate of change:** The average rate of change in $f$ from $x = 15$ to $x = 20$ is given by the change in $f$ divided by the change in $x$:

$$\frac{f(20) - f(15)}{20 - 15} = \frac{-7.9 - (-3.6)}{5} = -0.86.$$

Thus the average rate of change in $f$ is $-0.86$. To estimate the value of $f(19)$, we calculate $f(19)$ as $f(15)$ plus 4 times the average rate of change:

$$f(19) = f(15) + 4 \times -0.86 = -3.6 + 4 \times -0.86,$$

which equals $-7.04$, or about $-7.0$.

S-31. **When limiting values occur:** We expect $S$ to have a limiting value of 0. This is because the average speed $S$ gets closer and closer to 0 as the time $t$ required to travel 100 miles increases.

## 1.3  FUNCTIONS GIVEN BY GRAPHS

1. **Sketching a graph with given concavity:**

   (a)                                                           (b)

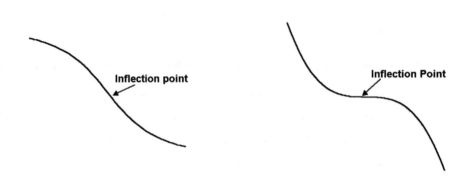

3. **Household debt**:

   (a) In practical terms $h(1975)$ is the average American household debt as a percentage of disposable income in 1975. From the graph, the value is about 60%.

   (b) The maximum value of about 130% occurs in about 2006.

5. **Supply and demand curves**:

   (a) The supply increases as the price increases.

   (b) The demand decreases as the price increases.

7. **Skirt length**:

   (a) As the ratio increases the hem get closer to the ankle, so a larger ratio indicates a longer skirt.

   (b) By the answer to Part (a), skirt lengths were the shortest when the ratio was the smallest. Locating the minimum point of the solid graph, we see that the ratio was the smallest in about 1969.

   (c) When skirt lengths reach to the ankle, the ratio is 1. According to the graph, the ratio never reached 1. So skirt lengths never reached to the ankle during this period.

9. **Unemployment**:

   (a) Here $U(1990)$ is the U.S. unemployment rate, as a percentage, in the year 1990. According to the graph, that rate is 5%.

(b) The decade of highest unemployment occurred where the graph was at its highest over a 10-year period, and that occurred from about 1930 to 1940. That was the decade of the Great Depression.

(c) The most recent date before 2010 when the graph reached the same height as in 2010 was around 1982 or 1983. Then employment was as high as in 2010.

11. **A stock market investment**

(a) According to the exercise our original investment made in 1970 was $10,000, so $v(1970) = \$10,000$. Our investment lost half its value in the 70's, so $v(1980) = \$5000$. The investment was worth $35,000 in 1990, so $v(1990) = \$35,000$. Since the stock remained stable after 1990, the value of the investment stays at the same level of $35,000. So $v(2010) = \$35,000$.

(b)

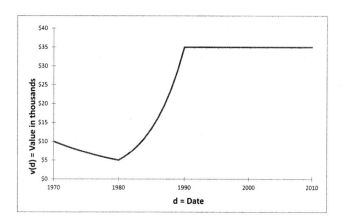

The graph should reflect the function values from Part (a), and it should increase from 1980 to 1990 and remain flat after that. There remains room for interpretation, and many graphs similar to the one above are acceptable.

(c) The stock was most rapidly increasing in the 1980's. For the graph above, 1985 is a good guess. (The answer will depend on what the graph in Part (b) looks like.)

13. **River flow**:

(a) Because the end of July is 7 months since the start of the year, the flow at that time is $F(7)$ in functional notation. According to the graph, the value is about 1500 cubic feet per second.

(b) The flow is at its greatest at $t = 6$, which corresponds to the end of June.

(c) The flow is increasing the fastest at $t = 5$, corresponding to an inflection point on the graph. The time is the end of May.

(d) The graph is practically level from $t = 0$ to $t = 2$, so the function is nearly constant there, and the average rate of change over this interval is about 0 cubic feet per second.

(e) Because the flow is measured near the river's headwaters in the Rocky Mountains, we expect any change in flow to come from melting snow primarily. This is consistent with almost no change in flow during the first two months of the year (as seen in Part (d)), a maximum increase in flow at the end of May (as seen in Part (c)), and a peak flow one month later (as seen in Part (b)).

15. **Cutting trees**:

(a) The graph shows the net stumpage value of a 60-year-old Douglas fir stand to be about $14,000 per acre.

(b) The graph shows a 110-year-old Douglas fir stand has net stumpage value of $40,000 per acre.

(c) When the costs involved in harvesting equal the commercial value, the net stumpage value is 0. Thus we need to know when $V$ is 0. The graph shows that a 30-year-old Douglas fir stand has a net stumpage value of $0 per acre.

(d) The net stumpage value seems to be increasing the fastest in trees about 60 years old.

(e) We would expect the trees to reach an age where they don't grow as much. When this happens the net stumpage value should level out. Here is one possible extended graph for the stumpage value of the Douglas fir.

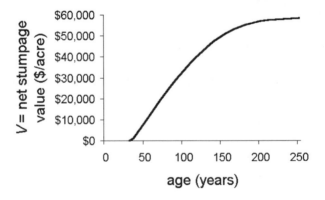

17. **Tornadoes in Oklahoma**:

(a) The most tornadoes were reported in 2011. There were about 74 tornadoes reported that year.

(b) The fewest tornadoes were reported in 2014. There were about 4 tornadoes reported then.

(c) In 2011 there were about 74 tornadoes reported, and in 2012 there were about 26 tornadoes reported. Hence the average rate of decrease was

$$\frac{\text{Decrease in } T}{\text{Years elapsed}} = \frac{74 - 26}{1} = 48 \text{ tornadoes per year.}$$

(d) In 2009 the number of tornadoes reported was about 25, and in 2011 the number of tornadoes reported was about 74. Hence the average rate of increase was

$$\frac{\text{Increase in } T}{\text{Years elapsed}} = \frac{74 - 25}{2} = 24.5 \text{ tornadoes per year.}$$

So there is an increase of about 24 to 25 tornadoes per year–the answer may vary depending on how the graph is interpreted.

(e) The number of tornadoes reported in 2009 was about 25, and the number of tornadoes reported in 2012 was about 26. Hence the average rate of change was

$$\frac{\text{Change in } T}{\text{Years elapsed}} = \frac{26 - 25}{3} = \frac{1}{3} \text{ tornado per year,}$$

which is close to 0. The answer may vary depending on how the graph is interpreted.

19. **Driving a car**: There are many possible stories. Common elements include that you start 6 miles from home and drive towards home, arriving there in about 12 minutes, staying at home for about 4 minutes, then driving about $2\frac{1}{2}$ miles away from home and staying there.

Additional features might be the velocity (rate of change) towards home as about 30 mph and the velocity away from home as about 15 mph.

21. **Photosynthesis**: In Parts (a) and (b) answers may vary somewhat.

(a) We want to find where the graph corresponding to 80 degrees crosses the horizontal axis. The crossing point at about 0.7 thousand foot-candles, or about 700 foot-candles.

(b) We want to find where the graph corresponding to 80 degrees crosses the graph corresponding to 40 degrees. The crossing point is at about 0.8 thousand foot-candles, or about 800 foot-candles.

(c) Among the three graphs, the graph corresponding to 80 degrees meets the vertical axis at the lowest point, so a temperature of 80 degrees will result in the largest emission of carbon dioxide in the dark.

(d) The graph corresponding to 40 degrees is the most level, so that temperature gives the net exchange that is least sensitive to light.

23. **Carbon dioxide concentrations**:

(a) The minimum of the graph occurs at about 2 p.m.

(b) The maximum concentration is attained over the entire interval from about 6 a.m. to about 9 a.m.

(c) Net absorption is indicated by the interval where the graph is decreasing. This is the period from about 9 a.m. to about 2 p.m.

(d) From 6 a.m. to 9 a.m. the graph is level, so the net carbon dioxide exchange is zero during that period.

25. **Profit from fertilizer**:

(a) The maximum value of yield occurs at about 100 on the horizontal axis, so about 100 pounds of nitrogen per acre should be applied to produce maximum crop yield.

(b) The profit can be read by measuring the difference between the yield and cost graphs.

(c) The maximum difference between the graphs occurs at about 85 on the horizontal axis, so about 85 pounds of nitrogen per acre should be applied to produce maximum profit. (In this part answers may vary somewhat.)

27. **Home equity**:

(a) The graph is concave up.

(b) Since the graph is increasing and concave up, this means that equity is increasing faster and faster, that is, increasing at an increasing rate.

(c) Since the mortgage is for 30 years, the amount of the mortgage is the home equity after 30 years, and this is about $335,000.

29. **Grain beetles**:

(a) The graph indicates that the beetles in the wheat stored at 32.3 degrees had a higher mortality rate than the other group. Consequently, it would be better to store the wheat at the higher temperature so that the beetles would not live as long.

(b) The graph indicates that the beetles feeding on the wheat had a higher mortality rate. Consequently, the beetles would be more likely to infest the maize.

31. **Span of life**: The following graph shows a sequence of survivorship curves which illustrates an increasing life span with a fixed maximum life span of 120 years.

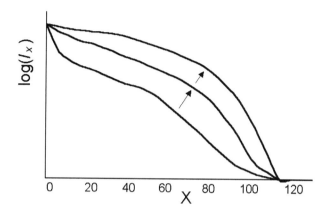

## Skill Building Exercises

S-1. **A function given by a graph**: The value of $f(1.8)$ is obtained from the graph by locating 1.8 on the horizontal axis, then moving vertically up to the point of the graph over 1.8, and then moving horizontally to locate the corresponding value on the vertical axis, Thus $f(1.8)$ is a little below 3.5, so about 3.3.

S-3. **A function given by a graph**: The value of $f(4.2)$ is obtained from the graph by locating 4.2 on the horizontal axis, then moving vertically up to the point of the graph over 4.2, and then moving horizontally to locate the corresponding value on the vertical axis, Thus $f(4.2)$ is about 0.5.

S-5. **A function given by a graph**: The largest value of $x$ for which $f(x) = 1.5$ is obtained from the graph by locating 1.5 on the vertical axis, then moving horizontally to the furthest point on the graph, and then moving vertically down to locate the corresponding value on the horizontal axis. Thus the smallest value of $x$ for which $f(x) = 1.5$ is about $x = 4.0$.

S-7. **A function given by a graph**: The graph reaches a maximum at its highest point, which is where $x = 2.4$ and $f(x)$ is about 4.0.

S-9. **A function given by a graph**: The graph is decreasing for values of $x$ from $x = 2.4$ to $x = 4.8$.

S-11. **A function given by a graph**: Between $x = 0$ and $x = 1.8$, the graph is concave up.

S-13. **A function given by a graph**: Between $x = 3.0$ and $x = 4.8$, the graph is concave up.

S-15. **A function given by a graph**: The graph achieves its maximum value at about $x = 0.5$ and $x = 2.5$.

S-17. **A function given by a graph**: The graph achieves its minimum value at about $x = 1.5$ and $x = 3.5$.

S-19. **A function given by a graph**: The inflection points of $f(x)$ occur at about $x = 1$, $x = 2$, and $x = 3$.

S-21. **A function given by a graph**: Between $x = 2$ and $x = 3$, the graph is concave down.

S-23. **A function given by a graph**: The graph is both increasing and concave up between $x = 1.5$ and $x = 2$ and between $x = 3.5$ and $x = 4$.

S-25. **Special points**: Points where concavity changes are called points of inflection.

S-27. **Sketching graphs**: Answers will vary. One solution has the same shape as the graph at the beginning of the Skill Building Exercises, but with the maximum at $x = 3$.

S-29. **Sketching graphs**: Answers will vary. One solution is

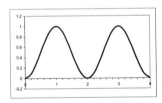

S-31. **Sketching graphs**: Answers will vary. One solution is

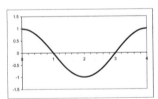

S-33. Answers will vary. One solution is

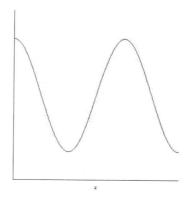

S-35. Answers will vary. One solution is

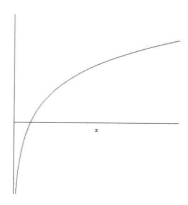

## 1.4  FUNCTIONS GIVEN BY WORDS

1. **Fixed and variable cost**: Now

$$C(0) = \text{Fixed cost } = 1500 \text{ dollars,}$$

and

$$
\begin{aligned}
C(1) &= \text{Fixed cost } + \text{ Variable costs due to 1 item} \\
&= 1500 + 3 \times 1 = 1503 \text{ dollars,} \\
C(2) &= \text{Fixed cost } + \text{ Variable costs due to 2 items} \\
&= 1500 + 3 \times 2 = 1506 \text{ dollars,} \\
C(3) &= \text{Fixed cost } + \text{ Variable costs due to 3 items} \\
&= 1500 + 3 \times 3 = 1509 \text{ dollars,}
\end{aligned}
$$

and so on. This suggests the formula $C = 1500 + 3n$.

3. **Fat in fast food:**

(a) There are 24 grams of fat in each hamburger, so for 3 hamburgers the total fat is $24 \times 3 = 72$ grams. Altogether, the amount of fat in an order of 3 hamburgers, 1 chicken sandwich, 2 orders of fries, and 2 orders of onion rings is

$$24 \times 3 + 13 \times 1 + 30 \times 2 + 25 \times 2 = 195 \text{ grams.}$$

(b) To find the total grams of fat $F$ in an order, as in part a we add the fat due to hamburgers, chicken sandwiches, and so forth. For example, $h$ hamburgers contribute $24h$ grams of fat. The resulting formula is

$$F = 24h + 13c + 30f + 25o.$$

5. **United States population growth:**

(a) Since $t$ is the number of years since 1960, the year 1963 is represented by $t = 3$. In functional notation, the population of the U.S. in 1963 is given by $N(3)$. To calculate its value, we use the fact that the population increases by 1.2% per year. Since $N(0) = 180$, in millions,

$$
\begin{aligned}
N(1) &= \text{Population in 1960} + 1.2\% \text{ growth} \\
&= 180 + 0.012 \times 180 = 182.16 \text{ million} \\
N(2) &= \text{Population in 1961} + 1.2\% \text{ increase} \\
&= 182.16 + 0.012 \times 182.16 = 184.35 \text{ million} \\
N(3) &= \text{Population in 1962} + 1.2\% \text{ growth} \\
&= 184.35 + 0.012 \times 184.35 = 186.56 \text{ million.}
\end{aligned}
$$

(b) The table is given below, calculating $N(4)$ and $N(5)$ as in Part (a) above.

| Year | $t$ | $N(t)$ |
|------|-----|--------|
| 1960 | 0 | 180.00 |
| 1961 | 1 | 182.16 |
| 1962 | 2 | 184.35 |
| 1963 | 3 | 186.56 |
| 1964 | 4 | 188.80 |
| 1965 | 5 | 191.06 |

(c) A graph is shown below.

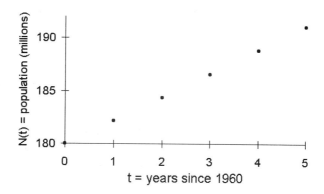

(d) The formula $180 \times 1.012^t$ can be shown to give the same values as those found in Part (b) by substituting $t = 0$, $t = 1, \ldots$ into the formula:

| $t$ | $180 \times 1.012^t$ |
|---|---|
| 0 | $180 \times 1.012^0 = 180.00$ |
| 1 | $180 \times 1.012^1 = 182.16$ |
| 2 | $180 \times 1.012^2 = 184.35$ |
| 3 | $180 \times 1.012^3 = 186.56$ |
| 4 | $180 \times 1.012^4 = 188.80$ |
| 5 | $180 \times 1.012^5 = 191.06$ |

(e) Using the formula and the fact that 2000 corresponds to $t = 40$, we find that the population in 2000 from this prediction is

$$N(40) = 180 \times 1.012^{40} = 290.06 \text{ million people.}$$

This estimate is about 9 million higher than the actual population.

7. **Altitude**:

(a) Let $t$ be the time in minutes since takeoff and $A$ the altitude in feet. Then

$$A(0) = \text{ Initial altitude } = 200 \text{ feet,}$$

and

$$
\begin{aligned}
A(1) &= \quad \text{Initial altitude } + \text{ Increase over 1 minute} \\
&= \quad 200 + 150 \times 1 = 350 \text{ feet,} \\
A(2) &= \quad \text{Initial altitude } + \text{ Increase over 2 minutes} \\
&= \quad 200 + 150 \times 2 = 500 \text{ feet,} \\
A(3) &= \quad \text{Initial altitude } + \text{ Increase over 3 minutes} \\
&= \quad 200 + 150 \times 3 = 650 \text{ feet,}
\end{aligned}
$$

and so on. This suggests the formula

$$A = \text{Initial altitude} + \text{Increase over } t \text{ minutes} = 200 + 150 \times t,$$

or $A = 200 + 150t$.

(b) In functional notation the altitude 90 seconds after takeoff is $A(1.5)$ because 90 seconds corresponds to 1.5 minutes. The value is $200 + 150 \times 1.5 = 425$ feet.

(c) Here is the graph.

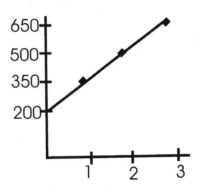

The altitude changes at a constant rate, and this is reflected in the fact that the graph is a straight line.

9. **A rental**:

(a) If we rent the car for 2 days and drive 100 miles, it will cost us

$$2 \text{ days rental} + \text{charge for 100 miles} = 49 \times 2 + 0.25 \times 100 = \$123.$$

(b) Let $d$ be the number of days we rent the car, $m$ the number of miles we drive the car, and $C$ the cost in dollars of renting the car. Then the formula that gives us the cost of renting a car is

$$C(d, m) = 49 \times \text{ days } + 0.25 \times \text{ miles } = 49d + 0.25m.$$

(c) Since we drove from Dallas to Austin and back then we traveled a total of 500 miles. We kept the car for one week, or 7 days. Hence in functional notation the cost of the rental car is $C(7,500)$. This is calculated as

$$C(7,500) = 49 \times 7 + 0.25 \times 500 = \$468.$$

11. **Preparing a letter, continued**:

(a) There are 2 pages of regular stationery and 1 of fancy letterhead stationery, so the total cost is $2 \times 0.03 + 1 \times 0.16 = 0.22$ dollar, or 22 cents.

(b) We know from Part (a) that the cost for stationery is $0.22$ dollar. The secretarial cost is $2 \times 9.25$ dollars, and the cost for the envelope is $0.50$ dollar. In total, the letter costs $0.22 + 2 \times 9.25 + 0.50 = 19.22$ dollars.

(c) Let $c$ be the cost in dollars of the stationery and $p$ the number of pages. There are $p - 1$ pages of regular stationery and 1 of fancy letterhead stationery. Then

$$
\begin{aligned}
c(p) &= \text{Cost for fancy letterhead stationery} + \text{Cost for regular stationery} \\
&= 0.16 + 0.03(p - 1).
\end{aligned}
$$

(d) Let $h$ be the number of hours spent typing the letter, let $p$ be the number of pages of the letter, and let $C$ be the cost of preparing and mailing the letter (measured in dollars). Then

$$
\begin{aligned}
C(h, p) &= \text{Secretarial cost} + \text{Cost for paper} + \text{Cost for envelope} \\
&= 9.25h + 0.16 + 0.03(p - 1) + 0.50.
\end{aligned}
$$

(This can also be written as $C(h, p) = 9.25h + 0.03(p - 1) + 0.66$ or as $C(h, p) = 9.25h + 0.03p + 0.63$.)

(e) Here $h = \dfrac{25}{60}$ and $p = 2$, so the cost is

$$
C\left(\frac{25}{60}, 2\right) = 9.25 \times \frac{25}{60} + 0.16 + 0.03(2 - 1) + 0.50 = 4.54 \text{ dollars.}
$$

13. **Stock turnover rate**:

(a) If 350 shirts are sold annually then the number of orders of 50 shirts is $\dfrac{350}{50} = 7$.

(b) The annual stock turnover rate is 7, the number computed in Part (a), because this is the number of times that the average inventory of 50 shirts needs to be replaced if 350 shirts are sold in a year.

(c) If 500 shirts were sold in a year then the annual stock turnover rate would be $\dfrac{500}{50} = 10$.

(d) Let $T$ be the annual stock turnover rate and $S$ the number of shirts sold in a year. Then

$$
T = \frac{\text{Number of shirts sold}}{\text{Average inventory}} = \frac{S}{50}.
$$

Thus the formula is $T = \dfrac{S}{50}$.

15. **Total cost**:

   (a) Let $N$ be the number of widgets produced in a month and $C$ the total cost in dollars. Because

   $$\text{Total cost} = \text{Variable cost} \times \text{Number of items} + \text{Fixed costs},$$

   we have $C = 15N + 9000$.

   (b) In functional notation the total cost is $C(250)$, and the value is $15 \times 250 + 9000 = 12{,}750$ dollars.

17. **More on revenue**:

   (a) We compute

   $$p(100) = 50 - 0.01 \times 100 = 49$$

   $$p(200) = 50 - 0.01 \times 200 = 48$$

   $$p(300) = 50 - 0.01 \times 300 = 47$$

   $$p(400) = 50 - 0.01 \times 100 = 46$$

   $$p(500) = 50 - 0.01 \times 500 = 45.$$

   These values agree with those in the table.

   (b) We have $R = pN$, so $R = (50 - 0.01N)N$ dollars.

   (c) In functional notation the total revenue is $R(450)$, and the value is

   $$(50 - 0.01 \times 450) \times 450 = 20{,}475 \text{ dollars.}$$

19. **Renting motel rooms**:

   (a) Since the group rents 2 extra rooms, we take off \$4 from the base price of each room, which means that we charge \$81 for each room. Since we rented 3 rooms altogether, we take in a total of $3 \times 81 = \$243$.

   (b) The formula that tells us how much to charge for each room is

   $$\text{Rate} = \text{Base price} - \$2 \text{ per extra room dollars.}$$

   Now if $n$ rooms are rented, then the number of extra rooms is $n - 1$. Since the base price is \$85, the formula can be written as

   $$\text{Rate} = 85 - 2 \times \text{number of extra rooms} = 85 - 2(n - 1) \text{ dollars.}$$

   This can also be written as Rate $= 87 - 2n$ dollars.

(c) The total revenue is the number of rooms times the rental per room:

$$R(n) = \text{Rooms rented} \times \text{Price per room} = n \times (85 - 2(n - 1)) \text{ dollars.}$$

This can also be written as $R(n) = n(87 - 2n)$ dollars.

(d) The total cost of renting 9 rooms is expressed in functional notation as $R(9)$. This is calculated as $R(9) = 9(85 - 2 \times (9 - 1)) = \$621$.

21. **Catering a dinner**:

(a)   i. If 50 people attend, then your cost is the rental fee plus the caterer's fee for each of the 50 people, or a total of $150 + 50 \times 10 = 650$ dollars. If 50 people attend, then to break even, each ticket should cost $\dfrac{650}{50} = 13$ dollars.

   ii. Let $n$ be the number of people attending and $C$ the amount in dollars you should charge per person. Your total cost is the rental fee plus the caterer's fee for each of the $n$ people, or a total of $150 + 10n$ dollars. Since $n$ people attend, then to break even, each ticket should cost

$$C = \frac{150 + 10n}{n} \text{ dollars.}$$

This can also be written as $C = \dfrac{150}{n} + 10$, which can be thought of as each person's share of the \$150 rental fee plus the \$10 caterer's fee.

   iii. If 65 people attend, then the amount to charge each is expressed in functional notation as $C(65)$. This is calculated as

$$C(65) = \frac{150}{65} + 10 = \$12.31 \text{ per ticket.}$$

(b) To make a profit of \$100, you should think of your cost as being \$100 more in determining your ticket price, so your new price would be

$$P = \frac{100 + 150 + 10n}{n} = \frac{250 + 10n}{n} \text{ dollars.}$$

This can also be written as $P = \dfrac{250}{n} + 10$ or as $P = \dfrac{100}{n} + \dfrac{150}{n} + 10$, which can be thought of as each person's share of the \$100 profit, plus each person's share of the \$150 rental fee, plus the \$10 caterer's fee.

23. **Production rate**:

(a) Let $k$ be the constant of proportionality. Then $t = kn$.

(b) Since $t$ is total number of items produced and $n$ is the number of employees, it follows that $k$ is the number of items produced per employee.

25. **Head and pressure:**

   (a) Because $p$ is proportional to $h$ with constant of proportionality 0.434, the equation is $p = 0.434h$.

   (b) The head of water at the mouth of the nozzle is $8 \times 12 = 96$ feet. The back pressure is the value of $p$ given by the equation in Part (a) with the head of $h = 96$ feet: $p = 0.434 \times 96 = 41.66$. Thus the back pressure is 41.66 pounds per square inch.

   (c) The head is the height of the nozzle above the pumper, and in this case that value is $185 - 40 = 145$ feet. The back pressure is $p = 0.434 \times 145 = 62.93$ pounds per square inch.

27. **Darcy's law:**

   (a) Because $V$ is proportional to $S$ with constant of proportionality $K$, the equation is $V = KS$.

   (b) The constant $K$ equals the permeability of sandstone, 0.041 meter per day. Also, $S$ is given as 0.03. We compute the velocity of the water flow using the equation in Part (a): $V = 0.041 \times 0.03 = 0.00123$. The units are found by multiplying the units for $K$ with those for $S$, and we have that the velocity is 0.00123 meter per day.

   (c) Now we take the constant $K$ to be 41 meters per day, but $S$ is still 0.03. The velocity of the water flow is $V = 41 \times 0.03 = 1.23$ meters per day.

29. **Loan origination fee:**

   (a) The fees for securing a mortgage for $322,000 are $2500 plus 2% of the mortgage amount. In this case, that is $2500 + 0.02 \times 322{,}000 = \$8940$.

   (b) The formula for the loan fee is $F = 2500 + 0.02M$.

31. **Research project:** Answers will vary.

## Skill Building Exercises

S-1. **A description:** If you have $5000 and spend half the balance each month, then the new balance will be

$$
\begin{aligned}
\text{New balance after 1 month} &= 5000 - \tfrac{1}{2} \times 5000 &= 2500 \\
\text{New balance after 2 months} &= 2500 - \tfrac{1}{2} \times 2500 &= 1250 \\
\text{New balance after 3 months} &= 1250 - \tfrac{1}{2} \times 1250 &= 625 \\
\text{New balance after 4 months} &= 625 - \tfrac{1}{2} \times 625 &= 312.5
\end{aligned}
$$

and so the balance left after 4 months is $312.50.

S-3. **A description**: We know that $f(0) = 5$ and that each time $x$ increases by 1, the value of $f$ triples, that is, it is three times its previous value. Therefore

$$
\begin{aligned}
f(1) &= 3 \times f(0) = 3 \times 5 = 15 \\
f(2) &= 3 \times f(1) = 3 \times 15 = 45 \\
f(3) &= 3 \times f(2) = 3 \times 45 = 135 \\
f(4) &= 3 \times f(3) = 3 \times 135 = 405
\end{aligned}
$$

so $f(4) = 405$. On the other hand, $5 \times 3^4$ is also $405$.

S-5. **Getting a formula**: If the man loses 67 strands of hair each time he showers, then if he showers $s$ times, he will lose $67 \times s$ strands of hair. Thus $N = 67s$.

S-7. **Getting a formula**: If you start with $500 and you add to that $37 each month, then the balance will be $500 plus $37 times the number of months. Thus the balance $B$ is given by $B = 500 + 37t$.

S-9. **Getting a formula**: You pay $249 for the iPod and the cost of downloaded songs is $0.99 per song, so the total cost you pay is $249 plus $0.99 times the number of songs. Thus $C = 249 + 0.99s$.

S-11. **Getting a formula**: Your total cost is the sum of the cost of the staples, the paper, and the pens. The cost of the staples is $1.46 times the number of boxes of staples, the cost of the paper is $3.50 times the number of reams of paper, and the cost of the pens if $2.40 times the number of boxes of pens. Thus $C = 1.46s + 3.50r + 2.40p$.

S-13. **Getting a formula**: To find the net profit, we find the revenue from the jewelry and subtract the cost of the silver and the turquoise. The revenue from the jewelry is $500 times the number of ounces, which is the total of the silver and turquoise ounces, so the revenue is $500(s + t)$. The cost of the materials is $300 times the number of ounces of silver plus the $21 times the number of ounces of turquoise, so the total cost is $300s + 21t$, and so $P = 500(s + t) - (300s + 21t)$, or $500(s + t) - 300s - 21t$.

S-15. **Constant of proportionality**: If $g(t) = 16t$, then $g$ is proportional to $t$ with constant of proportionality 16.

S-17. **Pizza**: For cheese pizzas of a fixed diameter, the weight is proportional to the thickness. For example, if you stack two pizzas of the same diameter the weight doubles, as does the thickness.

S-19. **A tire**: The circumference of a tire is proportional to its radius, since the circumference is $2\pi$ times the radius.

S-21. **A square box**: Since the volume of a cube is the third power of the length of a side, it is not proportional to the length of a side. For example, if the length of the sides doubles, then the volume of the cube is multiplied by 8, not 2.

S-23. **Wages**: Since the man makes $16 per hour, the monthly salary is 16 times the number of hours worked, so the monthly salary is proportional to the number of hours worked. Here the constant of proportionality is 16 dollars per hour.

S-25. **Sodas**: Since the price per soda is fixed, the cost of buying sodas is proportional to the number of sodas bought. Here the constant of proportionality is the price per soda.

## Chapter 1 Review Exercises

1. **Evaluating formulas**: To get the function value $M(9500, 0.01, 24)$, substitute $P = 9500$, $r = 0.01$, and $t = 24$ in the formula

$$M(P, r, t) = \frac{Pr(1 + r)^t}{(1 + r)^t - 1}.$$

The result is

$$\frac{9500 \times 0.01 \times (1 + 0.01)^{24}}{(1 + 0.01)^{24} - 1},$$

which equals 447.20.

2. **U.S. population**:

(a) Because 1790 corresponds to $t = 0$, the population in 1790 was $3.93 \times 1.03^0 = 3.93$ million.

(b) Because 1810 is 20 years after 1790, we take $t = 20$ and get that $N(20)$ is functional notation for the population in 1790.

(c) To find the population in 1810 we put $t = 20$ in the formula. The result is $3.93 \times 1.03^{20} = 7.10$, so the population in 1810 was 7.10 million according to the formula.

3. **Averages and average rate of change**:

(a) Because 5 is halfway between 4 and 6, we estimate $f(5)$ by

$$\frac{f(4) + f(6)}{2} = \frac{40.1 + 43.7}{2} = 41.9.$$

(b) The average rate of change is the change in $f$ divided by the change in $x$, and that is

$$\frac{f(6) - f(4)}{2} = \frac{43.7 - 40.1}{2} = 1.8.$$

4. **High school graduates**:

(a) Here $N(1989)$ represents the number, in millions, graduating from high school in 1989. According to the table, its value is 2.47 million.

(b) In functional notation the number of graduates in 1988 is $N(1988)$. We estimate its value by averaging:

$$\frac{N(1987) + N(1989)}{2} = \frac{2.65 + 2.47}{2} = 2.56.$$

Thus there were about 2.56 million graduating in 1988.

(c) The average rate of change is the change in $N$ divided by the change in $t$, and that is

$$\frac{N(1991) - N(1989)}{2} = \frac{2.29 - 2.47}{2} = -0.09 \text{ million per year.}$$

(d) To estimate the value of $N(1994)$, we calculate $N(1991)$ plus 3 years of change at the average rate found in the previous part. So, $N(1994)$ is estimated to be

$$N(1991) + 3 \times -0.09 = 2.29 - 0.27 = 2.02.$$

Our estimate for $N(1994)$ is 2.02 million.

5. **Increasing, decreasing, and concavity**:

(a) The function is increasing from 2006 to 2016.

(b) It is concave down from 2012 to 2016 and concave up from 2006 to 2012.

(c) There is an inflection point at $d = 2012$.

6. **Logistic population growth**:

(a) The population grows rapidly at first and then the growth slows. Eventually it levels off.

(b) The population reaches 300 in mid-2009.

(c) The population is increasing most rapidly in 2008.

(d) The point of most rapid population increase is an inflection point.

7. **Getting a formula**: The balance (in dollars) is the initial balance of \$780 minus \$39 times the number of withdrawals. Thus $B = 780 - 39t$.

8. **Cell phone charges**:

    (a) Let $t$ denote the number of text messages and $C$ the charge, in dollars.

    (b) The charge (in dollars) is the flat monthly rate of $39.95 plus $0.10 times the number of messages in excess of 100. That excess is $t - 100$, so the formula is $C = 39.95 + 0.1(t - 100)$.

    (c) In functional notation the cost if you have 450 messages is $C(450)$. The value is $39.95 + 0.1(450 - 100) = 74.95$ dollars.

    (d) If the number of text messages is less than 100 then the only charge is the flat monthly rate of $39.95. So the formula is $C = 39.95$.

9. **Cell phone charges again**:

    (a) Let $t$ denote the number of text messages, $m$ the number of minutes, and $C$ the charge, in dollars.

    (b) The charge (in dollars) is the flat monthly rate of $34.95, plus $0.35 times the number of minutes in excess of 4000, plus $0.10 times the number of messages in excess of 100. Thus the formula is $C = 34.95 + 0.35(m - 4000) + 0.1(t - 100)$.

    (c) The charges are $34.95 + 0.35(6000 - 4000) + 0.1(450 - 100) = 769.95$ dollars.

    (d) If the number of text messages is less than 100 then the only charges are the flat monthly rate of $34.95 plus $0.35 times the number of minutes in excess of 4000. So the formula is $C = 34.95 + 0.35(m - 4000)$.

    (e) We use the formula from Part (d). The charges are $34.95 + 0.35(4200 - 4000) = 104.95$ dollars.

10. **Practicing calculations**:

    (a) We calculate that $C(0) = 0.2 + 2.77e^{-0.37 \times 0} = 2.97$.

    (b) We calculate that $C(0) = \dfrac{12.36}{0.03 + 0.55^0} = 12$.

    (c) We calculate that $C(0) = \dfrac{0 - 1}{\sqrt{0 + 1}} = -1$.

    (d) We calculate that $C(0) = 5 \times 0.5^{0/5730} = 5$.

11. **Amortization**:

    (a) We calculate that

    $$M(5500, 0.01, 24) = \frac{5500 \times 0.01 \times (1 + 0.01)^{24}}{(1 + 0.01)^{24} - 1} = 258.90 \text{ dollars.}$$

    Your monthly payment if you borrow $5500 at a monthly rate of 1% for 24 months is $258.90.

(b) In functional notation the payment is $M(8000, 0.006, 36)$. The value is

$$\frac{8000 \times 0.006 \times (1 + 0.006)^{36}}{(1 + 0.006)^{36} - 1} = 247.75 \text{ dollars.}$$

12. **Using average rate of change**:

(a) The average rate of change is the change in $f$ divided by the change in $x$, and that is

$$\frac{f(3) - f(0)}{3} = \frac{55 - 50}{3} = 1.67.$$

(b) Again, the average rate of change is the change in $f$ divided by the change in $x$, and in this case that is

$$\frac{f(6) - f(3)}{3} = \frac{61 - 55}{3} = 2.$$

(c) Because 4 is 1 unit more than 3 and the average rate of change is 2, we estimate $f(4)$ by

$$f(3) + 1 \times 2 = 55 + 2 = 57.$$

13. **Timber values under Scribner scale**:

(a) The average rate of change from \$20 to \$24 is $\frac{81.60 - 68.00}{24 - 20} = 3.4$. The units here are dollar value per MBF Scribner divided by dollar value per cord. Continuing in this way, we get the following table, where the rate of change has the units just given.

| Interval | 20 to 24 | 24 to 28 | 28 to 36 |
|---|---|---|---|
| Rate of change | 3.4 | 3.4 | 3.4 |

Note that the change in the variable for the last interval is 8, not 4.

(b) No, the value per MBF Scribner should not have a limiting value: Its rate of change is a nonzero constant, so we expect it to increase at a constant rate.

(c) We use the average rate of change from \$24 to \$28 to estimate the value per MBF Scbribner when the value per cord is \$25. That estimate is $81.60 + 1 \times 3.4 = 85$ dollars per MBF Scbribner. Because \$85 is greater than \$71, if you are selling then \$25 per cord is a better value, but if you are buying then \$71 per MBF Scribner is a better value. Another way to do this is to note by inspecting the table that the value of \$25 per cord is greater than the value of \$71 per MBF Scribner: The value of \$25 per cord is greater than the value of \$24 per cord listed in the table, so it is higher than the equivalent value of \$81.60 per MBF Scribner listed in the table.

14. **Concavity**: If a graph is decreasing at an increasing rate then it is concave down. If it is decreasing at a decreasing rate then it is concave up.

15. **Longleaf pines**:

   (a) The height of the tree increases quickly at first, but the growth rate decreases as the tree ages. It makes sense for a young tree to grow more quickly than an older tree.

   (b) According to the graph the tree height for a 60-year-old tree is about 132 feet.

   (c) Yes, there is a limiting value, since the graph eventually levels off.

   (d) The graph is concave down. This means that the height is increasing at a decreasing rate, so each year the amount of growth decreases.

16. **Getting a formula**:

   (a) If we rent 3 rooms we get a discount of $2 \times 2 = 4$ dollars, so each room will cost $56 - 4 = 52$ dollars.

   (b) Because each room will cost $52 dollars, we will pay $3 \times 52 = 156$ dollars altogether.

   (c) If we rent $n$ rooms we get a discount of $2(n-1) = 2n-2$ dollars. If we let $R$ denote the rental cost in dollars per room then $R = 56 - 2(n-1)$ or $R = 58 - 2n$ dollars.

   (d) Let $C$ denote the total cost in dollars. Then $C = n \times (56 - 2(n-1))$ or $C = n \times (58 - 2n)$.

17. **A wedding reception**:

   (a) If you invite 100 guests then the cost is $3200 for the venue plus $31 times 50 (because 100 guests makes an excess of 50 over the number included). Thus the cost is $3200 + 31 \times 50 = 4750$ dollars.

   (b) If you invite $n$ guests then the cost is $3200 for the venue plus $31 times the number in excess of 50. That excess is $n - 100$, so if we let $C$ denote the cost in dollars then $C = 3200 + 31(n - 50)$ or $C = 1650 + 31n$.

   (c) We want to find $n$ so that $C = 5500$. By the formula from Part (b) this says that $1650 + 31n = 5500$. By trial and error (or by solving for $n$ using algebra) we find that we can invite 124 guests.

18. **Limiting values**:

   (a) No, not all tables show limiting values.

   (b) We can identify a limiting value from a table by checking whether the last few entries in the table show little change. If so, the limiting value is approximated by the trend established by the last few entries.

(c) No, not all graphs show limiting values.

(d) We can identify a limiting value from a graph by checking whether the last portion of the graph levels off. If so, the limiting value is approximated by that value where the graph is level.

## A FURTHER LOOK: AVERAGE RATES OF CHANGE WITH FORMULAS

1. **Calculating rates of change**: The average rate of change is

$$\frac{f(4) - f(2)}{4 - 2} = \frac{\frac{1}{4} - \frac{1}{2}}{2} = -\frac{1}{8} \text{ or } -0.125.$$

Rounding to two decimal places gives $-0.13$.

2. **Calculating rates of change**: The average rate of change is

$$\frac{f(2) - f(1)}{2 - 1} = \frac{5 - 2}{1} = 3.$$

3. **Calculating rates of change**: The average rate of change is

$$\frac{f(9) - f(4)}{9 - 4} = \frac{3 - 2}{5} = \frac{1}{5} \text{ or } 0.20.$$

4. **Average rates of change with variables**: The average rate of change is

$$\frac{f(3 + h) - f(3)}{(3 + h) - 3} = \frac{2(3 + h) + 1 - (2 \times 3 + 1)}{h} = \frac{2h}{h} = 2.$$

5. **Average rates of change with variables**: The average rate of change is

$$\frac{f(h) - f(0)}{h - 0} = \frac{h^2 - 0^2}{h} = \frac{h^2}{h} = h.$$

6. **Difference quotients**: The average rate of change from $x$ to $x + h$ is

$$\frac{f(x + h) - f(x)}{(x + h) - x} = \frac{3(x + h) + 1 - (3x + 1)}{h} = \frac{3x + 3h + 1 - 3x - 1}{h} = \frac{3h}{h} = 3.$$

7. **Difference quotients**: The average rate of change from $x$ to $x + h$ is

$$\frac{f(x+h) - f(x)}{(x+h) - x} = \frac{(x+h)^2 + (x+h) - (x^2 + x)}{h} = \frac{x^2 + 2xh + h^2 + x + h - x^2 - x}{h}$$

$$= \frac{2xh + h^2 + h}{h} = 2x + h + 1.$$

8. **Linear functions**: If $f(x) = mx + b$ then the average rate of change from $p$ to $q$ is

$$\frac{f(q) - f(p)}{q - p} = \frac{(mq + b) - (mp + b)}{q - p} = \frac{mq - mp}{q - p} = \frac{m(q - p)}{q - p} = m.$$

Thus the average rate of change is $m$, and this does not depend on either $p$ or $q$.

9. **The effect of adding a constant**: The average rate of change for $f$ is

$$\frac{f(b) - f(a)}{b - a},$$

and the average rate of change for $g$ is

$$\frac{g(b) - g(a)}{b - a} = \frac{(f(b) + c) - (f(a) + c)}{b - a} = \frac{f(b) - f(a)}{b - a}.$$

Thus the two rates are the same. This makes sense because the rate of change of a constant is $0$.

10. **A fish**: The average rate of growth over the first year is

$$\frac{L(1) - L(0)}{1 - 0} = \frac{(10 - \frac{1}{2}) - (10 - 1)}{1} = \frac{1}{2} \text{ or } 0.5 \text{ inch per year.}$$

11. **Radioactive decay**:

    (a) The average rate of change from $t = 0$ to $t = 2$ is

    $$\frac{A(2) - A(0)}{2 - 0} = \frac{\frac{20}{2^2} - \frac{20}{2^0}}{2} = \frac{5 - 20}{2} = \frac{-15}{2} \text{ or } -7.50 \text{ grams per minute.}$$

    (b) Physically the meaning of a negative rate of change is that the amount of the radioactive substance is decreasing.

12. **A derivative**: When $h$ is close to $0$, the average rate of change $2x + h$ is close to $2x$. Thus the derivative of $x^2$ is $2x$.

13. **A derivative**: When $h$ is close to $0$, the average rate of change $3x^2 + 3xh + h^2$ is close to $3x^2$, because the two terms involving $h$ get smaller and smaller as $h$ gets smaller and smaller. Thus the derivative of $x^3$ is $3x^2$.

# A FURTHER LOOK: AREAS ASSOCIATED WITH GRAPHS

1. **Rectangle**: The shaded region in the figure is a rectangle with base $6 - 3 = 3$. The height of the rectangle is the height of the graph of $f(x) = x^2$ at $x = 3$, so the height is $3^2 = 9$. Thus the area is

$$\text{Area of rectangle } = \text{ Base } \times \text{ Height} = 3 \times 9 = 27.$$

2. **Rectangle**: The shaded region in the figure is a rectangle with base $4 - 1 = 3$. The height of the rectangle is the height of the graph of $f(x) = x^2$ at $x = 4$, so the height is $4^2 = 16$. Thus the area is

$$\text{Area of rectangle } = \text{ Base } \times \text{ Height} = 3 \times 16 = 48.$$

3. **Triangle**: The shaded region in the figure is a triangle with base $8 - 2 = 6$. The height of the triangle is the height of the graph of $f(x) = x - 2$ at $x = 8$, so the height is $8 - 2 = 6$. Thus the area is

$$\text{Area of triangle } = \frac{1}{2} \text{ Base } \times \text{ Height} = \frac{1}{2} \times 6 \times 6 = 18.$$

4. **Triangle**: The shaded region in the figure is a triangle with base $5 - 1 = 4$. The height of the triangle is the height of the graph of $f(x) = 10 - 2x$ at $x = 1$, so the height is $10 - 2 \times 1 = 8$. Thus the area is

$$\text{Area of triangle } = \frac{1}{2} \text{ Base } \times \text{ Height} = \frac{1}{2} \times 4 \times 8 = 16.$$

5. **Lower sum**: The shaded region in the figure is made up of three rectangles. Each of the rectangles has base 1. The heights of the rectangles are determined by the graph. The left-hand rectangle has height $2^3 = 8$, so its area is $1 \times 8 = 8$. The middle rectangle has height $3^3 = 27$, so its area is $1 \times 27 = 27$. The right-hand rectangle has height $4^3 = 64$, so its area is $1 \times 64 = 64$. Therefore, the total area is $8 + 27 + 64 = 99$.

6. **Upper sum:** The shaded region in the figure is made up of three rectangles. Each of the rectangles has base 1. The heights of the rectangles are determined by the graph. The left-hand rectangle has height $3^3 = 27$, so its area is $1 \times 27 = 27$. The middle rectangle has height $4^3 = 64$, so its area is $1 \times 64 = 64$. The right-hand rectangle has height $5^3 = 125$, so its area is $1 \times 125 = 125$. Therefore, the total area is $27 + 64 + 125 = 216$.

7. **Trapezoid:** We think of the area of the shaded region as the difference between the areas of two triangles, one with base from $x = 0$ to $x = 3$ and the other with base from $x = 0$ to $x = 5$. The first of these triangles has base 3. Its height is the height of the graph of $f(x) = 4x$ at $x = 3$, so its height is $4 \times 3 = 12$. Thus the area of the first triangle is

$$\text{Area of triangle} \ = \frac{1}{2} \text{ Base } \times \text{ Height} = \frac{1}{2} \times 3 \times 12 = 18.$$

Similarly, the second triangle has base 5 and height $4 \times 5 = 20$, so its area is $\frac{1}{2} \times 5 \times 20 = 50$. The area of the shaded region is the difference between these two areas, which equals $50 - 18 = 32$. This area can be found in other ways, one of which is to use the formula for the area of a trapezoid.

8. **Trapezoid:** We think of the area of the shaded region as the difference between the areas of two triangles, one with base from $x = 4$ to $x = 6$ and the other with base from $x = 2$ to $x = 6$. The first of these triangles has base 2. Its height is the height of the graph of $f(x) = 18 - 3x$ at $x = 4$, so its height is $18 - 3 \times 4 = 6$. Thus the area of the first triangle is

$$\text{Area of triangle} \ = \frac{1}{2} \text{ Base } \times \text{ Height} = \frac{1}{2} \times 2 \times 6 = 6.$$

Similarly, the second triangle has base 4 and height $18 - 3 \times 2 = 12$, so its area is $\frac{1}{2} \times 4 \times 12 = 24$. The area of the shaded region is the difference between these two areas, which equals $24 - 6 = 18$. This area can be found in other ways, one of which is to use the formula for the area of a trapezoid.

9. **Describing an area:** The shaded region in the figure below represents the region indicated in the exercise. (The right-hand edge of the region is part of the line $x = 4$.)

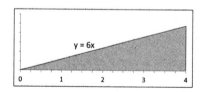

The region is a triangle with base 4. The height of the triangle is the height of the graph of $f(x) = 6x$ at $x = 4$, so the height is $6 \times 4 = 24$. Thus the area is

$$\text{Area of triangle } = \frac{1}{2} \text{ Base } \times \text{ Height} = \frac{1}{2} \times 4 \times 24 = 48.$$

10. **Describing an area**: The shaded region in the figure below represents the region indicated in the exercise. (The left-hand edge of the region is part of the line $x = 4$, and the right-hand edge is part of the line $x = 8$.)

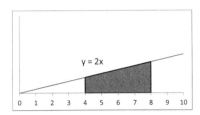

We think of the area of the region as the difference between the areas of two triangles, one with base from $x = 0$ to $x = 4$ and the other with base from $x = 0$ to $x = 8$. The first of these triangles has base 4. Its height is the height of the graph of $f(x) = 2x$ at $x = 4$, so its height is $2 \times 4 = 8$. Thus the area of the first triangle is

$$\text{Area of triangle } = \frac{1}{2} \text{ Base } \times \text{ Height} = \frac{1}{2} \times 4 \times 8 = 16.$$

Similarly, the second triangle has base 8 and height $2 \times 8 = 16$, so its area is $\frac{1}{2} \times 8 \times 16 = 64$. The area of the shaded region is the difference between these two areas, which equals $64 - 16 = 48$. This area can be found in other ways, one of which is to use the formula for the area of a trapezoid.

11. **An upper sum**: The shaded region in the figure is made up of two rectangles. Each of the rectangles has base 2. The heights of the rectangles are determined by the graph. The left-hand rectangle has height $40 - 1^2 = 39$, so its area is $2 \times 39 = 78$. The right-hand rectangle has height $40 - 3^2 = 31$, so its area is $2 \times 31 = 62$. Therefore, the total area is $78 + 62 = 140$.

12. **A lower sum**: The shaded region in the figure is made up of two rectangles. Each of the rectangles has base 2. The heights of the rectangles are determined by the graph. The left-hand rectangle has height $40 - 3^2 = 31$, so its area is $2 \times 31 = 62$. The right-hand rectangle has height $40 - 5^2 = 15$, so its area is $2 \times 15 = 30$. Therefore, the total area is $62 + 30 = 92$.

## A FURTHER LOOK: DEFINITION OF A FUNCTION

1. We calculate $f(2) = \dfrac{2+3}{2-1} = 5$.

2. We calculate $f(9) = 9 + \sqrt{9} = 12$.

3. We replace $x$ in the formula by $x + 2$: $f(x+2) = (x+2) + 1 = x + 3$.

4. We replace $x$ in the formula by $x^2$: $f(x^2) = x^2 - \left(x^2\right)^2 = x^2 - x^4$.

5. Because division by 0 is not allowed, the domain of $f$ is all numbers for which the denominator is not 0. Thus, the domain is all real numbers except 1.

6. Because division by 0 is not allowed, the domain of $f$ is all numbers for which the denominator is not 0. Thus, the domain is all real numbers except 0 and 3.

7. Because we cannot take the square root of a negative number, the domain of $f$ is all numbers for which $x - 4$ is nonnegative. Thus, the domain is all real numbers greater than or equal to 4.

8. Because we can solve the equation to get $y = x^{1/3}$, the equation does determine $y$ as a function of $x$.

9. The equation does not determine $y$ as a function of $x$. For example, if $x = 1$ there are two possible values for $y$: $y = 2^{1/4}$ and $y = -2^{1/4}$.

10. We calculate $f(0) = 5 \times 0.5^{0/4} = 5$ and $f(5) = 5 \times 0.5^{5/4} = 2.10$.

# Solution Guide for Chapter 2: Graphical and Tabular Analysis

## 2.1  TABLES AND TRENDS

1. **A savings account**:

   (a) Here is a partial table of values:

   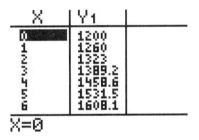

   (b) We want to find what value of $t$ in the formula first gives the function value $B = 1500$ or higher. From the table we see that, in terms of whole numbers for $t$, it is not until $t$ reaches 5 that the function $B$ first reaches 1500 or higher. Thus, you will need to wait about 5 years until you have enough money.

3. **Paying off a credit card**: We are interested in the number of payments, so we require the answer to be a whole number of months. We first make a table of values for $B$ with a table starting value of 100 and a table increment of 10. This table is on the left below, and we see that the function $B$ first takes a value below 100 somewhere close to $t = 160$. We change the table starting value to 155 and the table increment to 1 to get the table on the right below.

| X | Y1 |
|---|---|
| 100 | 643.38 |
| 110 | 469.58 |
| 120 | 342.72 |
| 130 | 250.14 |
| 140 | 182.57 |
| 150 | 133.25 |
| 160 | 97.251 |

X=100

| X | Y1 |
|---|---|
| 155 | 113.84 |
| 156 | 110.31 |
| 157 | 106.89 |
| 158 | 103.57 |
| 159 | 100.36 |
| 160 | 97.251 |
| 161 | 94.237 |

X=155

From the table on the right above we see that, in terms of whole numbers for $t$, it is not until $t$ reaches 160 that the function $B$ first takes a value below 100. Thus, you will have to make 160 payments to get the balance below \$100. (This corresponds to 13 years and 4 months of payments.)

5. **Economic efficiency**: To find the production level at which marginal cost and marginal benefit are the same, we want to find the value of $n$ for which $C(n) = B(n)$. First we make the following table of values of the two functions $C$ and $B$.

| X | Y1 | Y2 |
|---|---|---|
| 3 | 8 | 18.225 |
| 4 | 9 | 16.403 |
| 5 | 10 | 14.762 |
| 6 | 11 | 13.286 |
| 7 | 12 | 11.957 |
| 8 | 13 | 10.762 |
| 9 | 14 | 9.6855 |

X=9

From the table we see that the functions agree (to one decimal place) when the variable is 7. Thus the production level at which marginal cost and marginal benefit are the same is $n = 7$ items.

7. **Spache Readability Formula**:

(a) We are assuming that $W = 200$ and $U = 10$. Therefore, for the passages being considered the formula for the grade level as a function of $S$ is

$$\text{Grade level} = 0.121 \times \frac{200}{S} + 0.082 \times 10 + 0.659.$$

The table of values below was made using this formula. This table suggests that the grade level decreases as the number of sentences increases, and this observation is confirmed by inspecting the formula: Increasing $S$ causes the denominator of the fraction to be larger, so the fraction (and hence the function) is smaller.

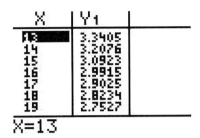

X=13

Therefore, for the passages being considered, as the number of sentences increases the grade level decreases.

(b) There can't be more sentences than words, so in the passages being considered the maximum number of sentences is 200 (the number of words).

(c) To find the minimum number of sentences that would put this passage at the second-grade level, we want to find the smallest value of $S$ for which the formula in part a gives 2 when rounded down. From the table of values above we see that 16 sentences would put the passage at a second-grade level, but no smaller number of sentences would do so. (Recall from part a that the function is decreasing.)

9. **Harvard Step Test**: Here is a table of values for $E$ as a function of $P$:

| X | Y₁ |
|---|---|
| 150 | 100 |
| 160 | 93.75 |
| 170 | 88.235 |
| 180 | 83.333 |
| 190 | 78.947 |
| 200 | 75 |
| 210 | 71.429 |

X=150

(a) From the table above, we see that the index $E$ decreases with increasing values of the variable $P$. This should hold for all values of $P$ because increasing the denominator in the fraction defining $E$ makes the fraction smaller. This means that a person having a larger total pulse count after the exercise than another person has a lower physical efficiency index and hence is not as physically fit.

(b) The index of someone with a total pulse count of 200 is expressed in functional notation by $E(200)$. From the table we generated we find the value to be 75.

(c) We saw in Part (b) that the index of someone with a total pulse count of 200 is 75. This is in the interval from 65 to 79 of the table in the text, and that table indicates that the physical condition of such a person is high average.

(d) According to the table in the text, for an excellent rating the index must be 90 or above. From the table we generated we see that this transition occurs between $P = 160$ and $P = 170$. Here is a table examining the function values more closely:

| X | Y1 |
|-----|--------|
| 161 | 93.168 |
| 162 | 92.593 |
| 163 | 92.025 |
| 164 | 91.463 |
| 165 | 90.909 |
| 166 | 90.361 |
| 167 | 89.82 |

X=161

We see that the index is 90 or above when the total pulse count is 166 or lower.

11. **Later public high school enrollment**: Here are two tables of values for $N$ as a function of $t$:

| X | Y1 |
|---|--------|
| 0 | 13.37 |
| 1 | 13.797 |
| 2 | 14.158 |
| 3 | 14.453 |
| 4 | 14.682 |
| 5 | 14.845 |
| 6 | 14.942 |

X=0

| X | Y1 |
|----|--------|
| 7 | 14.973 |
| 8 | 14.938 |
| 9 | 14.837 |
| 10 | 14.67 |
| 11 | 14.437 |
| 12 | 14.138 |
| 13 | 13.773 |

X=7

(a) From the second table we generated we find the value of $N(10)$ to be 14.67 million students. This is the number (in millions) of students enrolled in U.S. public high schools in the year 2010 (because that is 10 years after 2000).

(b) We want to consider the model from 2000 to 2010, which corresponds to the range from $t = 0$ to $t = 10$. Examining the tables given above shows that the maximum value of the function occurs when $t = 7$. Thus the enrollment was the largest in 2007 (7 years after 2000). From the second table we see that the largest enrollment was 14.97 million students.

(c) Here 2004 corresponds to $t = 4$, and we noted in Part (a) that 2010 corresponds to $t = 10$. From the first table above we find that $N(4) = 14.682$, and we found in Part (a) that $N(10) = 14.67$. The average yearly rate of change is then

$$\frac{N(10) - N(4)}{10 - 4} = \frac{14.67 - 14.682}{6} = 0.00 \text{ million students per year.}$$

Thus the average yearly rate of change from 2004 to 2010 is 0.00 million students per year. This relatively small result is misleading because, as we saw in Part (b), the enrollment actually increased significantly over this period (reaching 14.97 million students in 2007) before it decreased to a level near the value in 2004.

13. **Competition**: This question can be answered by trial and error, but we will solve it by finding two formulas and making a table. Let $n$ be the number of races run, $F = F(n)$ the time in seconds it takes the first friend to run a mile, and $S = S(n)$ the time in seconds it takes the second friend. We have

$$\text{Time taken } = \text{ Initial time taken } - \text{ Decrease in time } \times \text{ Number of races.}$$

Now for the first friend the initial time taken is 7 minutes, or 420 seconds, and the decrease in time is 13 seconds. This gives the formula $F = 420 - 13n$. Similar reasoning gives the formula $S = 440 - 16n$ for the second friend. Here is a partial table of values of these two functions:

| X | Y₁ | Y₂ |
|---|---|---|
| 1 | 407 | 424 |
| 2 | 394 | 408 |
| 3 | 381 | 392 |
| 4 | 368 | 376 |
| 5 | 355 | 360 |
| 6 | 342 | 344 |
| 7 | 329 | 328 |

X=1

From the table we see that the first race in which the second friend beats the first is the seventh race.

15. **Counting when order matters**: Here is a table of values for the factorial function:

| X | Y₁ | |
|---|---|---|
| 1 | 1 | |
| 2 | 2 | |
| 3 | 6 | |
| 4 | 24 | |
| 5 | 120 | |
| 6 | 720 | |
| 7 | 5040 | |

X=7

(a) The number of ways you can arrange 5 people in a line is 5!. From the table we see that the value is 120. (This can also be computed directly: $5! = 5 \times 4 \times 3 \times 2 \times 1 = 120$.) Thus there are 120 ways to arrange 5 people in a line.

(b) From the table we see that the factorial function first exceeds 1000 when the variable is 7, so 7 (or more) people will result in more than 1000 possible arrangements for a line.

(c) The number of guesses is the same as the number of ways to arrange 4 people in a line because there are 4 different digits given. Thus the number is 4!, which is 24. Hence we would need at most 24 guesses.

(d) The number of shufflings is the same as the number of ways to arrange 52 people in a line. Thus the number is 52!. The calculator gives that $52! = 8.07 \times 10^{67}$. Thus there are $8.07 \times 10^{67}$ possible shufflings of a deck of cards.

17. **APR and EAR:**

(a) We would expect the EAR to be larger if interest is compounded more often because once interest is compounded it accrues additional interest.

(b) We proceed by making a table of values for

$$\text{EAR} = \left(1 + \frac{\text{APR}}{n}\right)^n - 1 = \left(1 + \frac{0.1}{n}\right)^n - 1$$

The table below on the left shows that when $n = 1$ (yearly compounding), the EAR is exactly 10%. The table on the right has been extended to include $n = 12$, monthly compounding, and it shows the EAR to be 10.471%.

| X | Y1 |
|---|---|
| 1 | .1 |
| 2 | .1025 |
| 3 | .10337 |
| 4 | .10381 |
| 5 | .10408 |
| 6 | .10426 |
| 7 | .10439 |

X=1

| X | Y1 |
|---|---|
| 6 | .10426 |
| 7 | .10439 |
| 8 | .10449 |
| 9 | .10456 |
| 10 | .10462 |
| 11 | .10467 |
| 12 | .10471 |

X=12

The table below includes $n = 365$ (daily compounding), and it shows an EAR of 10.516%.

| X | Y1 |
|---|---|
| 360 | .10516 |
| 361 | .10516 |
| 362 | .10516 |
| 363 | .10516 |
| 364 | .10516 |
| 365 | .10516 |
| 366 | .10516 |

X=365

(c) On a loan of \$5000 compounded monthly, after one year the interest owed would be

$$5000 \times \text{ monthly EAR } = 5000 \times 0.10471 = 523.55 \text{ dollars.}$$

The total amount owed is the loan amount plus the interest, which is $5000 + 523.55 = 5523.55$ dollars.

If we compound daily, after one year the interest owed would be

$$5000 \times \text{ daily EAR } = 5000 \times 0.10516 = 525.80 \text{ dollars.}$$

That gives a total amount owed of \$5525.80.

(d) The EAR when we compound continuously is

$$r = e^{\text{APR}} - 1 = e^{0.10} - 1 = 0.10517.$$

This is $10.517\%$, which is only $0.0046$ percentage point (less than $0.05$ percentage point) higher than monthly compounding.

19. **An amortization table for continuous compounding:**

(a) The monthly interest rate for this exercise is $r = \dfrac{\text{APR}}{12} = \dfrac{0.09}{12} = 0.0075$, we borrowed $P = \$3500$, and $t = 24$ months is the life of the loan. Our monthly payment for compounding continuously would be

$$M = \frac{3500(e^{0.0075} - 1)}{1 - e^{-0.0075 \times 24}} = \$159.95.$$

This is only 5 cents more than the payment for compounding monthly.

(b) The amortization table is a table of values for

$$B = \frac{P\left(e^{rt} - e^{rk}\right)}{e^{rt} - 1} = \frac{3500 \times \left(e^{0.0075 \times 24} - e^{0.0075 \times k}\right)}{e^{0.0075 \times 24} - 1}.$$

The amortization table is below. Comparing the entries here with those in Part (c) of Exercise 18 above, we see that the current entries are larger, but the difference is less than a dollar.

| Number of payments made | Amount still owed | | Number of payments made | Amount still owed |
|---|---|---|---|---|
| 0 | 3500.00 | | 13 | 1682.51 |
| 1 | 3366.40 | | 14 | 1535.23 |
| 2 | 3231.79 | | 15 | 1386.83 |
| 3 | 3096.17 | | 16 | 1237.32 |
| 4 | 2959.53 | | 17 | 1086.69 |
| 5 | 2821.85 | | 18 | 934.92 |
| 6 | 2683.15 | | 19 | 782.01 |
| 7 | 2543.40 | | 20 | 627.94 |
| 8 | 2402.59 | | 21 | 472.72 |
| 9 | 2260.73 | | 22 | 316.33 |
| 10 | 2117.80 | | 23 | 158.76 |
| 11 | 1973.79 | | 24 | 0 |
| 12 | 1828.70 | | | |

21. **Inventory**:

(a) The exercise tells us that $N = 36$, $c = 850$, and $f = 230$. So inventory expense is given by

$$E(Q) = \left(\frac{Q}{2}\right) \times 850 + \left(\frac{36}{Q}\right) \times 230.$$

(b) For 3 cars, the yearly inventory expense is

$$E(3) = \left(\frac{3}{2}\right) \times 850 + \left(\frac{36}{3}\right) \times 230 = \$4035.$$

(c) We can find this by generating a table of values for the inventory cost function given in Part (a). As we scroll down the table we see that the cost decreases until we get to \$3770, which corresponds to ordering 4 cars at a time. The table is shown below for $Q = 1$ to $Q = 7$ cars per order.

(d) Since we expect to sell 36 cars this year and we order 4 cars at a time to minimize our inventory cost, then we will place $\dfrac{36}{4} = 9$ orders to Detroit this year.

(e) The average rate of increase in yearly inventory cost from ordering four cars to ordering six cars is

$$\frac{\text{Change in cost}}{\text{Change in cars}} = \frac{E(6) - E(4)}{6 - 4} = \frac{3930 - 3770}{2} = \$80 \text{ per year per additional car.}$$

23. **Falling with a parachute**:

(a) The velocity 2 seconds into the fall is expressed in functional notation as $v(2)$. Its value is

$$v(2) = 20(1 - 0.2^2) = 19.2 \text{ feet per second.}$$

(b) The average change in velocity during the first second of the fall is

$$\frac{\text{Change in velocity}}{\text{Seconds}} = \frac{v(1) - v(0)}{1} = \frac{16 - 0}{1} = 16 \text{ feet per second per second.}$$

The average velocity from the fifth to sixth second is

$$\frac{\text{Change in velocity}}{\text{Seconds}} = \frac{v(6) - v(5)}{1} = \frac{19.999 - 19.994}{1} = 0.005 \text{ foot per second per second.}$$

The velocity is increasing as the seconds go by (as can be seen from a table of values), but the average rate of increase is decreasing.

(c) We make a table of values for velocity. The table below, which shows velocity from the $t = 2$ through $t = 8$, indicates that velocity levels off at about 20 feet per second.

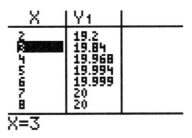

(d) Now 99% of terminal velocity is 99% of 20 feet per second, which is $0.99 \times 20 = 19.8$ feet per second. The table above shows that this occurs approximately 3 seconds into the fall.

In Example 2.1 it took 25 seconds to reach 99% of terminal velocity, while in this exercise it took only 3 seconds. This indicates that objects (such as parachutes) for which air resistance is large will reach terminal velocity faster. Thus, we would expect the feather to reach terminal velocity more quickly than a cannonball.

25. **Profit**:

(a) Assume that the total cost $C$ is measured in dollars. Because

$$\text{Total cost} = \text{Variable cost} \times \text{Number of items} + \text{Fixed costs},$$

we have $C = 50N + 150$.

(b) Assume that the total revenue $R$ is measured in dollars. Because

$$\text{Total revenue} = \text{Selling price} \times \text{Number of items},$$

we have $R = 65N$.

(c) Assume that the profit $P$ is measured in dollars. Because

$$\text{Profit} = \text{Total revenue} - \text{Total cost},$$

we have $P = R - C$. Using the formulas from Parts (a) and (b), we obtain

$$P = 65N - (50N + 150).$$

(This can also be written as $P = 15N - 150$.)

(d) We want to find for what value of $N$ we have $P(N) = 0$. Examining a table of values for $P$ shows that this occurs when $N = 10$. Thus a break-even point occurs at a production level of 10 widgets per month.

27. **A precocious child and her blocks**:

(a) If we think of going around the rectangle and adding up the sides as we encounter them, we see that the perimeter is $h + w + h + w$. That is two $h$'s and two $w$'s, so $P = 2h + 2w$ inches.

(b) We substitute $w = \dfrac{64}{h}$ in the above formula and get

$$P = 2h + 2 \times \frac{64}{h} \text{ inches.}$$

(c) To solve this exercise we first generate a table for $P(h)$. This is shown below for heights of 5 through 11 blocks. The smallest value is 32, and this happens when the height is 8. If the height is 8 then the width is $w = \dfrac{64}{8} = 8$ inches. So the blocks should be arranged in an 8 by 8 square to get a minimum perimeter.

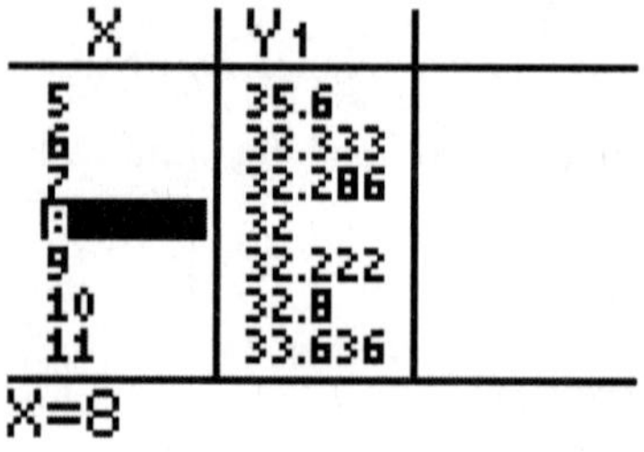

| X | Y1 |
|---|---|
| 5 | 35.6 |
| 6 | 33.333 |
| 7 | 32.286 |
| 8 | 32 |
| 9 | 32.222 |
| 10 | 32.8 |
| 11 | 33.636 |

X=8

(d) In this case the perimeter would still be $P = 2h + 2w$, and then $P = 2h + 2 \times \dfrac{60}{h}$ (measuring $P$ in inches). We make a table of values as before. The table below shows a minimum perimeter of 31 inches, and this occurs when the height is 8 inches. However, a height of 8 inches gives a width of $w = \dfrac{60}{8} = 7.5$ inches, which cannot be accomplished without cutting the blocks into pieces.

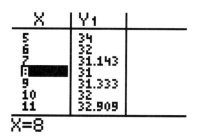

We know that $hw = 60$ and so $h$ must divide evenly into 60. This means that $h$ can only be one of 1, 2, 3, 4, 5, 6, 10, 12, 15, 20, 30, or 60. If we look at these values in the table we see that a height of 6 or 10 leads us to a perimeter of 32, which is the minimum perimeter among these values. If the height is 6, then the width must be 10, and so the child should lay the blocks in a 6 by 10 pattern if she wants to get a minimum perimeter.

29. **Growth in length of haddock:**

(a) The value of $L(4)$ is

$$L(4) = 53 - 42.82 \times 0.82^4 = 33.64 \text{ centimeters.}$$

(This can also be found from a table of values for $L$.) This means that a haddock that is 4 years old is approximately 33.64 centimeters long.

(b) Using a table for the function $L$ gives the values $L(5) = 37.12$, $L(10) = 47.11$, $L(15) = 50.82$, and $L(20) = 52.19$. Now the average yearly rate of growth in length from age 5 years to age 10 years is

$$\frac{L(10) - L(5)}{10 - 5} = \frac{47.11 - 37.12}{5} = 2.00 \text{ centimeters per year.}$$

The average yearly rate of growth in length from age 15 years to age 20 years is

$$\frac{L(20) - L(15)}{20 - 15} = \frac{52.19 - 50.82}{5} = 0.27 \text{ centimeter per year.}$$

The growth rate is lower over the later period, and this suggests that haddock grow more rapidly when they are young than when they are older.

(c) Scanning down a table on the calculator indicates that the function $L$ increases to a limiting value of 53 as $t$ gets larger and larger. Thus the longest haddock is 53 centimeters long.

31. **California earthquakes**: Here is a table of values for the function $p$:

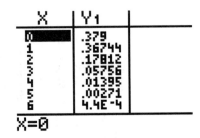

(a) We are given that $n = 3$, and from the table we see that the function value is $p(3) = 0.058$. Thus the probability is 0.058 or 5.8%.

(b) Scanning down the table suggests that the function values approach 0 as the variable $n$ increases. Thus the limiting value is 0. This means that it is very unlikely that a California home will be affected by a large number of earthquakes over a 10-year period.

(c) We are given that $n = 0$, and from the table we see that the function value is $p(0) = 0.379$. Thus the probability of a California home being affected by no major earthquakes over a 10-year period is 0.379 or 37.9%.

(d) According to the hint, we have

Probability of no major earthquake occurring
+ Probability of at least one major earthquake occurring $= 1$,

which can be rearranged to give

Probability of at least one major earthquake occurring
$= 1 -$ Probability of no major earthquake occurring.

We saw in Part (c) that the probability on the right has the value 0.379. Thus the probability of at least one major earthquake occurring is $1 - 0.379 = 0.621$ or 62.1%.

33. **Research project**: Answers will vary.

## Skill Building Exercises

S-1. **Making tables and comparing functions**: To make a table for $f(x) = x^2 - 1$ showing function values for $x = 4, 6, 8, \ldots$, we enter the function as $\mathsf{Y1} = X^2 - 1$ and use a table starting value of 4 and a table increment value of 2, resulting in the following table.

| X | Y1 |
|---|---|
| 4 | 15 |
| 6 | 35 |
| 8 | 63 |
| 10 | 99 |
| 12 | 143 |
| 14 | 195 |
| 16 | 255 |

X=4

S-3. **Making tables and comparing functions**: To make a table for $f(x) = 16 - x^3$ showing function values for $x = 3, 7, 11, \ldots$, we enter the function as $\mathsf{Y1} = 16 - X^3$ and use a table starting value of 3 and a table increment value of 4, resulting in the following table.

| X | Y1 |
|---|---|
| 3 | -11 |
| 7 | -327 |
| 11 | -1315 |
| 15 | -3359 |
| 19 | -6843 |
| 23 | -12151 |
| 27 | -19667 |

X=3

S-5. **Making tables and comparing functions**: To make a table for $f(x) = 2^x + x^2$ showing function values for $x = 0.1, 0.2, 0.3, \ldots$, we enter the function as $\mathsf{Y1} = 2^X + X^2$ and use a table starting value of 0.1 and a table increment value of 0.1, resulting in the following table.

| X | Y1 |
|---|---|
| .1 | 1.0818 |
| .2 | 1.1887 |
| .3 | 1.3211 |
| .4 | 1.4795 |
| .5 | 1.6642 |
| .6 | 1.8757 |
| .7 | 2.1145 |

X=.1

S-7. **Making tables and comparing functions**: To make a table for $f(x) = \sqrt{x} - x/30$ showing function values for $x = 5, 10, 15, \ldots$, we enter the function as $Y1 = \sqrt{(X)} - X/30$ and use a table starting value of $5$ and a table increment value of $5$, resulting in the following table.

| X | Y1 |
|---|-----|
| 5 | 2.0694 |
| 10 | 2.8289 |
| 15 | 3.373 |
| 20 | 3.8055 |
| 25 | 4.1667 |
| 30 | 4.4772 |
| 35 | 4.7494 |

X=5

S-9. **Making tables and comparing functions**: To make a table for $f(x) = (1 + 2^x)/(1 + 3^x)$ showing function values for $x = 1, 2, 3, \ldots$, we enter the function as $Y1 = (1 + 2^X)/(1 + 3^X)$ and use a table starting value of $1$ and a table increment value of $1$, resulting in the following table.

| X | Y1 |
|---|-----|
| 1 | .75 |
| 2 | .5 |
| 3 | .32143 |
| 4 | .20732 |
| 5 | .13525 |
| 6 | .08904 |
| 7 | .05896 |

X=1

S-11. **Finding limiting values**: We enter the function as $Y1 = (2X - 1)/(X + 1)$ and use a table starting value of $0$ and a table increment of $100$, resulting in the following table.

| X | Y1 |
|---|-----|
| 0 | -1 |
| 100 | 1.9703 |
| 200 | 1.9851 |
| 300 | 1.99 |
| 400 | 1.9925 |
| 500 | 1.994 |
| 600 | 1.995 |

X=0

The table suggests a limiting value of 2.

S-13. **Finding limiting values**: We enter the function as $Y1 = \sqrt{X+1} - \sqrt{X}$ and use a table starting value of 0 and a table increment of 100, resulting in the following table.

| X | Y1 |
|---|---|
| 0 | 1 |
| 100 | .04988 |
| 200 | .03531 |
| 300 | .02884 |
| 400 | .02498 |
| 500 | .02235 |
| 600 | .0204 |

X=0

The table suggests a limiting value of 0.

S-15. **Finding limiting values**: We enter the function as $Y1 = (4X^2 - 1)/(7X^2 + 1)$ and use a table starting value of 0 and a table increment value of 20, resulting in the following table.

| X | Y1 |
|---|---|
| 0 | -1 |
| 20 | .57087 |
| 40 | .57129 |
| 60 | .57137 |
| 80 | .57139 |
| 100 | .57141 |
| 120 | .57141 |

X=0

The table suggests a limiting value of about 0.5714.

S-17. **Finding limiting values**: We enter the function as $Y1 = (1 + 1/X)^X$ and use a table starting value of 0 and a table increment value of 10,000, resulting in the following table.

| X | Y1 |
|---|---|
| 0 | ERROR |
| 10000 | 2.7181 |
| 20000 | 2.7182 |
| 30000 | 2.7182 |
| 40000 | 2.7182 |
| 50000 | 2.7183 |
| 60000 | 2.7183 |

Y1=2.71825918189

The table suggests a limiting value of about 2.72.

S-19. **Finding limiting values**: We enter the function as $\mathsf{Y1} = X(2^{(1/X)} - 1)$ and use a table starting value of $0$ and a table increment value of $1000$, resulting in the following table.

| X | Y1 |
|------|--------|
| 0 | ERROR |
| 1000 | .69339 |
| 2000 | .69327 |
| 3000 | .69323 |
| 4000 | .69321 |
| 5000 | .6932 |
| 6000 | .69319 |

$$Y1 = .6931872198$$

The table suggests a limiting value of about $0.69$.

S-21. **Finding limiting values**: We enter the function as $\mathsf{Y1} = \sqrt{(9X^2 + X)} - 3X$ and use a table starting value of $0$ and a table increment value of $10$, resulting in the following table.

| X | Y1 |
|------|--------|
| 0 | 0 |
| 10 | .16621 |
| 20 | .16644 |
| 30 | .16651 |
| 40 | .16655 |
| 50 | .16657 |
| 60 | .16658 |

$$Y1 = .16658957754$$

The table suggests a limiting value of about $0.17$.

S-23. **Finding limiting values**: We enter the function as $\mathsf{Y1} = \sqrt{(X^2 + X + 1)} - X$ and use a table starting value of $0$ and a table increment value of $100$, resulting in the following table.

| X | Y1 |
|------|--------|
| 0 | 1 |
| 100 | .50373 |
| 200 | .50187 |
| 300 | .50125 |
| 400 | .50094 |
| 500 | .50075 |
| 600 | .50062 |

$$Y1 = .50062447928$$

The table suggests a limiting value of about $0.50$.

S-25. **Finding maxima and minima**: To find the minimum value of $f$, we make a table. We enter the function as $\mathsf{Y1} = X^2 - 8X + 21$ and use a table starting value of 0 and a table increment value of 1, resulting in the following table.

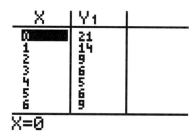

The table shows that $f$ has a minimum value of 5 at $x = 4$, at least for values of $x$ up to 6. Scrolling down the table through $x = 20$, we see that the function increases, so indeed the minimum value of $f$ is 5 at $x = 4$.

S-27. **Finding maxima and minima**: To find the maximum value of $f$, we make a table. We enter the function as $\mathsf{Y1} = 9X^2 - 2^X + 1$ and use a table starting value of 0 and a table increment value of 1. Scrolling farther down the table yields the following table.

| X | Y1 |
|---|---|
| 4 | 129 |
| 5 | 194 |
| 6 | 261 |
| 7 | 314 |
| 8 | 321 |
| 9 | 218 |
| 10 | -123 |

X=8

The table shows that $f$ reaches a maximum of 321 at $x = 8$.

S-29. **Finding maxima and minima**: To find the maximum value of $f$, we make a table. We enter the function as $\mathsf{Y1} = (X^2)/(2^X)$ and use a table starting value of 0 and a table increment value of 1. This yields the following table:

| X | Y1 |
|---|---|
| 0 | 0 |
| 1 | .5 |
| 2 | 1 |
| 3 | 1.125 |
| 4 | 1 |
| 5 | .78125 |
| 6 | .5625 |

Y1=1.125

The table shows that $f$ has a maximum value of 1.13 at $x = 3$; scrolling farther down the table shows that the function decreases, so indeed this is the maximum value.

S-31. **Finding maxima and minima**: To find the maximum value of $f$, we make a table. We enter the function as $Y1 = 2^X - X^3 + 300$ and use a table starting value of 0 and a table increment value of 1. Scrolling farther down the table yields the following table.

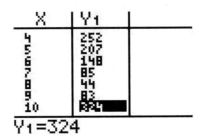

The table shows that $f$ has a maximum value of 324 at $x = 10$.

## 2.2  GRAPHS

1. **Continuous compounding**:

   (a) We show the graph with a horizontal span of 0 to 20 and a vertical span of 0 to 14,000. In choosing the vertical span we were guided by the table on the left below.

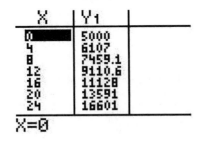

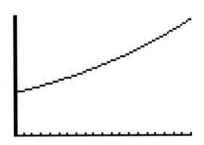

   (b) The graph is concave up. The value of the investment is increasing at an increasing rate, so it grows more rapidly in later years than in earlier years.

3. **Equity**:

   (a) We are assuming that $P = 200,000$, $r = 0.06/12 = 0.005$, and $t = 360$, so the formula for $E$ as a function of $k$ is $E(k) = \dfrac{200,000(1.005^k - 1)}{1.005^{360} - 1}$.

(b) We show the graph with a horizontal span of 0 to 360 and a vertical span of 0 to 200,000. In choosing the vertical span we were guided by the table on the left below.

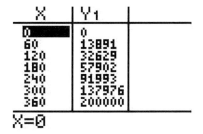    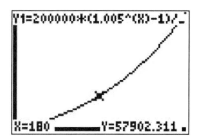

(c) Halfway through the term of the mortgage corresponds to 180 payments. From the graph (or the table), we see that $E(180)$ is about 57,902 dollars. This amount is less than half of the mortgage amount, so we do not have half-ownership in the home halfway through the term of the mortgage.

5. **Adult weight from puppy weight**:

(a) We are assuming that $w = 5$, so the formula for $W$ as a function of $a$ is $W = 52 \times \dfrac{5}{a}$ or $W = \dfrac{260}{a}$.

(b) We show the graph with a horizontal span of 0 to 16 and a vertical span of 0 to 100. In choosing the vertical span we were guided by the table below.

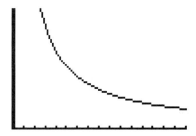

(c) The graph above shows that the adult weight $W$ decreases as the age $a$ at which the puppy weighs 5 pounds increases. Therefore, a weight of 5 pounds at an early age indicates a larger adult weight than a weight of 5 pounds at a later age. This conclusion makes sense in terms of the way animals grow.

(d) The graph is concave up. The expected adult weight decreases at a decreasing rate as the age at which 5 pounds is reached increases.

7. **Weekly cost**:

   (a) The weekly cost if there are 3 employees is $2500 + 350 \times 3 = 3550$ dollars.

   (b) Let $n$ be the number of employees and $C$ the weekly cost in dollars. Because

   $$\text{Weekly cost} = \text{Fixed cost} + \text{Cost per employee} \times \text{Number of employees},$$

   we have $C = 2500 + 350n$.

   (c) We show the graph with a horizontal span of 0 to 10 and a vertical span of 2500 to 7000. In choosing this vertical span we were guided by the table below.

   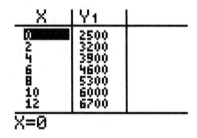 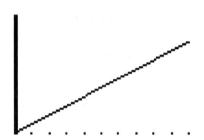

   (d) We want to find what value of the variable $n$ gives the function value $C = 4250$. Looking at a table of values or the graph shows that this occurs when $n = 5$. Thus if there are 5 employees the weekly cost will be \$4250.

9. **Resale value**:

   (a) The resale value in the year 2013 is

   $$
   \begin{aligned}
   \text{Resale value} \quad &= \quad \text{Value in the year 2012} - \text{Decrease over 1 year} \\
   &= \quad 18{,}000 - 1700 \\
   &= \quad 16{,}300 \text{ dollars.}
   \end{aligned}
   $$

   Thus the resale value in the year 2013 is \$16,300.

   (b) Because

   $$\text{Resale value} = \text{Value in the year 2012} - \text{Decrease each year} \times \text{Number of years since 2012},$$

   we have $V = 18{,}000 - 1700t$.

   (c) We show the graph with a horizontal span of 0 to 4 and a vertical span of 10,000 to 20,000. In choosing this vertical span we were guided by the table below.

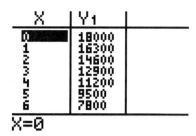 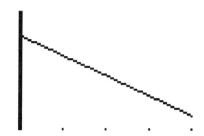

(d) Since 2015 is 3 years since 2012, the resale value in that year is $V(3)$ in functional notation. The value is $18,000 - 1700 \times 3 = 12,900$ dollars. (This can also be found from a table of values or the graph.)

11. **Baking a potato**:

(a) The first step is to make a table of values to select a vertical span. As suggested, we use a horizontal span of 0 to 120 minutes. This table with 20 minute increments is shown below. It suggests a vertical span of 0 to 420. We used this window setting to make the graph below, which shows time in minutes on the horizontal axis and temperature on the vertical axis.

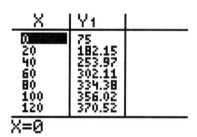 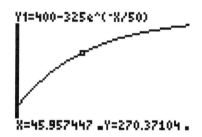

(b) The initial temperature of the potato is its temperature when placed in the oven. That is the value of $P$ when $t$ is zero:

$$P(0) = 400 - 325e^{-0/50} = 75 \text{ degrees.}$$

(c) The change in the potato's temperature in the first 30 minutes is

$$
\begin{aligned}
\text{Temperature at 30 minutes } - \text{ Initial temperature} \quad &= \quad P(30) - P(0) \\
&= \quad 221.636 - 75 \\
&= \quad 146.636 \text{ degrees.}
\end{aligned}
$$

The change in the potato's temperature in the second 30 minutes is

$$\text{Temperature at 60 minutes } - \text{ Temperature at 30 minutes } = P(60) - P(30)$$
$$= 302.112 - 221.636$$
$$= 80.476 \text{ degrees.}$$

The potato's temperature rose the most in the first 30 minutes.

The average rate of change during the first 30 minutes of baking is

$$\frac{\text{Change in temp}}{\text{Minutes elapsed}} = \frac{P(30) - P(0)}{30} = \frac{146.636}{30} = 4.89 \text{ degrees per minute.}$$

The average rate of change during the second 30 minutes of baking is

$$\frac{\text{Change in temp}}{\text{Minutes elapsed}} = \frac{P(60) - P(30)}{30} = \frac{80.476}{30} = 2.68 \text{ degrees per minute.}$$

We used three decimal accuracy for the various values of $P$ to ensure that our final answer would still be accurate to two decimal places after division.

(d) The graph is concave down, and this tells us that, although the temperature is rising, the rate of increase in temperature is decreasing. In other words, the temperature is not rising as fast at later times as it was at first. Note that in Part (c) the average increase from 0 to 30 minutes was 4.89 degrees per minute, whereas the average increase from 30 to 60 minutes was only 2.68 degrees per minute.

(e) The potato will reach a temperature of 270 degrees after about 46 minutes. This can be seen by tracing the graph, as we have done in the graph above.

(f) If the potato is left in the oven a long time, its temperature will match that of the oven. To see what happens to the potato after a long time, we made a new table with a table starting value of 0 and a table increment value of 120. It appears that the limiting value is 400. We see that the temperature of the oven is about 400 degrees.

| X | Y1 |
|---|---|
| 0 | 75 |
| 120 | 370.52 |
| 240 | 397.33 |
| 360 | 399.76 |
| 480 | 399.98 |
| 600 | 400 |
| 720 | 400 |

X=0

13. **Population growth**:

(a) We show the graph with a horizontal span of 0 to 25 and a vertical span of $-35$ to 35. In choosing this vertical span we were guided by the table below.

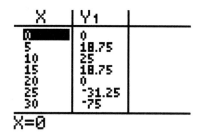 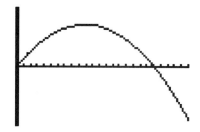

(b) The growth over a week if the population at the beginning is 4 thousand animals is expressed as $G(4)$ in functional notation. From a table of values or the graph we find that the value is 16 thousand animals.

(c) From a table of values or the graph we find that the value of $G(22)$ is $-11$ thousand animals. This means that if the population at the beginning of a week is 22 thousand animals then the population will decrease by 11 thousand over that week.

(d) Tracing the graph shows that the function increases from $n = 0$ to about $n = 10$. Over this interval the graph is concave down. This means that, if the population at the beginning of the week is at most 10 thousand, for larger initial populations the growth over a week is larger, but the rate of increase in this growth actually decreases.

15. **The economic order quantity model**:

(a)   i. Since the demand is $N = 400$ units per year and the carrying cost is $h = 24$ dollars per year, the function we want is

$$Q(c) = \sqrt{\frac{800c}{24}} \text{ items per order.}$$

Since we do not expect the fixed ordering cost to exceed \$25, we use 0 to 25 as the horizontal span. The table below suggests a vertical span of 0 to 30. The graph is shown below. The horizontal axis corresponds to fixed order cost, and the vertical span is number of items per order.

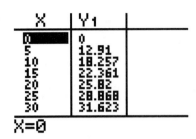 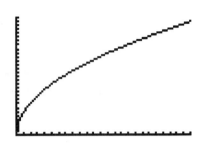

ii. If we evaluate the function above at 6 (using a table of values or the graph), we get that the number of items we need to order at a time is 14.14, or around 14.

iii. From the graph we can see that as the fixed order cost increases so does the number of items we need to order at a time.

(b) i. The exercise tells us that $N = 400$ and $c = 14$. So the function we need is

$$Q = \sqrt{\frac{2 \times 400 \times 14}{h}} = \sqrt{\frac{11200}{h}} \text{ items per order.}$$

Since $h$ ranges from 0 to 25 dollars, we use that for the horizontal span. We used the table of values below to get the vertical span. We used 0 to 150 to make the graph below. The horizontal axis corresponds to carrying cost, and the vertical axis shows number of items per order.

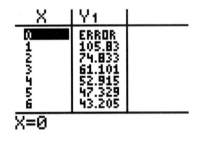 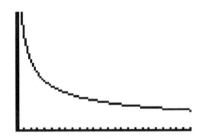

ii. If we evaluate the above function at 15 we get 27.33. So the optimum order size for a carrying cost of $15 is about 27 items.

iii. From the graph we can see that an increase in the carrying cost decreases the number of items ordered.

iv. Thinking of $Q$ as a function of $h$ alone and evaluating the function at 15 and 18, we see that $Q(15) = 27.325$ and $Q(18) = 24.944$. So the average rate of change from $15 to $18 is

$$\frac{\text{Change in number}}{\text{Change in carrying cost}} = \frac{Q(18) - Q(15)}{3} = -0.79 \text{ item per dollar.}$$

Here again we used three decimal places to ensure that our answer is still accurate to two decimal places after division.

v. The graph is concave up, and this tells us that the rate of decrease in the number of items ordered is decreasing as the carrying cost increases.

17. **An annuity**:

(a) The monthly interest rate is $r = 0.01$ and we want a monthly withdrawal of $M = \$200$, so the function is

$$P(t) = 200 \times \frac{1}{0.01} \times \left(1 - \frac{1}{(1 + 0.01)^t}\right) \text{ dollars.}$$

Note that $t$ is measured in months.

We use a horizontal span of 0 to 500 months and we chose a vertical span of 0 to 25,000 dollars from the table below. The horizontal axis is number of months of withdrawal, and the vertical axis is dollars invested.

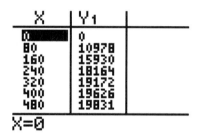
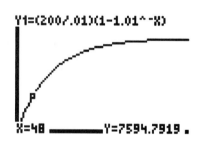

(b) We use the graph to evaluate the function at $t = 48$ (4 years is 48 months) and find that we need to invest \$7594.79 so that our child can withdraw \$200 per month for 4 years.

(c) We evaluate the function at $t = 120$ (10 years is 120 months) and find that we need to invest \$13,940.10 so that our child can withdraw \$200 per month for 10 years.

(d) The graph levels off and appears to have a limiting value. If we trace out toward the tail end of the graph, we see that an investment of \$20,000 will be sufficient to establish a \$200 per month perpetuity.

19. **Artificial gravity**:

(a)  i. We want an acceleration of $a = 9.8$ meters per second per second, so we use the function

$$N = \frac{30}{\pi} \times \sqrt{\frac{9.8}{r}} \text{ rotations per minute.}$$

ii. We use a horizontal span of 10 to 200 meters, and from the table below choose a vertical span of 0 to 10 revolutions per minute. The horizontal axis corresponds to the radius of the station, and the vertical axis to revolutions per minute.

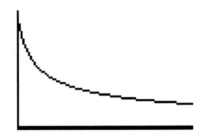

iii. As the distance $r$ increases, the number of rotations decreases, although more and more slowly. In practical terms, the larger the space station, the fewer rotations per minute are needed to produce artificial gravity.

iv. If we evaluate the function using the graph, or a table of values, we see that in order to produce Earth gravity for a space station of radius $r = 150$ meters, we need $N = 2.44$ rotations per minute.

(b)  i. Now we are looking at a space station of radius $r = 150$ meters. Then gravity as a function of revolutions per minute is given by

$$N = \frac{30}{\pi} \times \sqrt{\frac{a}{150}} \text{ revolutions per minute.}$$

ii. We want the horizontal span to be from 2.45 meters per second per second to 9.8 meters per second per second. The table of values below suggests a vertical span of 1 to 3 revolutions per minute. The horizontal axis shows acceleration due to gravity, and the vertical axis shows revolutions per second.

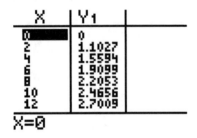
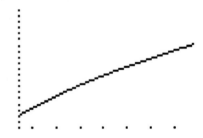

iii. As the desired acceleration increases, the number of rotations needed also increases, although more and more slowly.

21. **More on plant growth**:

   (a) i. We are assuming a rainfall of $R = 100$ millimeters, so our function, measured in kilograms per hectare, is

$$Y = -55.12 - 0.01535N - 0.00056N^2 + 3.946 \times 100$$
$$= 339.48 - 0.01535N - 0.00056N^2.$$

   ii. We use a horizontal span of 0 to 800 kilograms per hectare and from the table of values below, a vertical span of $-50$ to 400 kilograms per hectare. The horizontal axis corresponds to initial biomass and the vertical axis corresponds to growth in biomass.

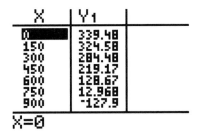  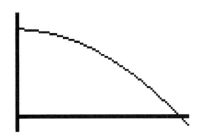

   iii. As the amount of initial plant biomass, $N$, increases, the amount of growth, $Y$, decreases. Practically, there is a limit to how much biomass the land can support, and each kilogram present leaves less room for more plants to grow.

   (b) i. We are now looking at a rainfall of $R = 80$ millimeters, so our function now, measured in kilograms per hectare, is

$$Y = -55.12 - 0.01535N - 0.00056N^2 + 3.946 \times 80 = 260.56 - 0.01535N - 0.00056N^2.$$

   ii. We add this new graph to the one above using the same horizontal and vertical spans as before.

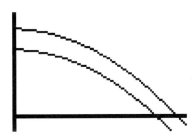

   iii. We see that lowered rainfall decreases biomass growth. Yes, this is quite consistent with the prediction made in the previous exercise.

23. **Magazine circulation**:

(a) We use a horizontal span from 0 to 6 years, and we used the table of values shown below to choose a vertical span from 0 to 55 thousand circulation. The horizontal axis is years since 2006, and the vertical axis is circulation in thousands.

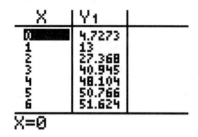 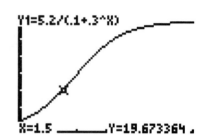

(b) Since 18 months is 1.5 years and $t$ is measured in years, $C(1.5)$ expresses in functional notation the circulation of the magazine 18 months after it was started. The value of $C(1.5)$ can be found using the graph, as shown in the figure above on the right. We see that 18 months after it started the magazine has a circulation of 19.67 thousand, or 19,670 magazines.

(c) The graph is concave up for about the first 2 years. (Answers will vary here.) In practical terms, the circulation was increasing more and more quickly during the first two years.

(d) The circulation was increasing the fastest where the graph is steepest. The graph appears to be steepest near $t = 2$, or after 2 years. (Answers will vary here but should be consistent with the answer to Part (c).)

(e) To see the limiting value for $C$, which is in practical terms the level of market saturation, we increase the horizontal span to 20 years and trace the tail end of the graph. We see from the graph below that the limiting value is about 52 thousand magazines. (This can also be seen by scanning down a table.)

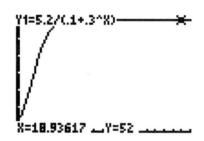

25. **Buffalo**:

(a) We show the graph with a horizontal span of 0 to 30 and a vertical span of 0 to 330.

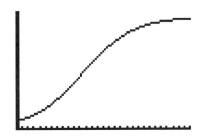

(b) The year 2002 corresponds to $t = 0$. From a table of values or the graph we find that the value of the function at $t = 0$ is 21. Thus there were 21 buffalo in the herd in 2002.

(c) We want to find what value of the variable $t$ gives the function value $N = 300$. This can be done by tracing the graph, and the value is between $t = 24$ and $t = 25$. Since $t$ is the number of years since 2002, the number of buffalo will first exceed 300 in the year 2026.

(d) Scrolling down a table on the calculator (or tracing the graph), we see that the limiting value of the function is 315. Thus the population will eventually increase to 315 buffalo.

(e) Tracing the graph shows that it is concave up from about $t = 0$ to about $t = 11$. It is concave down afterwards. This means that the herd is growing at an increasing rate from 2002 to 2013 and that the rate of growth begins to decrease after that time.

27. **Research project**: Answers will vary.

## Skill Building Exercises

S-1. **Graphs and function values**: To get the value of $f(3)$, we enter the function as $Y1 = 2 - X^2$ and use the standard view, that is, the window with a horizontal span from $-10$ to $10$ and the same vertical span. Locating the point on the graph corresponding to $X = 3$, we see from the graph below that $f(3) = -7$.

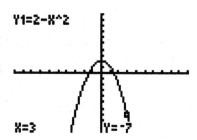

S-3. **Graphs and function values**: To get the value of $f(3)$, we enter the function as $Y1 = (X^2 + 2^X)/(X + 10)$ and use the standard view, that is, the window with a horizontal span from $-10$ to $10$ and the same vertical span. Locating the point on the graph corresponding to $X = 3$, we see from the graph below that $f(3) = 1.31$.

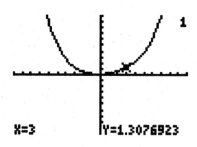

S-5. **Finding a window**: To find an appropriate window setup that will show a good graph of $y = \dfrac{x^3}{500}$ with a horizontal span of $-3$ to $3$, first we enter the function as $Y1 = (X^3)/500$, and then we make a table with a table starting value of $-3$ and a table increment value of $1$. Such a table is shown in the figure on the left below. The table of values shows that a vertical span from $-0.06$ to $0.06$ will display the graph, as shown on the right below.

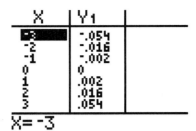

S-7. **Finding a window**: To find an appropriate window setup that will show a good graph of $y = \dfrac{x^4 + 1}{x^2 + 1}$ with a horizontal span of 0 to 300, first we enter the function as $Y1 = (X^4 + 1)/(X^2 + 1)$, and then we make a table with a table starting value of 0 and a table increment value of 50. Such a table is shown in the figure on the left below. The table of values shows that a vertical span from 0 to 90,000 will display the graph, as shown on the right below.

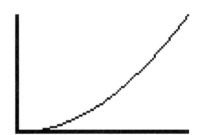

S-9. **Finding a window**: To find an appropriate window setup that will show a good graph of $y = \dfrac{1}{x^2 + 1}$ with a horizontal span of $-2$ to 2, first we enter the function as $Y1 = 1/(X^2 + 1)$, and then we make a table with a table starting value of $-2$ and a table increment value of 1. Such a table is shown in the figure on the left below. The table of values shows that a vertical span from 0 to 1 will display the graph, as shown on the right below.

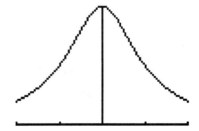

S-11. **Finding windows and making graphs**: To find an appropriate window setup that will show a good graph of $y = 1/x$ with a horizontal span of 1 to 10, first we enter the function as Y1 $= 1/X$, and then we make a table with a table starting value of 1 and a table increment value of 3. Such a table is shown in the figure on the left below. The table of values shows that a vertical span from 0 to 1 will display the graph, as shown on the right below.

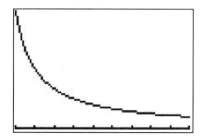

S-13. **Finding windows and making graphs**: To find an appropriate window setup that will show a good graph of $y = x^2 - x$ with a horizontal span of 0 to 10, first we enter the function as Y1 $= X^2 - X$, and then we make a table with a table starting value of 0 and a table increment value of 2. Such a table is shown in the figure on the left below. The table of values shows that a vertical span from 0 to 90 will display the graph, as shown on the right below.

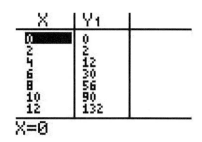
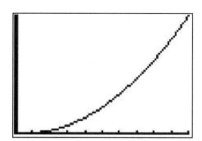

S-15. **Finding windows and making graphs**: To find an appropriate window setup that will show a good graph of $y = 3x/(50 + x)$ with a horizontal span of 0 to 1000, first we enter the function as Y1 $= 3X/(50 + X)$, and then we make a table with a table starting value of 0 and a table increment value of 20. Such a table is shown in the figure on the left below. The table of values shows that a vertical span from 0 to 2.5 will display the graph, as shown on the right below.

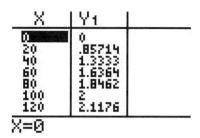 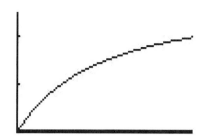

S-17. **Finding windows and making graphs**: To find an appropriate window setup that will show a good graph of $y = -180 + 100x - 4x^2$ with a horizontal span of 0 to 20, first we enter the function as $Y1 = -180 + 100X - 4X^2$, and then we make a table with a table starting value of 0 and a table increment value of 4. Such a table is shown in the figure on the left below. The table of values shows that a vertical span from $-300$ to $500$ will display the graph, as shown on the right below.

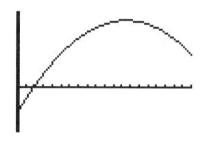

S-19. **Finding windows and making graphs**: To find an appropriate window setup that will show a good graph of $y = 120{,}000(e^{0.005} - 1)/(1 - e^{-0.06x})$ with a horizontal span of 5 to 30, first we enter the function as $Y1 = 120000(e^{(0.005)} - 1)/(1 - e^{(-0.06X)})$, and then we make a table with a table starting value of 5 and a table increment value of 5. Such a table is shown in the figure on the left below. The table of values shows that a vertical span from 0 to 2500 will display the graph, as shown on the right below.

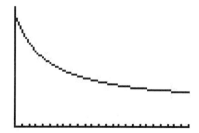

S-21. **Finding windows and making graphs**: To find an appropriate window setup that will show a good graph of $y = (30/\pi)\sqrt{9.8/x}$ with a horizontal span of 10 to 200, first we enter the function as $\mathsf{Y1} = (30/\pi)\sqrt{(9.8/X)}$, and then we make a table with a table starting value of 0 and a table increment value of 30. Such a table is shown in the figure on the left below. The table of values shows that a vertical span from 0 to 10 will display the graph, as shown on the right below.

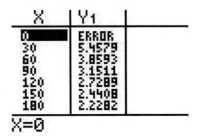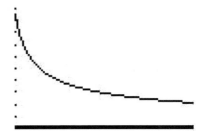

S-23. **Finding windows and making graphs**: To find an appropriate window setup that will show a good graph of $y = x/(1 + x^2)$ with a horizontal span of 0 to 20, first we enter the function as $\mathsf{Y1} = X/(1 + X^2)$, and then we make a table with a table starting value of 0 and a table increment value of 4. Such a table is shown in the figure on the left below. The table of values shows that a vertical span from 0 to 0.6 will display the graph, as shown on the right below.

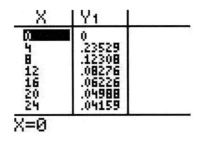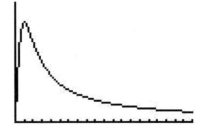

S-25. **Finding windows and making graphs**: To find an appropriate window setup that will show a good graph of $y = x/(1 + 1.1^x)$ with a horizontal span of 0 to 20, first we enter the function as $\mathsf{Y1} = X/(1 + 1.1^X)$, and then we make a table with a table starting value of 0 and a table increment value of 4. Such a table is shown in the figure on the left below. The table of values shows that a vertical span from 0 to 3 will display the graph, as shown on the right below.

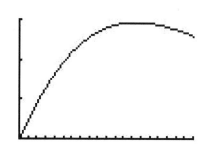

S-27. **Finding windows and making graphs**: To find an appropriate window setup that will show a good graph of $y = x/(\sqrt{x} + \sqrt{20 - x})^2$ with a horizontal span of 0 to 20, first we enter the function as $\mathsf{Y1} = X/(\sqrt{(X)} + \sqrt{(20 - X)})^2$, and then we make a table with a table starting value of 0 and a table increment value of 4. Such a table is shown in the figure on the left below. The table of values shows that a vertical span from 0 to 1.2 will display the graph, as shown on the right below.

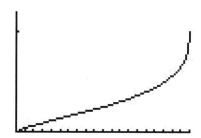

S-29. **Finding windows and making graphs**: To find an appropriate window setup that will show a good graph of $y = (x - 10)/(x + 10)$ with a horizontal span of 0 to 20, first we enter the function as $\mathsf{Y1} = (X - 10)/(X + 10)$, and then we make a table with a table starting value of 0 and a table increment value of 4. Such a table is shown in the figure on the left below. The table of values shows that a vertical span from $-1.2$ to 0.4 will display the graph, as shown on the right below.

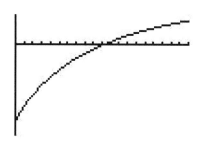

## 2.3  SOLVING LINEAR EQUATIONS

1. **Sweater sales**:

   (a) Initially there are 2000 sweaters on hand. Each week the number of sweaters decreases by 400. Therefore,

   $$\text{Number of sweaters remaining} = 2000 - 400 \times \text{Number of weeks}$$

   or $s = 2000 - 400w$.

   (b) Our inventory will be depleted when $s = 0$, so we want to solve the equation $2000 - 400w = 0$. We solve it as follows:

   $$
   \begin{aligned}
   2000 - 400w &= 0 \\
   -400w &= -2000 \qquad \textbf{Subtract 2000 from both sides.} \\
   w &= \frac{-2000}{-400} \qquad \textbf{Divide both sides by } -400. \\
   w &= 5.
   \end{aligned}
   $$

   Therefore, our inventory will be depleted after 5 weeks.

3. **Gross domestic product**:

   (a) Because $P$ is calculated as the sum of $C$, $I$, $G$, and $E$, we have $P = C + I + G + E$.

   (b) We are given that $C = 11{,}930$, $I = 2852$, $G = 3175$, and $E = -538$, all in billions of dollars.

       i. Recall that $E$ is exports minus imports. Because $E$ is negative, imports were larger.

       ii. From the given values we have

   $$P = 11{,}930 + 2852 + 3175 - 538 = 17{,}419,$$

   so the gross domestic product was 17,419 billion dollars.

   (c) We solve the equation $P = C + I + G + E$ for $E$ by subtracting $C + I + G$ from both sides. The result is $E = P - C - I - G$.

   (d) We are given the same values for $C$, $I$, and $G$ as in Part (b), and we know that $P = 18{,}184$ (in billions of dollars). The equation from Part (c) gives

   $$E = 18{,}184 - 11{,}930 - 2852 - 3175 = 227,$$

   so the net export of goods and services is 227 billion dollars.

5. **Break-even point**:

(a) Because

$$\text{Cost} \ = \ \text{Fixed costs} \ + \text{Variable cost} \ \times \ \text{Number of T-shirts,}$$

we have $C = 3500 + 8t$.

(b) Because

$$\text{Revenue} \ = \text{Selling price} \ \times \ \text{Number of T-shirts,}$$

we have $R = 10t$.

(c) Because the break-even point occurs when cost equals revenue, we want to find $t$ so that $C = R$. Thus, we want to solve the equation $3500 + 8t = 10t$. We solve it as follows:

$$
\begin{aligned}
3500 + 8t \ &= \ 10t \\
3500 \ &= \ 2t \qquad \textbf{Subtract } 8t \textbf{ from both sides.} \\
\frac{3500}{2} \ &= \ t \qquad \textbf{Divide both sides by } t. \\
1750 \ &= \ t.
\end{aligned}
$$

Therefore, the break-even point occurs when 1750 T-shirts are produced and sold in a month.

7. **Resale value**:

(a) The formula gives $V(3) = 12.5 - 1.1 \times 3 = 9.2$ thousand dollars. This means that the resale value of the boat will be $9.2$ thousand dollars, or $9200, at the start of the year 2014 (because that is 3 years after the start of 2011).

(b) We want to find what value of the variable $t$ gives the value 7 for the function $V$. This can be done either by examining a table of values (or a graph) for the function or by solving the linear equation $12.5 - 1.1t = 7$. We take the second approach:

$$
\begin{aligned}
12.5 - 1.1t \ &= \ 7 \\
-1.1t \ &= \ 7 - 12.5 \qquad \textbf{Subtract } 12.5 \textbf{ from both sides.} \\
t \ &= \ \frac{7 - 12.5}{-1.1} \qquad \textbf{Divide both sides by } -1.1. \\
t \ &= \ 5.
\end{aligned}
$$

Thus the resale value will be 7 thousand dollars at the start of the year 2016 (because that is 5 years after the start of 2011).

(c) We solve the equation $V = 12.5 - 1.1t$ for $t$ as follows:

$$
\begin{aligned}
V &= 12.5 - 1.1t \\
V - 12.5 &= -1.1t \qquad \textbf{Subtract } 12.5 \textbf{ from both sides.} \\
\frac{V - 12.5}{-1.1} &= t \qquad \textbf{Divide both sides by } -1.1.
\end{aligned}
$$

Thus the formula is $t = \dfrac{V - 12.5}{-1.1}$. This can also be written as $t = \dfrac{12.5 - V}{1.1}$ or as $t = 11.36 - 0.91V$.

(d) When we use the value $V = 4.8$ in the answer to Part (c), we obtain $t = \dfrac{4.8 - 12.5}{-1.1} = 7$. Thus the resale value will be 4.8 thousand dollars at the start of the year 2018 (because that is 7 years after the start of 2011).

9. **Gas mileage**:

(a) The exercise tells us that we have $g = 12$ gallons of gas in the tank and that our car gets $m = 24$ miles per gallon. So we can travel

$$
d = gm = 12 \times 24 = 288 \text{ miles.}
$$

(b) To solve for $m$, we divide both sides by $g$:

$$
\begin{aligned}
d &= gm \\
\frac{d}{g} &= m \text{ miles per gallon.}
\end{aligned}
$$

   i. This equation tells us that the mileage our car gets is the distance traveled divided by the amount of gas used.

   ii. If we drive $d = 335$ miles using $g = 13$ gallons of gas, then our car got

$$
m = \frac{d}{g} = \frac{335}{13} = 25.77 \text{ miles per gallon.}
$$

(c)  i. Since $d = 425$ miles, we have $425 = gm$, so $g = \dfrac{425}{m}$.

   ii. To choose a window size we need first to pick a reasonable range for gas mileage. We choose 0 to 30 for the horizontal span. The table below suggests a vertical span from 0 to 75. The horizontal axis corresponds to gas mileage, and the vertical axis corresponds to tank capacity.

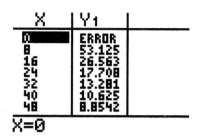
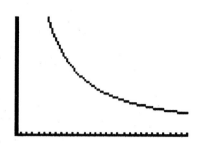

The graph is not a straight line.

11. **Supply and demand**: We have two equations for the price $P$: $P = 1.9S - 0.7$ and $P = 2.8 - 0.6D$. Combining them, since each equals $P$, we have

$$1.9S - 0.7 = 2.8 - 0.6D.$$

At the equilibrium price $D = S$, so we can substitute $S$ for $D$ on the right-hand side of the equation and solve for $S$. Here are the steps:

$$
\begin{aligned}
1.9S - 0.7 &= 2.8 - 0.6S \\
1.9S + 0.6S &= 2.8 + 0.7 \qquad \textbf{Add } 0.7 \textbf{ and } 0.6S \textbf{ to both sides.} \\
2.5S &= 3.5 \qquad \textbf{Combine terms.} \\
S &= \frac{3.5}{2.5} \qquad \textbf{Divide both sides by } 2.5. \\
S &= 1.4.
\end{aligned}
$$

Thus the equilibrium price occurs at the quantity 1.4 billion bushels; using the formula $P = 1.9S - 0.7$ shows that that price is $1.9 \times 1.4 - 0.7 = 1.96$ dollars per bushel.

13. **Fire engine pressure**:

(a) We are given that $NP = 80$ and $K = 0.51$. Because the length of the hose is 80 feet, we have $L = \dfrac{80}{50} = 1.6$. Thus $EP = 80(1.1 + 0.51 \times 1.6) = 153.28$. The engine pressure is 153.28 psi.

(b) We are given that $K = 0.73$ and $EP = 150$. Because the length of the hose is 45 feet, we have $L = \dfrac{45}{50} = 0.9$. Thus $150 = NP(1.1 + 0.73 \times 0.9)$. Dividing both sides by $1.1 + 0.73 \times 0.9$ gives $NP = \dfrac{150}{1.1 + 0.73 \times 0.9} = 85.37$. The nozzle pressure is 85.37 psi.

(c) To solve the equation $EP = NP(1.1 + K \times L)$ for $NP$, we divide both sides by $1.1 + K \times L$. The result is $NP = \dfrac{EP}{1.1 + K \times L}$.

(d) We are given that $EP = 160$ and $NP = 90$. Because the length of the hose is 190 feet, we have $L = \dfrac{190}{50} = 3.8$. Thus $160 = 90(1.1 + K \times 3.8)$. We solve for $K$ as follows:

$$
\begin{aligned}
160 &= 90(1.1 + 3.8K) \\
\frac{160}{90} &= 1.1 + 3.8K \qquad \textbf{Divide both sides by } 90. \\
\frac{160}{90} - 1.1 &= 3.8K \qquad \textbf{Subtract 1.1 from both sides.} \\
\frac{\frac{160}{90} - 1.1}{3.8} &= K \qquad \textbf{Divide both sides by } 3.8. \\
0.18 &= K.
\end{aligned}
$$

Thus the "K" factor is $0.18$.

(e) We solve for $K$ as follows:

$$
\begin{aligned}
EP &= NP(1.1 + K \times L) \\
\frac{EP}{NP} &= 1.1 + K \times L \qquad \textbf{Divide both sides by } NP. \\
\frac{EP}{NP} - 1.1 &= K \times L \qquad \textbf{Subtract 1.1 from both sides.} \\
\frac{\frac{EP}{NP} - 1.1}{L} &= K \qquad \textbf{Divide both sides by } L.
\end{aligned}
$$

Thus $K = \dfrac{\frac{EP}{NP} - 1.1}{L}$.

15. **Net profit**:

(a) Our gross profit is the number of dolls sold times the selling price. Then we have to subtract the cost of making the dolls, which is the number of dolls sold times the cost of a doll. We also have to subtract the cost of rent. We find that

$$
\begin{aligned}
\text{Net profit} &= \text{Selling price} \times \#\,\text{sold} - \text{Cost of a doll} \times \#\,\text{sold} - \text{Rent} \\
p &= dn - cn - R.
\end{aligned}
$$

This can also be written as $P = (d - c)n - R$.

(b) The exercise tells us that the rent is $R = \$1280$, it costs $c = \$2$ each to make the dolls, the selling price is $d = \$6.85$, and we sell $n = 826$ dolls. So our net profit would be

$$
p = dn - cn - R = 6.85 \times 826 - 2 \times 826 - 1280 = \$2726.10.
$$

(c) We have

$$p = dn - cn - R$$

$$p + cn = dn - R \qquad \textbf{Add } cn \textbf{ to both sides.}$$

$$p + cn + R = dn \qquad \textbf{Add } R \textbf{ to both sides.}$$

$$\frac{p + cn + R}{n} = d \qquad \textbf{Divide both sides by } n.$$

Thus the formula is $d = \dfrac{p + cn + R}{n}$. This can also be written as $d = \dfrac{p + R}{n} + c$.

(d) The required net profit is $p = \$4000$, the rent is $R = \$1200$, the cost per doll is $c = \$2$, and we expect to sell $n = 700$ dolls. So under these conditions the price we need to get for each doll is

$$d = \frac{p + cn + R}{n} = \frac{4000 + 700 \times 2 + 1200}{700} = \$9.43.$$

17. **Temperature conversion**:

(a) Here $K(30)$ is the temperature in kelvins when the temperature on the Celsius scale is 30 degrees. The value is $30 + 273.15 = 303.15$ kelvins.

(b) To express the Celsius temperature $C$ in terms of kelvins $K$ we solve the given equation relating the two:

$$K = C + 273.15$$

$$K - 273.15 = C \quad \textbf{Subtract 273.15 from each side.}$$

Thus we have

$$C = K - 273.15.$$

(c) We use the expression from Part (b) to replace $C$ in $F = 1.8C + 32$. The result is

$$F = 1.8(K - 273.15) + 32.$$

This can also be written as $F = 1.8K - 459.67$.

(d) If the temperature is $K = 310$ kelvins, then by Part (c) the Fahrenheit temperature is

$$F = 1.8(310 - 273.15) + 32 = 98.33 \text{ degrees.}$$

19. **Running ants**:

(a) The ambient temperature is $T = 30$ degrees Celsius, so in functional notation $S(30)$ is the speed of the ants. This is calculated as

$$S(30) = 0.2 \times 30 - 2.7 = 3.3 \text{ centimeters per second.}$$

(b) We have

$$S = 0.2T - 2.7$$

$$S + 2.7 = 0.2T \quad \textbf{Add 2.7 to each side.}$$

$$\frac{S + 2.7}{0.2} = T \quad \textbf{Divide each side by 0.2.}$$

Thus the formula is $T = \dfrac{S + 2.7}{0.2}$.

(c) The speed is $S = 3$ centimeters per second, so the ambient temperature is

$$T = \frac{S + 2.7}{0.2} = \frac{3 + 2.7}{0.2} = 28.5 \text{ degrees Celsius.}$$

21. **Profit**:

(a) Assume that the total cost $C$ is in dollars. Because

$$\text{Total cost } = \text{ Variable cost } \times \text{ Number of items } + \text{ Fixed costs,}$$

we have $C = 55N + 200$.

(b) Assume that the total revenue $R$ is in dollars. Because

$$\text{Total revenue } = \text{Selling price } \times \text{ Number of items,}$$

we have $R = 58N$.

(c) Assume that the profit $P$ is in dollars. Because

$$\text{Profit } = \text{ Total revenue } - \text{ Total cost,}$$

we have $P = R - C$. Using the formulas from Parts (a) and (b), we obtain

$$P = 58N - (55N + 200).$$

Because

$$58N - (55N + 200) = 58N - 55N - 200 = 3N - 200,$$

this can also be written as $P = 3N - 200$. The latter is the formula we use in Part (d).

(d) We want to find what value of the variable $N$ gives the function value $P = 0$. Thus we need to solve the equation $3N - 200 = 0$. We solve it as follows:

$$3N - 200 = 0$$

$$3N = 200 \quad \textbf{Add 200 to both sides.}$$

$$N = \frac{200}{3} \quad \textbf{Divide both sides by 3.}$$

$$N = 66.67.$$

Thus the break-even point occurs at a production level of about 66.67 thousand widgets per month. This is 66,670 widgets per month.

23. **Plant growth**:

(a) We have

$$Y = -55.12 - 0.01535N - 0.00056N^2 + 3.946R$$

$Y + 55.12 = -0.01535N - 0.00056N^2 + 3.946R$ **Add 55.12 to each side.**

$Y + 55.12 + 0.01535N = -0.00056N^2 + 3.946R$ **Add $0.01535N$ to each side.**

$Y + 55.12 + 0.01535N + 0.00056N^2 = 3.946R$ **Add $0.00056N^2$ to each side.**

$$\frac{Y + 55.12 + 0.01535N + 0.00056N^2}{3.946} = R$$ **Divide both sides by 3.946.**

Thus the formula is

$$R = \frac{Y + 55.12 + 0.01535N + 0.00056N^2}{3.946}.$$

(b) If there is no plant growth, then $Y = 0$, and so

$$R = \frac{55.12 + 0.01535N + 0.00056N^2}{3.946} \text{ millimeters.}$$

(c) We use a horizontal span of 0 to 800 kilograms per hectare. Using the table below, we choose a vertical span of 0 to 120 millimeters. The horizontal axis corresponds to initial plant biomass, and the vertical axis corresponds to rainfall.

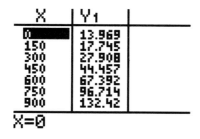

(d) As $N$ increases, $R$ also increases. In practical terms, the greater the initial plant biomass, the more rainfall it needs just to maintain its size.

(e) If the initial plant biomass is $N = 400$ kilograms per hectare, then we can evaluate the function using the graph to get $R = 38.23$ millimeters.

(f) If the initial plant biomass $N$ is 500 kilograms per hectare, then, by using the graph, we find that the corresponding zero isocline rainfall $R$ needed to sustain that biomass is $R(500) = 51.39$ millimeters of rain. Unfortunately, there are only 40 millimeters of rain, so the plants will die back.

25. **Competition between populations:**

(a) We need to solve the equation $5(1 - m - n) = 0$ for $n$. We solve it as follows:

$$\begin{aligned} 5(1 - m - n) &= 0 \\ 1 - m - n &= 0 \qquad \textbf{Divide both sides by } 5. \\ 1 - m &= n \qquad \textbf{Add } n \textbf{ to both sides.} \end{aligned}$$

Thus $n = 1 - m$ describes this isocline.

(b) We need to solve the equation $6(1 - 0.7m - 1.2n) = 0$ for $n$. We solve it as follows:

$$\begin{aligned} 6(1 - 0.7m - 1.2n) &= 0 \\ 1 - 0.7m - 1.2n &= 0 \qquad \textbf{Divide both sides by } 6. \\ 1 - 0.7m &= 1.2n \qquad \textbf{Add } 1.2n \textbf{ to both sides.} \\ \frac{1 - 0.7m}{1.2} &= n \qquad \textbf{Divide both sides by } 1.2. \end{aligned}$$

Thus $n = \dfrac{1 - 0.7m}{1.2}$ describes this isocline.

(c) By Parts (a) and (b), the per capita growth rates will both be zero if $n = 1 - m$ and $n = \dfrac{1 - 0.7m}{1.2}$. This gives the equation $1 - m = \dfrac{1 - 0.7m}{1.2}$. We solve this equation as follows:

$$\begin{aligned} 1 - m &= \frac{1 - 0.7m}{1.2} \\ 1.2(1 - m) &= 1 - 0.7m \qquad \textbf{Multiply both sides by } 1.2. \\ 1.2 - 1.2m &= 1 - 0.7m \qquad \textbf{Expand.} \\ 1.2 - 1 &= -0.7m + 1.2m \qquad \textbf{Add } 1.2m \textbf{ and subtract } 1 \textbf{ on both sides.} \\ 0.2 &= 0.5m \qquad \textbf{Combine terms.} \\ \frac{0.2}{0.5} &= m \qquad \textbf{Divide both sides by } 0.5. \\ 0.4 &= m. \end{aligned}$$

Thus $m = 0.4$, and from the equation in Part (a) we find that $n = 1 - m = 1 - 0.4 = 0.6$. Hence the equilibrium point occurs at $m = 0.4, n = 0.6$ thousand animals.

27. **Research project:** Answers will vary.

## Skill Building Exercises

**S-1. Identifying linear equations**: This equation is not linear because of the term $x^2$.

**S-3. Identifying linear equations**: This equation is not linear because of the term $\dfrac{1}{x}$ (or because of the term $\dfrac{1}{x+1}$).

**S-5. Identifying linear equations**: This equation is linear because when the variable $x$ appears it is only divided by a constant.

**S-7. Solving linear equations**: To solve $3x + 7 = x + 21$ for $x$, first we gather the terms involving the variable to the left side of the equation and those not involving the variable to the right side, then we factor out the variable, and then finally we divide by the coefficient of the variable.

$$
\begin{aligned}
3x + 7 &= x + 21 \\
3x - x &= 21 - 7 \\
2x &= 14 \\
x &= 7.
\end{aligned}
$$

**S-9. Solving linear equations**: To solve $3 - 5x = 23 + 4x$ for $x$, we follow the procedure described above, just as for Exercise S-7:

$$
\begin{aligned}
3 - 5x &= 23 + 4x \\
-5x - 4x &= 23 - 3 \\
-9x &= 20 \\
x &= \frac{20}{-9} = -\frac{20}{9}.
\end{aligned}
$$

**S-11. Solving linear equations**: To solve $12x + 4 = 55x + 42$ for $x$, we follow the procedure described above, just as for Exercise S-7:

$$
\begin{aligned}
12x + 4 &= 55x + 42 \\
12x - 55x &= 42 - 4 \\
-43x &= 38 \\
x &= \frac{38}{-43} = -\frac{38}{43}.
\end{aligned}
$$

**S-13. Solving linear equations**: To solve for $cx + d = 12$ for $x$, we follow the procedure described above, just as for Exercise S-7:

$$
\begin{aligned}
cx + d &= 12 \\
cx &= 12 - d \\
x &= \frac{12 - d}{c}.
\end{aligned}
$$

**S-15. Solving linear equations**: To solve $2k + m = 5k + n$ for $k$, we follow the procedure described above, just as for Exercise S-7, only in this case the variable is named $k$ instead of $x$:

$$
\begin{aligned}
2k + m &= 5k + n \\
2k - 5k &= n - m \\
-3k &= n - m \\
k &= \frac{n - m}{-3} = -\frac{n - m}{3}.
\end{aligned}
$$

This expression can also be written as $k = \dfrac{m - n}{3}$.

**S-17. Reversing roles of variables**: To reverse the roles of $F$ and $C$ in the equation, we solve $F = (9/5)C + 32$ for $C$. To do so, first we gather the terms involving $C$ to the left side of the equation and those not involving $C$ to the right side, then we factor out $C$, and then finally we divide by the coefficient of $C$. It is convenient to begin by reversing the sides of the equation:

$$
\begin{aligned}
(9/5)C + 32 &= F \\
(9/5)C &= F - 32 \\
C &= (5/9)(F - 32).
\end{aligned}
$$

**S-19. Reversing roles of variables**: To reverse the roles of $T$ and $S$ in the equation, proceed just as in Exercise S-17, only in this case the variable is named $S$:

$$
\begin{aligned}
5S + 15 &= T \\
5S &= T - 15 \\
S &= \frac{T - 15}{5}.
\end{aligned}
$$

**S-21. Reversing roles of variables**: To reverse the roles of $w$ and $v$ in the equation, proceed just as in Exercise S-17, only in this case the variable is named $v$:

$$
\begin{aligned}
uv + t &= w \\
uv &= w - t \\
v &= \frac{w - t}{u}.
\end{aligned}
$$

**S-23. Reversing roles of variables**: To reverse the roles of $a$ and $b$ in the equation, proceed just as in Exercise S-17, only in this case the variable is named $b$:

$$
\begin{aligned}
cb + db - e &= a \\
cb + db &= a + e \\
(c + d)b &= a + e \\
b &= \frac{a + e}{c + d}.
\end{aligned}
$$

**S-25. Reversing roles of variables**: To reverse the roles of $a$ and $b$ in the equation, proceed just as in Exercise S-17, only in this case the variable is named $b$:

$$
\begin{aligned}
bc + c &= a \\
bc &= a - c \\
b &= \frac{a - c}{c}.
\end{aligned}
$$

**S-27. Reversing roles of variables**: To reverse the roles of $a$ and $b$ in the equation, proceed just as in Exercise S-17, only in this case the variable is named $b$:

$$
\begin{aligned}
bc + ba &= a \\
b(c + a) &= a \\
b &= \frac{a}{c + a}.
\end{aligned}
$$

**S-29. Reversing roles of variables**: To reverse the roles of $a$ and $b$ in the equation, proceed just as in Exercise S-17, only in this case the variable is named $b$:

$$
\begin{aligned}
bc + bd &= a \\
b(c + d) &= a \\
b &= \frac{a}{c + d}.
\end{aligned}
$$

## 2.4 SOLVING NONLINEAR EQUATIONS

**PLEASE NOTE:** The exercises in this section involve solving nonlinear equations. There are many ways to do this, so it is important that students, graders, and instructors be aware of the various ways to solve equations. For example, some people prefer to use tables and get better and better approximations by refining the table; some will graph and zoom in repeatedly; other will use two graphs and find their intersection; others may graph and trace; and still others will rearrange the expression and find the appropriate root of an equation.

1. **Doubling time**:

    (a) The initial balance of the investment is the value of $B$ when $t = 0$. That value is $B(0) = 5000 \times 1.0075^0 = 5000$ dollars. Therefore, the initial balance was \$5000.

    (b) The balance doubles when $B(t)$ takes the value $2 \times 5000 = 10{,}000$ dollars. Thus, we want to solve the equation $5000 \times 1.0075^t = 10{,}000$. We do this using the crossing-graphs method. After consulting a table of values, we use a horizontal span of 0 to 100 months and a vertical span of 0 to 11,000 dollars. In the figure on the left below, we have graphed the balance function along with the constant function 10,000 (thick line). We see that they intersect when $t = 93$ (rounded to the nearest whole number of months). Thus, it takes about 93 months for the balance to double.

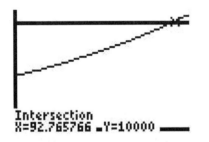

```
Intersection
X=92.765766  Y=10000 ▬▬▬
```

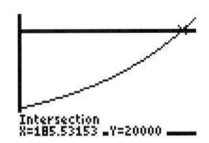

```
Intersection
X=185.53153  Y=20000 ▬▬▬
```

    (c) The balance doubles again when $B(t)$ takes the value 20,000 dollars. Thus, we want to solve the equation $5000 \times 1.0075^t = 20{,}000$. We do this using the crossing-graphs method. After consulting a table of values, we use a horizontal span of 0 to 200 months and a vertical span of 0 to 23,000 dollars. In the figure on the right above, we have graphed the balance function along with the constant function 20,000 (thick line). We see that they intersect when $t = 185.53153\ldots$ To find the answer, we subtract from this number the answer to part b:

$$185.53153\ldots - 93$$

equals 93 (rounded to the nearest whole number of months). Thus, it takes about 93 months for the balance to double again, the same time it took to double in the first place.

3. **The economic order quantity model**:

(a) Because the cost for each order is $c = 30$ dollars and the weekly holding cost is $h = 8$ dollars, the function we want is $Q = \sqrt{\dfrac{2 \times 30N}{8}}$.

(b) We use a horizontal span of 0 to 50 bikes per week and, after consulting a table of values, we use a vertical span of 0 to 25. In the figure on the left below, we have graphed the function.

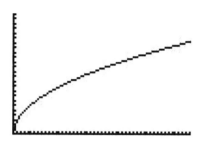
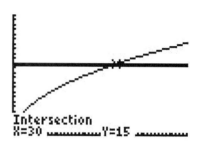

(c) We want to solve the equation $\sqrt{\dfrac{2 \times 30N}{8}} = 15$. We do this using the crossing-graphs method. The figure on the right above shows that the graph from part b intersects with the graph of the constant function 15 when $N = 30$. Thus, selling 30 bikes per week corresponds to an optimal order quantity of 15 bikes per week.

(d) Because you are to sell 30 bikes per week and order 15 bikes per week, you must place 2 orders per week to maintain your inventory.

5. **Engine displacement**:

(a) This engine has 4 cylinders, so $n = 4$. The bore is $B = 3.92$ inches, and the stroke is $S = 3.11$ inches. Therefore, the displacement for this engine is

$$D = 4\pi \times 3.11(3.92/2)^2 = 150.14 \text{ cubic inches.}$$

(b) This engine has $n = 6$ cylinders and a stroke of $S = 3.01$ inches. Therefore, the displacement for this engine as a function of the bore is

$$D = 6\pi \times 3.01(B/2)^2.$$

We want to find the bore $B$ so that the total displacement $D$ is 175 cubic inches. That is, we want to solve the equation

$$6\pi \times 3.01(B/2)^2 = 175.$$

We do this using the crossing-graphs method. After consulting a table of values, we use a horizontal span of 0 to 4 inches and a vertical span of 0 to 250 cubic inches. In the figure below, we have graphed the displacement function along with the constant function 175 (thick line). We see that they intersect when $B = 3.51$. Thus the proper bore for this engine is 3.51 inches.

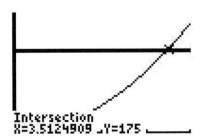

7. **Admiralty Coefficient**: We are assuming that the horsepower is $H = 6000$ and that the displacement is $D = 14{,}800$ tons, so the formula for $A$ as a function of $S$ is

$$A = \frac{14{,}800^{2/3} S^3}{6000}.$$

We want to find the value of $S$ so that $A = 427$. That is, we want to solve the equation

$$\frac{14{,}800^{2/3} S^3}{6000} = 427.$$

We do this using the crossing-graphs method. After consulting a table of values, we use a horizontal span of 0 to 17 knots and a vertical span of 0 to 500. In the figure below, we have graphed $A$ along with the constant function 427 (thick line). We see that they intersect when $S = 16.20$ knots. Thus the speed of a typical Victory Ship was 16.20 knots.

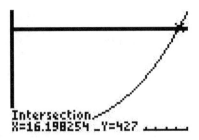

9. **A population of foxes**:

(a) The number of foxes introduced is the value of $N$ when $t$ is zero:

$$N(0) = \frac{37.5}{0.25 + 0.76^0} = 30 \text{ foxes.}$$

(b) The exercise gave no information about the horizontal span. We have chosen to look at the fox population over a 25-year period. The following table of values led us to choose a vertical span of 0 to 160 foxes. The horizontal axis is time in years, and the vertical axis is number of foxes.

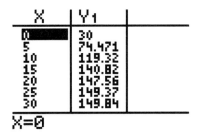 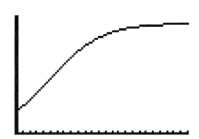

The population of foxes grows rapidly at first, and then the growth slows down until there is almost no growth at all.

(c) The fox population reaches 100 individuals when $N(t) = 100$. Thus we need to solve the equation

$$\frac{37.5}{0.25 + 0.76^t} = 100$$

for $t$. We do this using the crossing graphs method. In the figure below we have added the graph of 100 (thick line) and calculated the crossing point. We read from the bottom of the screen that the fox population will reach 100 individuals after 7.58 years.

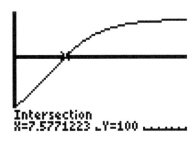

11. **Revisiting the floating ball**: We need to make a graph with a horizontal span of $-2$ to 7. We used the table of values on the left below to choose a vertical span of $-3300$ to 2100. The graph is on the right below.

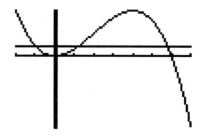

In the figures below we have found the other two solutions, $d = -0.98$ foot and $d = 5.80$ feet. The negative solution would have the ball floating above the surface of the water. The positive solution is larger than the diameter of the ball, and so the ball would be completely submerged.

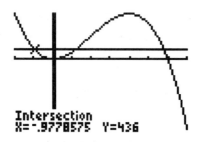

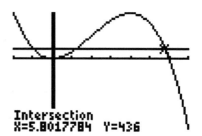

13. **Falling with a parachute**: The man has fallen 140 feet when

$$12.5(0.2^t - 1) + 20t = 140.$$

Thus we need to solve this equation. We use the crossing graphs method. Even with a parachute, it doesn't take long to fall 140 feet—surely less than 20 seconds. Thus we take a horizontal span of 0 to 20. The distance he falls increases from 0 up to 140, and so we use a slightly larger vertical span of 0 to 160. Below we have graphed $12.5(0.2^t - 1) + 20t$, and we have added the graph of 140 (thick line). We get the crossing point at $t = 7.62$. Thus the man falls 140 feet in 7.62 seconds.

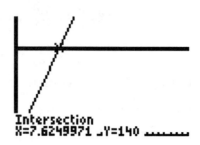

15. **Reaction rates**:

(a) The exercise suggests a horizontal span of 0 to 100. The table of values on the left below led us to choose a vertical span of $-5$ to 25. The graph is on the right below.

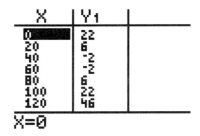 

(b) The reaction rate when the concentration is 15 moles per cubic meter is represented by $R(15)$ in functional notation. Using a table of values or the graph gives that the value is 9.25 moles per cubic meter per second.

(c) We want to find what two values of the variable $x$ give the value 0 for the function $R$. That is, we want to find the two solutions of the equation $0.01x^2 - x + 22 = 0$. We do this using the single-graph method. As the graphs below indicate, the solutions are $x = 32.68$ and $x = 67.32$. Thus the reaction is in equilibrium at the concentrations of 32.68 moles per cubic meter and 67.32 moles per cubic meter.

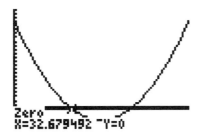 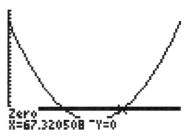

17. **Van der Waals equation**: The pressure is $p = 100$ atm, and the temperature is $T = 500$ kelvins. Putting these values into the equation, we have

$$100 = \frac{0.082 \times 500}{V - 0.043} - \frac{3.592}{V^2}.$$

We want to know the value of $V$. That is, we need to solve the equation above.

For the graph of $\dfrac{0.082 \times 500}{V - 0.043} - \dfrac{3.592}{V^2}$ the exercise suggests we use a horizontal span from 0.1 to 1 liter. The table of values below leads us to choose a vertical span of 0 to 400 atm. We have added the graph of 100 (thick line) to the picture and calculated the intersection point at $V = 0.37$. Thus one mole of carbon dioxide at this temperature and pressure will occupy a volume of 0.37 liter.

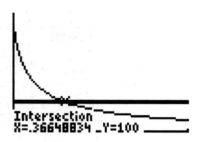

19. **Radiocarbon dating**: Since the half-life of carbon-14 is $H = 5.77$ thousand years, the formula that gives the fraction of the original amount present is

$$\frac{A}{A_0} = 0.5^{t/5.77}.$$

(Here $t$ is measured in thousands of years.) We want to know then this value is $\frac{1}{3}$. That is, we want to solve the equation

$$\frac{1}{3} = 0.5^{t/5.77}.$$

We use the crossing graphs method to do that. Since half of the carbon-14 will be gone in 5.77 thousand years, and half of that will be gone in another 5.77 thousand years, we know that there will be only a quarter of the original amount of carbon-14 left after $2 \times 5.77 = 11.54$ thousand years. Thus the fraction reaches $\frac{1}{3}$ sometime before 12 thousand years. Hence to graph $\frac{A}{A_0}$ we use a horizontal span of 0 to 12. The fraction starts at 1 and decreases. Thus we use a vertical span of 0 to 1. In the picture below, we have added the graph of $\frac{1}{3}$ and calculated the intersection point at $t = 9.15$. Thus the tree died about 9.15 thousand years ago.

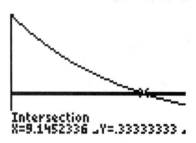

21. **Grazing kangaroos**:

    (a)    i. The exercise suggests a horizontal span of 0 to 2000 pounds per acre. The table of values below for $G$ led us to choose a vertical span of $-1$ to 3. The graph shows available vegetation biomass on the horizontal axis and daily intake on the vertical axis.

ii. The graph is concave down. Since the graph is increasing, the kangaroo eats more when the vegetation biomass is greater. Small changes in $V$ at the lower biomass levels cause the kangaroo to eat much more, while small changes in $V$ at higher levels do not change the eating habits of the kangaroo very much.

iii. The minimal vegetation biomass is the amount that will cause the kangaroo to consume zero pounds. Thus we want to find the value of $V$ where the graph crosses the horizontal axis. In the left-hand picture below, we have calculated where the graph crosses the horizontal axis as $V = 163.08$ pounds per acre.

iv. The satiation level is the limit of the amount the kangaroo will eat. That is where the curve levels off and becomes horizontal. Tracing toward the end of the graph, we see that $G$ values approach 2.5. So the satiation level for the western grey kangaroo is a daily intake of about 2.5 pounds.

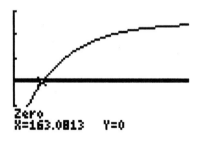

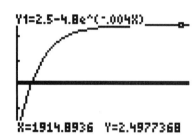

(b)   i. In the figure below, we have added the graph of $R$.

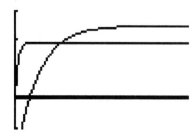

ii. The red kangaroo has a minimal vegetation biomass level of $V = 0$ pounds per

acre, since that is where the second graph crosses the horizontal axis. Consequently, the red kangaroo is more efficient at grazing than the grey kangaroo.

23. **Growth rate**:

(a) As suggested in the exercise, we use a horizontal span of 0 to 1000 pounds per
acre. The table below leads us to choose a vertical span of $-1$ to $1$. The horizontal
axis on the graph below corresponds to vegetation biomass and the vertical axis
to the per capita growth rate.

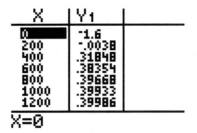 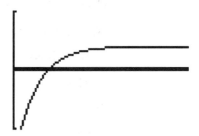

(b) The per capita growth rate is zero where the graph crosses the horizontal axis. This
point is calculated in the graph below and shows that $r = 0$ when $V = 201.18$. The
population size is stable when the vegetation level is 201.18 pounds per acre.

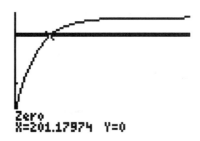

25. **Breaking even**:

(a) Assume that $C$ is measured in dollars. Because

$$\text{Total cost} = \text{Variable cost} \times \text{Number} + \text{Fixed costs,}$$

we have

$$C = 65N + 700.$$

(b) Assume that $R$ is measured in dollars. Because

$$\text{Total revenue} = \text{Price} \times \text{Number,}$$

we have $R = pN$, and thus

$$R = (75 - 0.02N)N.$$

(c) Assume that $P$ is measured in dollars. Because

$$\text{Profit} = \text{Total revenue} - \text{Total cost,}$$

we have $P = R - C$. Using Parts (a) and (b) gives

$$P = (75 - 0.02N)N - (65N + 700).$$

(This can also be written as $P = -0.02n^2 + 10N - 700$.)

(d) We want to find what two values of the variable $N$ give the value 0 for the function $P$. That is, we want to find the two solutions of the equation $(75 - 0.02N)N - (65N + 700) = 0$. We do this using the single-graph method. As the graphs below indicate, the solutions are $N = 84.17$ and $N = 415.83$. (We used a horizontal span of 0 to 500, as suggested in the exercise, and a vertical span of $-800$ to 650.) Thus the two break-even points are 84.17 thousand widgets per month and 415.83 thousand widgets per month.

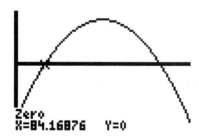
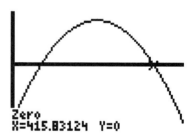

27. **Water flea**: We are given that $N_0 = 50$, so the relation is

$$e^{0.44t} = \frac{N}{50}\left(\frac{228 - 50}{228 - N}\right)^{4.46}.$$

We want to find what value of $t$ corresponds to the value 125 for $N$. That is, we want to find the solution of the equation

$$e^{0.44t} = \frac{125}{50}\left(\frac{228 - 50}{228 - 125}\right)^{4.46}.$$

Note that the right-hand side of this equation is a constant. We solve this equation using the crossing-graphs method, putting the function $e^{0.44t}$ in Y1 and the constant $\frac{125}{50}\left(\frac{228 - 50}{228 - 125}\right)^{4.46}$ in Y2. As the graph below indicates, the solution is $t = 7.63$. (After examining a table of values, we used a horizontal span of 0 to 8 and a vertical span of 0 to 40.) Thus it takes 7.63 days for the population to grow to 125.

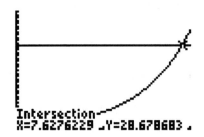

29. **Radius of a shock wave:**

(a) We are given that $E = 10^{15}$, and we want to find what value of the variable $t$ gives the value 4000 for the function $R$. That is, we want to find the solution of the equation $4.16(10^{15})^{0.2}t^{0.4} = 4000$. We do this using the crossing-graphs method. As the graph on the left below indicates, the solution is $t = 0.91$. (From the statement about the period over which the relation is valid, we expect the time to be around 1 second or less. After examining a table of values, we used a horizontal span of 0 to 1 and a vertical span of 0 to 4500.) Thus 0.91 second is required for the shock wave to reach a point 40 meters away.

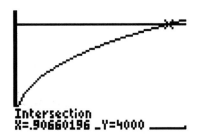
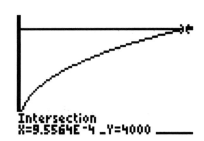

(b) We are given that $E = 9 \times 10^{20}$, and we want to find what value of the variable $t$ gives the value 4000 for the function $R$. That is, we want to find the solution of the equation $4.16(9 \times 10^{20})^{0.2}t^{0.4} = 4000$. We do this using the crossing-graphs method. As the graph on the right above indicates, the solution is $t = 9.6 \times 10^{-4} = 0.00096$. (Because the energy is much greater here than in Part (a), we expect the time required to be much smaller. After examining a table of values, we used a horizontal span of 0 to 0.001 and a vertical span of 0 to 4500.) Thus 0.00096 second is required for the shock wave to reach a point 40 meters away.

(c) We are given that $t = 1.2$, and we want to find what value of the variable $E$ gives the value 5000 for the function $R$. That is, we want to find the solution of the equation $4.16E^{0.2}1.2^{0.4} = 5000$. We do this using the crossing-graphs method. As the graph below indicates, the solution is $E = 1.74 \times 10^{15}$. (Because the time required in Part (a) is close to the time required here, we expect the energy to be on the order of $10^{15}$ here also. After examining a table of values, we used a horizontal span of 0 to $2 \times 10^{15}$ and a vertical span of 0 to 5500.) Thus an energy of $1.74 \times 10^{15}$ ergs was released by the explosion.

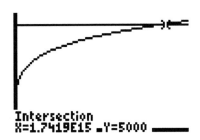

31. **Home equity**: Our home equity formula with $P = 425{,}000$, $r = 0.09/12 = 0.0075$, and $t = 30 \times 12 = 360$ months becomes

$$E = 425{,}000 \times \frac{(1 + 0.0075)^k - 1}{(1 + 0.0075)^{360} - 1}.$$

We wish to solve this for $k$ when $E = \dfrac{425{,}000}{2} = 212{,}500$. Using crossing-graphs as shown in the graph below (using a horizontal span of 0 to 360 and a vertical span of 0 to 425,000.)

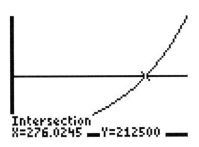

we see that the equity is half the original mortgage after 276 months, or 23 years.

## Skill Building Exercises

S-1. **An investment**: We want to find for what $t$ we have $B = 2000$, so we want to solve the equation $2000 = 500e^{0.05t}$.

S-3. **Argon-39**: The initial amount of argon-39 is 10 grams, so half of the initial amount is 5. Thus, we want to find for what $t$ we have $A = 5$. This means that we want to solve the equation $5 = 10 \times 0.5^{t/269}$.

S-5. **The crossing-graphs method**: To solve $\dfrac{20}{1+2^x} = x$ using the crossing-graphs method, we enter the $\dfrac{20}{1+2^X}$ as Y1 and $X$ as Y2. Using the standard viewing window, we obtain the graph below, which shows that the solution is $x = 2.69$.

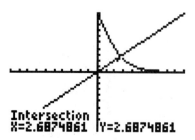

S-7. **The crossing-graphs method**: To solve $3^x + x = 2^x + 1$ using the crossing-graphs method, we enter the $3^X + X$ as Y1 and $2^X + 1$ as Y2. Using a horizontal span of $-2$ to $2$ and a vertical span of $0$ to $5$, we obtain the graph below, which shows that the solution is $x = 0.59$.

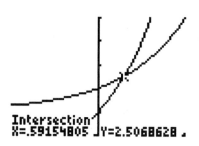

S-9. **The crossing-graphs method**: To solve $\dfrac{5}{x^2 + x + 1} = 1$ using the crossing-graphs method, we enter the $5/(X^2 + X + 1)$ as Y1 and $1$ as Y2. Using a horizontal span of $-3$ to $3$ and a vertical span of $0$ to $7$, we obtain the graphs below, which show that the solutions are $x = -2.56$ and $x = 1.56$.

S-11. **The crossing-graphs method**: To solve $x + \sqrt{x+1} = \sqrt{x+2}$ using the crossing-graphs method, we enter the $X + \sqrt{(X+1)}$ as Y1 and $\sqrt{(X+2)}$ as Y2. Using a horizontal span of $-2$ to $2$ and a vertical span of $-2$ to $2$, we obtain the graph below, which shows that the solution is $x = 0.37$.

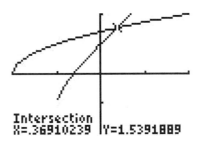

S-13. **The crossing-graphs method**: To solve $2^x + 3^x = 4^x$ using the crossing-graphs method, we enter the $2^X + 3^X$ as Y1 and $4^X$ as Y2. Using a horizontal span of $0$ to $2$ and a vertical span of $0$ to $15$, we obtain the graph below, which shows that the solution is $x = 1.51$.

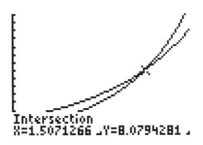

S-15. **The crossing-graphs method**: To solve $4 + x = (1+x)(1+x^2)$ using the crossing-graphs method, we enter the $1 + X$ as Y1 and $(1+X)(1+X^2)$ as Y2. Using a horizontal span of $0$ to $2$ and a vertical span of $0$ to $10$, we obtain the graph below, which shows that the solution is $x = 1.17$.

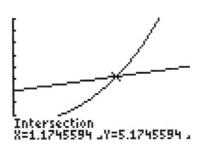

S-17. **The single-graph method**: To solve $\dfrac{20}{1+2^x} - x = 0$ using the single-graph method, we enter the function as Y1. Using the standard viewing window, we obtain the graph below, which shows that the solution is $x = 2.69$.

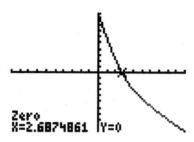

S-19. **The single-graph method**: To solve $3^x - 8 = 0$ using the single-graph method, we enter the function as Y1. Using a horizontal span of 0 to 3 and a vertical span from $-10$ to 10, we obtain the graph below, which shows that the solution is $x = 1.89$.

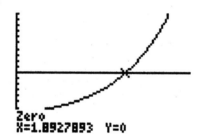

S-21. **The single-graph method**: To solve $x^3 - x - 5 = 0$ using the single-graph method, we enter the function as Y1. Using a horizontal span of 0 to 3 and a vertical span from $-10$ to 10, we obtain the graph below, which shows that the solution is $x = 1.90$.

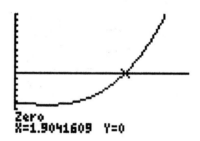

S-23. **The single-graph method**: To solve $\sqrt{x^2 + 1} = 2x$ using the single-graph method, we move the $2x$ to the left-hand side to obtain $\sqrt{x^2 + 1} - 2x = 0$ and we enter this function as Y1. Using a horizontal span of 0 to 1 and a vertical span from $-2$ to 2, we obtain the graph below, which shows that the solution is $x = 0.58$.

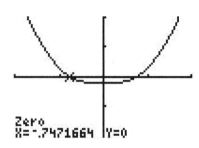

**S-25. The single-graph method**: To solve $\dfrac{x^4}{x^2+1} - 0.2 = 0$ using the single-graph method, we enter the function as Y1. Using a horizontal span of $-2$ to $2$ and a vertical span of $-2$ to $2$, we obtain the graphs below, which show solutions at $x = -0.75$ and $x = 0.75$.

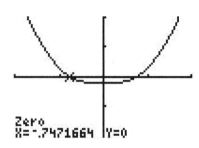 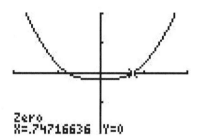

**S-27. The single-graph method**: To solve $x^4 + x^5 = x^6$ using the single-graph method, we move the $x^6$ to the left-hand side to obtain $x^4 + x^5 - x^6 = 0$ and enter this function as Y1. Using a horizontal span of $-1$ to $2$ and a vertical span of $-1$ to $2$, we obtain three graphs below, which show solutions at $x = -0.62$, $x = 0$, and $x = 1.62$.

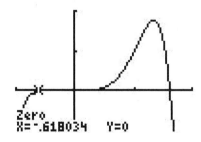

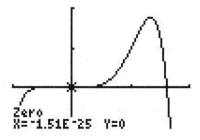 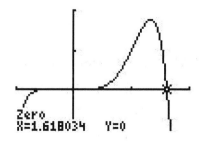

## 2.5  INEQUALITIES

1.  **Paying off a credit card**:

    (a) We use a horizontal span of 0 to 100 monthly payments. After consulting a table of values, we use a vertical span of 0 to 8000 dollars. In the figure on the left below, we have graphed the function $B$ along with the target values, which correspond to the constant function 1000 and the constant function 2000 (both using thick lines).

        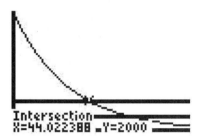

    (b) We want to solve the inequality

    $$1000 \leq 8000 \left(1.02 \times 0.95\right)^t \leq 2000.$$

    The inequality is satisfied when the graph of $B$ is at or between the lines. To find the intersection points we use the crossing-graphs method. We see from the figure on the right above that the first crossing point occurs when $t = 44.02$. A similar calculation shows that the second crossing point occurs when $t = 66.03$. Therefore, the function $B$ is in the desired range when $44.02 \leq t \leq 66.03$.

    (c) You can resume charging after the first payment that lowers the balance below \$1000. Because the number of payments made is a whole number, the answer corresponds to the first whole number greater than the first crossing point found in part b, which was $t = 44.02$. Thus, you can resume charging after 45 monthly payments. (An alternative to using the graph would be to make a table of values of $B$ using integer values for the variable.)

3.  **Dangers of smoking**:

    (a) We use a horizontal span of 0 to 60 minutes. After consulting a table of values, we use a vertical span of 0 to 0.5 nanogram per deciliter. In the figure on the left below, we have graphed the function $C$ along with the target value, which corresponds to the constant function 0.3 (thick line).

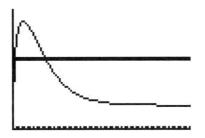 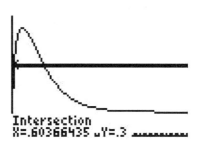

(b) We want to solve the inequality

$$0.1 + 0.3t^{0.6}e^{-0.17t} \geq 0.3.$$

The inequality is satisfied when the graph of $C$ is at or above the line. To find the intersection points we use the crossing-graphs method. We see from the figure on the right above that the first crossing point occurs when $t = 0.60$ minute. A similar calculation shows that the second crossing point occurs when $t = 10.78$ minutes. Thus, the concentration of cyanide is 0.3 nanogram per deciliter or higher from 0.60 minute to 10.78 minutes.

5. **Yellow-bellied sunbirds**:

(a) We use a horizontal span of 0 to 15 days. After consulting a table of values, we use a vertical span of 0 to 10 grams. In the figure on the left below, we have graphed the function $M$ along with the target value, which corresponds to the constant function 7.5 (thick line).

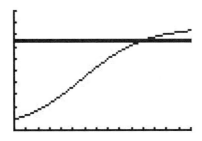 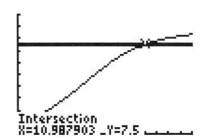

(b) We want to solve the inequality

$$\frac{8.6}{1 + 8.55e^{-0.37t}} \leq 7.5.$$

The inequality is satisfied when the graph of $M$ is at or below the line. To find the intersection point we use the crossing-graphs method. We see from the figure on the right above that the crossing point occurs when $t = 10.99$. Therefore, the hatchlings require parental care from hatching to 10.99 (practically 11) days.

7. **Estimating wave height:**

   (a) We use a horizontal span of 0 to 25 miles per hour. After consulting a table of values, we use a vertical span of 0 to 15 feet. The graph of $h$ is on the left below.

   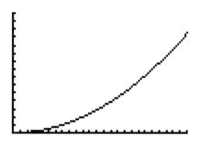 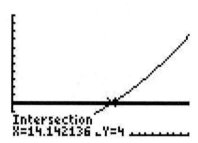

   (b) We want to solve the inequality

   $$0.02w^2 \leq 4.$$

   To do this we add to the figure from part a the graph of the constant function 4 (thick line) to get the figure on the right above. The inequality is satisfied when the graph of $h$ is at or below the line. To find the intersection point we use the crossing-graphs method. We see from the figure that the graphs intersect when $w = 14.14$. Therefore, winds from 0 to 14.14 miles per hour will give safe sailing for this boat.

9. **Adult weight from puppy weight:**

   (a) We are assuming that $w = 12$, so $W = 52 \times \dfrac{12}{a}$ or $W = \dfrac{624}{a}$. We use a horizontal span of 2 to 16 weeks. After consulting a table of values, we use a vertical span of 0 to 300 pounds. The graph of $W$ is on the left below.

   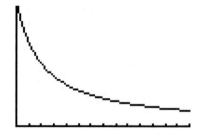 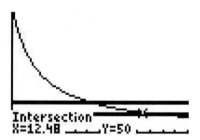

   (b) We want to solve the inequality

   $$50 \leq \frac{624}{a} \leq 75.$$

To do this we add to the figure from part a two graphs: the graph of the constant function 50 (thick line) and the graph of the constant function 75 (thick line). The result is the figure on the right above. The inequality is satisfied when the graph of $W$ is between the lines. To find the intersection points we use the crossing-graphs method. We see from the figure that the lower crossing point occurs when $a = 12.48$. A similar calculation shows that the upper crossing point occurs when $a = 8.32$. Therefore, the age range between 8.32 and 12.48 weeks corresponds to this adult weight.

11. **Drug concentration**: We want to solve the inequality

$$2.65(e^{-0.2t} - e^{-2t}) \geq 1.5.$$

After consulting a table of values, we use a horizontal span of 0 to 3.5 hours and a vertical span of 0 to 2 milligrams per liter for a graph of $C$. The graph is below.

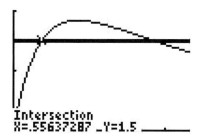

To the figure we have added the graph of the constant function 1.5 (thick line). The inequality is satisfied when the graph of $C$ is at or above the line. To find the intersection points we use the crossing-graphs method. We see from the figure that the first crossing point occurs when $t = 0.56$. A similar calculation shows that the second crossing point occurs when $t = 2.81$. Therefore, the drug concentration level is at least 1.5 milligrams per liter from 0.56 hour to 2.81 hours after the drug is administered.

13. **Mosteller formula for body surface area**:

   (a) We are assuming that $w = 80$, so the formula for $B$ as a function of $h$ is $B = \dfrac{\sqrt{80h}}{60}$.

   (b) We want to solve the inequality

$$1.9 \leq \frac{\sqrt{80h}}{60} \leq 2.$$

After consulting a table of values, we use a horizontal span of 0 to 200 centimeters and a vertical span of 0 to 2.5 square meters for a graph of $B$. The graph is below.

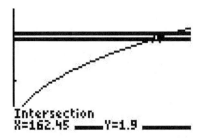

The inequality is satisfied when the graph of $B$ is between the lines. To find the intersection points we use the crossing-graphs method. We see from the figure that the first crossing point occurs when $h = 162.45$. A similar calculation shows that the second crossing point occurs when $h = 180$. Therefore, a height range from 162.45 to 180 centimeters will result in a body surface area range from 1.9 to 2 square meters.

15. **A train**: We want to solve the inequality

$$\sqrt{(30t)^2 + (5 - 90t)^2} \le 2.$$

A table of values shows that the function takes the value 5 when the variable is 0 and when it is 0.1 and that the function increases after that point. Therefore we use a horizontal span of 0 to 0.1 hour and a vertical span of 0 to 5 miles for a graph of $D$. The graph is below.

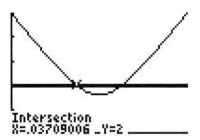

To the figure we have added the graph of the constant function 2 (thick line). The inequality is satisfied when the graph of $D$ is at or below the line. To find the intersection points we use the crossing-graphs method. We see from the figure that the first crossing point occurs when $t = 0.0370\cdots$. A similar calculation shows that the second crossing point occurs when $t = 0.0629\cdots$. These times are in hours. We convert them to minutes and round to the nearest minute: $0.0370\cdots \times 60$ is about 2, and $0.0629\cdots \times 60$ is about 4. Therefore, the train is too near the home from 2 to 4 minutes after midnight, or from 12:02 to 12:04 a.m.

17. **Two savings accounts**:

    (a) We use a horizontal span of 0 to 25 years. After consulting a table of values, we use a vertical span of 0 to 1200 dollars. The graphs of $B_1$ and $B_2$ (thick line) are in the figure on the left below.

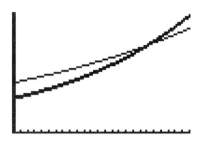 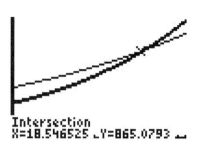

    (b) To find the intersection point we use the crossing-graphs method. We see from the figure on the right above that the crossing point occurs when $t = 18.55$. Therefore, the two account balances are the same after 18.55 years.

    (c) From the figure on the right above we see that the graph of $B_1$ is above that of $B_2$ until $t = 18.55$. After that point the graph of $B_2$ is above that of $B_1$. Therefore, the first balance (corresponding to $B_1$) is greater than the second balance (corresponding to $B_2$) up to 18.55 years. After that the second balance is greater.

19. **Fireworks**: We are assuming that the initial velocity is $v_0 = 50$ feet per second. We want to solve the inequality

$$50t - 16t^2 \geq 30.$$

After consulting a table of values, we use a horizontal span of 0 to 3 seconds and a vertical span of 0 to 40 feet for a graph of $H$. The graph is below.

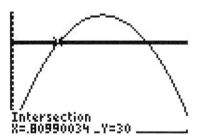

To the figure we have added the graph of the constant function 30 (thick line). The inequality is satisfied when the graph of $H$ is at or above the line. To find the intersection points we use the crossing-graphs method. We see from the figure that the first crossing point occurs when $t = 0.81$. A similar calculation shows that the second crossing point occurs when $t = 2.32$. Therefore, the device can be viewed during the time span from 0.81 to 2.32 seconds.

21. **More on logistic growth with a threshold:**

(a) We want to solve the inequality

$$-0.76n \left(1 - \frac{n}{8}\right) \left(1 - \frac{n}{50}\right) \geq 20.$$

To do this we add to the graphs from the preceding exercise the graph of the constant function 20 (thick line) to get the figure on the left below. (As in that exercise we use a horizontal span of 0 to 60 thousand and a vertical span of $-10$ to 30 thousand per year.) The inequality is satisfied when the graph of $G$ is at or above the line. To find the intersection points we use the crossing-graphs method. We see from the figure that the first crossing point occurs when $n = 24.84$ thousand. A similar calculation shows that the second crossing point occurs when $n = 43.01$ thousand. Therefore, the growth rate is at least 20 thousand per year for population levels between 24.84 thousand and 43.01 thousand.

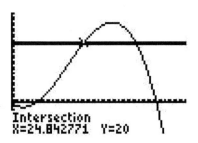
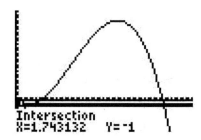

(b) Now we want to solve the inequality

$$-0.76n \left(1 - \frac{n}{8}\right) \left(1 - \frac{n}{50}\right) \leq -1.$$

We add to the graphs from the preceding exercise the graph of the constant function $-1$ (thick line) to get the figure on the right above. The inequality is satisfied when the graph of $G$ is at or below the line. To find the intersection points we use the crossing-graphs method. We see from the figure that the first crossing point occurs when $n = 1.74$ thousand. Similar calculations show that the second crossing point occurs when $n = 6.01$ thousand and that the third crossing point occurs when $n = 50.25$ thousand. Therefore, the growth rate is $-1$ thousand per year or less for population levels between 1.74 thousand and 6.01 thousand and for population levels of 50.25 thousand or more.

## Skill Building Exercises

S-1. **An investment**: We want to find for what $t$ we have $700 \leq B \leq 750$, so we want to solve the inequality $700 \leq 500e^{0.05t} \leq 750$.

S-3. **Two investments**: We want to find for what $t$ we have $B_2 \geq B_1$, so we want to solve the inequality $200 + 50t \geq 250 \times 1.07^t$.

S-5. **Solving inequalities**: To solve the inequality $x^3 + x \geq 4$, we enter $X^3 + X$ as Y1 and 4 as Y2 (thick graph). Using a horizontal span of 0 to 2 and a vertical span of 0 to 10, we obtain the graph below. The inequality is satisfied when the graph of Y1 is at or above the graph of Y2. Using the crossing-graphs method, we see from the figure that the crossing point occurs when $x = 1.38$. Therefore, the inequality holds when $x \geq 1.38$.

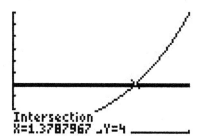

S-7. **Solving inequalities**: To solve the inequality $xe^x \geq 8$, we enter $Xe^X$ as Y1 and 8 as Y2 (thick graph). Using a horizontal span of 0 to 2 and a vertical span of 0 to 15, we obtain the graph below. The inequality is satisfied when the graph of Y1 is at or above the graph of Y2. Using the crossing-graphs method, we see from the figure that the crossing point occurs when $x = 1.61$. Therefore, the inequality holds when $x \geq 1.61$.

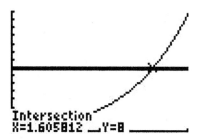

S-9. **Solving inequalities**: To solve the inequality $x \geq 1 + \sqrt{x}$, we enter $X$ as Y1 and $1 + \sqrt{X}$ as Y2 (thick graph). Using a horizontal span of 0 to 3 and a vertical span of 0 to 3, we obtain the graph below. The inequality is satisfied when the graph of Y1 is at or above the graph of Y2. Using the crossing-graphs method, we see from the figure that the crossing point occurs when $x = 2.62$. Therefore, the inequality holds when $x \geq 2.62$.

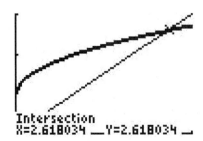

S-11. **Solving inequalities**: To solve the inequality $5 - x^2 \geq 2 + x^4$, we enter $5 - X^2$ as Y1 and $2 + X^4$ as Y2 (thick graph). Using a horizontal span of $-2$ to 2 and a vertical span of 0 to 20, we obtain the graph below. The inequality is satisfied when the graph of Y1 is at or above the graph of Y2. Using the crossing-graphs method, we see from the figure that the crossing point on the right occurs when $x = 1.14$. By symmetry, or by the crossing-graphs method, the crossing point on the left occurs when $x = -1.14$. Therefore, the inequality holds when $-1.14 \leq x \leq 1.14$.

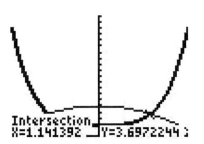

S-13. **Solving inequalities**: To solve the inequality $x^2 + 3x \leq x^3$, we enter $X^2 + 3X$ as Y1 and $X^3$ as Y2 (thick graph). Using a horizontal span of $-1.5$ to 2.5 and a vertical span of $-3.5$ to 16, we obtain the graph on the left below. The inequality is satisfied when the graph of Y1 is at or below the graph of Y2. Using the crossing-graphs method, we see from the figure on the right below that the crossing point on the far right occurs when $x = 2.30$. A similar calculation shows that the crossing point on the far left occurs when $x = -1.30$, and the middle crossing point occurs when $x = 0$. Therefore, the inequality holds when $-1.30 \leq x \leq 0$ and when $x \geq 2.30$.

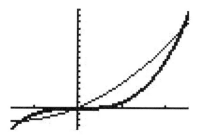
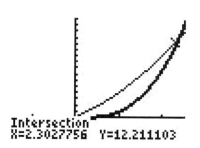

S-15. **Solving inequalities**: To solve the inequality $x^2 + 4 \geq x$, we enter $X^2 + 4$ as Y1 and $X$ as Y2 (thick graph). Using a horizontal span of $-2$ to 2 and a vertical span of $-2$ to 8, we obtain the graph below. The inequality is satisfied when the graph of Y1 is at or above the graph of Y2, and we see from the figure that that is always the case. Therefore, the inequality is true for all values of $x$.

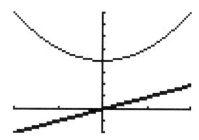

S-17. **Solving inequalities**: To solve the inequality $6 \leq x^3 + x \leq 20$, we enter $X^3 + X$ as Y1, 6 as Y2 (thick graph), and 20 as Y3 (thick graph). Using a horizontal span of 0 to 3 and a vertical span of 0 to 30, we obtain the graph below. The inequality is satisfied when the graph of Y1 is between the graphs of Y2 and Y3. Using the crossing-graphs method, we see from the figure that the crossing point on the left occurs when $x = 1.63$. A similar calculation shows that the crossing point on the right occurs when $x = 2.59$. Therefore, the inequality holds when $1.63 \leq x \leq 2.59$.

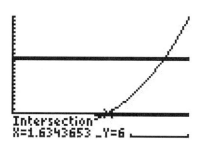

S-19. **Solving inequalities**: To solve the inequality $2^x \le x + 4 \le 3^x$, we enter $X + 4$ as Y1, $2^X$ as Y2 (thick graph), and $3^X$ as Y3 (thick graph). Using a horizontal span of 0 to 3.2 and a vertical span of 0 to 35, we obtain the graph below. The inequality is satisfied when the graph of Y1 is between the graphs of Y2 and Y3. Using the crossing-graphs method, we see from the figure that the crossing point on the left occurs when $x = 1.56$. A similar calculation shows that the crossing point on the right occurs when $x = 2.76$. Therefore, the inequality holds when $1.56 \le x \le 2.76$.

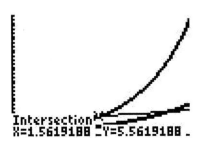

## 2.6  OPTIMIZATION

1. **World crude oil production**:

   (a) We use a horizontal span of 0 to 40 years. After consulting a table of values, we use a vertical span of 0 to 35 billion barrels per year. The graph is on the left below.

   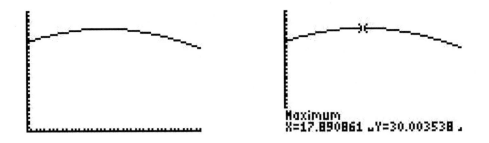

   (b) To find the peak production, we locate the maximum of the graph. In the figure on the right above, we see that the peak occurs when $t = 17.89$ and $P = 30.00$. Thus, the model predicts that a peak in world crude oil production will occur 17.89 years after 2000, or late 2017.

   (c) By the solution of part b, the maximum crude oil production predicted by this model is 30.00 billion barrels per year.

3. **Builder's Old Measurement:**

(a) We are assuming that $L = 150$, so

$$T = \frac{(150 - 0.6W)W^2}{188}.$$

We use a horizontal span of 0 to 250 feet. After consulting a table of values, we use a vertical span of 0 to 8000 feet. The graph is on the left below.

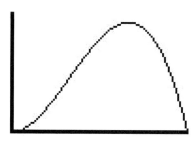 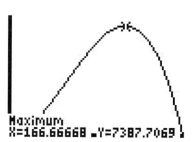

Maximum
X=166.66668 _Y=7387.7069

(b) To find the maximum tonnage, we locate the peak of the graph. In the figure on the right above, we see that the peak occurs when $W = 166.67$ and $T = 7387.71$. Thus the maximum tonnage for a ship of this length is 7387.71.

(c) In the figure on the right above, we see that the beam width of the ship giving maximum tonnage is $W = 166.67$ feet. The length is 150 feet, so the ship is wider than it is long.

(d) We are assuming that $W$ is at most 75 feet. In the figure on the right above, we see that $T$ is increasing to the left of the maximum at $W = 166.67$, so the maximum tonnage if $W$ is at most 75 feet occurs at the width $W = 75$. Using the graph or a table of values shows that $T = 3141.62$ when $W = 75$. Therefore, the maximum tonnage of a ship whose width is no more than half its length is 3141.62.

5. **The cannon at a different angle:**

(a) We can look at the table of values below to get both horizontal and vertical spans. The cannonball follows the graph of $y$ until it hits the ground, which corresponds to the horizontal axis. Thus we are only interested in the graph where it is positive. The table below shows that the cannonball strikes the ground just short of 4 miles, so we use 0 to 4 as a horizontal span. The table also shows that the cannonball doesn't get higher than about 1.5 miles. We use 0 to 2 miles for the vertical span.

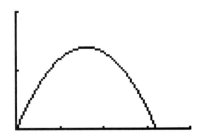

(b) The cannonball strikes the ground where the graph crosses the horizontal axis. In the left-hand figure below, we have expanded the vertical span a little so that the crossing point is clearly visible, and the calculator shows that this crossing point occurs at $x = 3.22$. Thus the cannonball travels 3.22 miles downrange.

(c) The cannonball reaches its maximum height where the graph reaches a peak. In the right-hand picture below, we see that this occurs when $x = 1.61$ and $y = 1.39$. Thus the cannonball reaches a maximum height of 1.39 miles at 1.61 miles downrange.

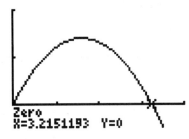
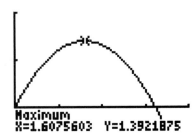

7. **Marine fishery**:

(a) The exercise suggests a horizontal span of 0 to 1.5. The table of values on the left below led us to choose a vertical span of $-0.1$ to 0.1. The graph is on the right below.

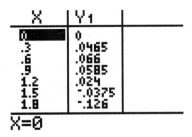
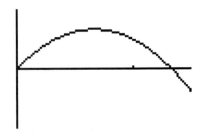

(b) The growth rate if the population size is 0.24 million tons is represented by $G(0.24)$ in functional notation. From a table of values or the graph we see that the value is 0.04 million tons per year.

(c) From a table of values or the graph we see that $G(1.42) = -0.02$. This means that if the population size is 1.42 million tons then the growth rate is $-0.02$ million tons per year, so the population is decreasing at a rate of 0.02 million tons per year.

(d) To find where the largest growth rate occurs, we locate the peak of the graph. In the figure below, we see that the peak occurs when $n = 0.67$ and $G = 0.07$. Thus the growth rate is the largest at a population size of 0.67 million tons.

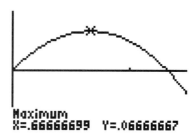

Maximum
X=.66666699   Y=.06666667

9. **Forming a pen**:

(a) We use two sides that are $W$ feet long and one that is $L$ feet long, so the total amount that we need is $2W + L$ feet of fence.

(b) The area of a rectangle is the width times the length, and because that area is to be 100 square feet we have $WL = 100$.

(c) To solve $WL = 100$ for $L$ we divide both sides by $W$. The result is $L = \dfrac{100}{W}$.

(d) By Part (a) the total amount of fence needed is $2W + L$ feet, and by the equation in Part (c) this gives $F = 2W + \dfrac{100}{W}$.

(e) After examining a table of values, we used a horizontal span of 1 to 15 and a vertical span of 0 to 110. The graph is on the left below.

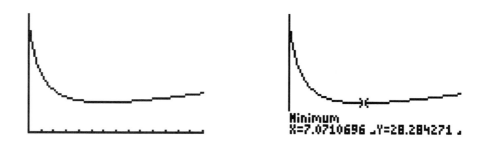

Minimum
X=7.0710696  ₌Y=28.284271 ₌

(f) In the graph on the right above we have located the minimum point, and we see that it occurs where $W = 7.07$ feet. From the equation in Part (c) we find that the corresponding length is $L = \dfrac{100}{7.07} = 14.14$. Thus using 7.07 feet perpendicular to the building and 14.14 feet parallel requires a minimum amount of fence.

11. **Maximum sales growth:**

(a) First we find a formula for unattained sales. Now

$$\text{Unattained sales} = \text{Limit} - \text{Sales level},$$

and we are told that the limit is 4 thousand dollars. Thus the formula for unattained sales is $4 - s$ thousand dollars. Now $G$ is proportional to the product of the sales level $s$ and the unattained sales, and we are told that the constant of proportionality is 0.3. Using the formula for unattained sales, we see that the equation is $G = 0.3s(4 - s)$.

(b) After examining a table of values, we used a horizontal span of 0 to 4 and a vertical span of 0 to 1.5. The graph is on the left below.

 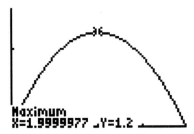

(c) In the graph on the right above we have located the maximum point, and we see that it occurs where $s = 2.00$. Thus the growth rate is as large as possible at a sales level of 2 thousand dollars.

(d) We choose a limit of 6 thousand dollars, so the new equation is $G = 0.3s(6 - s)$. In the graph on the left below we have located the maximum point, and we see that it occurs where $s = 3.00$. (After examining a table of values, we used a horizontal span of 0 to 6 and a vertical span of 0 to 3.) Thus the growth rate is as large as possible at a sales level of 3 thousand dollars. In each of the cases we have examined, the sales level for maximum growth is half of the limit on sales.

If we change the constant of proportionality, say to 0.5, and keep a limit of 6 thousand dollars, then the formula for $G$ changes to $G = 0.5s(6 - s)$. In the graph on the right below we have located the maximum point, and we see that it still occurs where $s = 3.00$. (After examining a table of values, we used a horizontal span of 0 to 6 and a vertical span of 0 to 5.) Thus the growth rate is again as large as possible at a sales level of 3 thousand dollars. This is still half of the limit on sales, which suggests that the relationship doesn't change if the proportionality constant is changed.

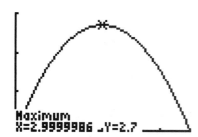
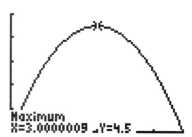

(e) Because the sales level for maximum growth is always half of the limit on sales, we can calculate this limit from the data by taking twice the sales level for maximum growth—that is, by doubling the sales level where the growth rate changes from increasing to decreasing.

13. **An aluminum can, continued**:

   (a) If the radius is 1 inch then the height is

   $$h(1) = \frac{15}{\pi} = 4.77 \text{ inches.}$$

   The area is

   $$A(1) = 2\pi + 30 = 36.28 \text{ square inches.}$$

   (b) If the radius is 5 inches then the height is

   $$h(5) = \frac{15}{\pi 5^2} = 0.19 \text{ inch.}$$

   The area is

   $$A(5) = 2\pi 5^2 + \frac{30}{5} = 163.08 \text{ square inches.}$$

   (c)   i. The aluminum will have a radius of only a few inches, and so we use a horizontal span of 0 to 4 inches. The table below suggests a vertical span of 0 to 50. In the graph below, the radius is on the horizontal axis, and the surface area is on the vertical axis. The graph shows the surface area needed to make a can holding 15 cubic inches as a function of the radius. From the graph it is evident that, as the radius increases, first the amount of aluminum needed decreases to a minimum, and then it increases.

| X | Y1 |
|---|---|
| 1 | 36.283 |
| 2 | 40.133 |
| 3 | 66.549 |
| 4 | 108.03 |
| 5 | 163.08 |
| 6 | 231.19 |
| 7 | 312.16 |

X=7

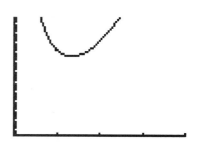

ii. The can that uses the minimum amount of aluminum is represented by the point at the bottom of the graph. In the figure below, we have located that point at $r = 1.34$ and $A = 33.67$. Thus we get a can of minimum surface area, 33.67 square inches, if we use a radius of 1.34 inches.

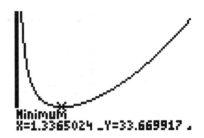

iii. We know that the radius should be about 1.34 to minimize the amount of aluminum used, so the height in that case would be

$$h(1.34) = \frac{15}{\pi 1.34^2} = 2.66 \text{ inches.}$$

15. **Profit**:

(a) Assume that $C$ is measured in dollars. Because

$$\text{Total cost} = \text{Variable cost} \times \text{Number} + \text{Fixed costs,}$$

we have

$$C = 60N + 600.$$

(b) Assume that $R$ is measured in dollars. Because

$$\text{Total revenue} = \text{Price} \times \text{Number,}$$

we have $R = pN$, and thus, using $p = 70 - 0.03N$, we find

$$R = (70 - 0.03N)N.$$

(c) Assume that $P$ is measured in dollars. Because

$$\text{Profit} = \text{Total revenue} - \text{Total cost,}$$

we have $P = R - C$. Using Parts (a) and (b) gives

$$P = (70 - 0.03N)N - (60N + 600).$$

(This can also be written as $P = -0.03N^2 + 10N - 600$.)

(d) We graph the function $P$. In the graph below we used a horizontal span of 0 to 300, as suggested in the exercise, and a vertical span of $-600$ to 300. The graph below shows the maximum, and we see that it occurs at $N = 166.67$. Thus profit is maximized at a production level of 166.67 thousand widgets per month.

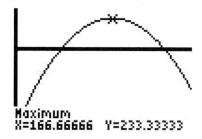

Maximum
X=166.66666  Y=233.33333

17. **Laying phone cable, continued**:

(a) Now

$$\text{Total cost } = \text{ Cost for cable on shore } + \text{ Cost for cable under water.}$$

Also,

$$\begin{aligned} \text{Cost for cable on shore } &= 300 \times \text{ Length of cable on shore} \\ &= 300L, \end{aligned}$$

and

$$\begin{aligned} \text{Cost for cable under water } &= 500 \times \text{ Length of cable under water} \\ &= 500W. \end{aligned}$$

Thus

$$C = 300L + 500W.$$

(b) We use the formula for $W$ in terms of $L$ from Part (a) of Exercise 16: Because $C = 300L + 500W$, that formula for $W$ gives

$$C = 300L + 500\sqrt{1 + (5 - L)^2}.$$

(c) Again, we use the formula for $W$ in terms of $L$ from Part (a) of Exercise 16. Because $L = 1$, the amount of cable under water is

$$W = \sqrt{1 + (5 - 1)^2} = 4.12 \text{ miles.}$$

By Part (b) of this exercise, the total cost of the project is

$$C = 300 \times 1 + 500\sqrt{1 + (5 - 1)^2} = 2361.55 \text{ dollars.}$$

Here is the picture:

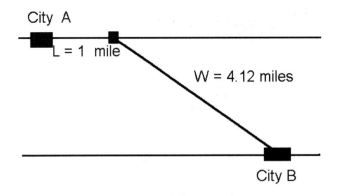

(d) Because $L = 3$, the amount of cable under water is

$$W = \sqrt{1 + (5 - 3)^2} = 2.24 \text{ miles.}$$

The total cost is

$$C = 300 \times 3 + 500\sqrt{1 + (5 - 3)^2} = 2018.03 \text{ dollars.}$$

Here is the picture:

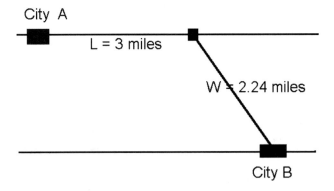

(e) Because City B is only 5 miles downriver from City A, the value of $L$ will be no more than 5. Thus we use a horizontal span of 0 to 5. The table below led us to choose a vertical span from 1000 to 3000. The horizontal axis is miles of cable laid on shore, and the vertical axis is cost in dollars. The graph shows how much it costs to run cable from City A to City B as a function of miles of cable on shore. From the graph it is evident that, as the amount of cable on shore increases, first the cost decreases to a minimum, and then it increases.

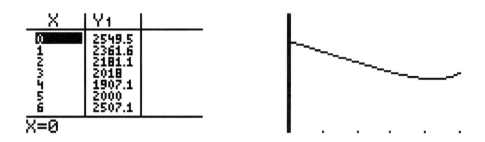

(f) In the picture below, we have located the minimum at $L = 4.25$ and $C = 1900$. Thus the minimum cost plan uses $4.25$ miles on shore (and costs \$1900).

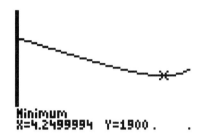

(g) Because $L = 4.25$ miles,

$$W = \sqrt{1 + (5 - 4.25)^2} = 1.25 \text{ miles.}$$

Here is the picture:

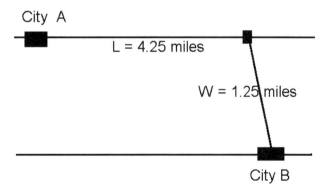

(h) The effect of the increase on the formulas from Parts (a) and (b) of this exercise is to replace $300$ by $700$. The new formula is

$$C = 700L + 500\sqrt{1 + (5 - L)^2}.$$

Because the land cost of \$700 per mile is greater than the water cost and the water route is the shortest, the least cost project would lay all the cable under water.

This can also be seen by making a graph (shown below) of the new cost function. (We used a vertical span of $1500$ to $4500$.) As expected (see Parts (c) and (d) of

Exercise 16), this function is always increasing, so the least cost is at the beginning, when $L = 0$.

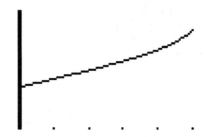

19. **Growth of fish biomass**:

(a) If a plaice weighs three pounds, then $w = 3$, and we want to know the age. That is, we want to solve the equation $w = 3$, that is

$$6.32(1 - 0.93e^{-0.095t})^3 = 3.$$

The table of values below for $w$ suggests a horizontal span of 0 to 20 years and a vertical span of 0 to 5 pounds. In the right-hand picture, we have plotted the graph of $w$, along with that of the constant function 3, and calculated the intersection point. This shows that a 3-pound plaice is about 15.18 years old.

| X | Y₁ |
|---|---|
| 3 | .17171 |
| 6 | .67332 |
| 9 | 1.396 |
| 12 | 2.1917 |
| 15 | 2.957 |
| 18 | 3.6372 |
| 21 | 4.2123 |

X=21

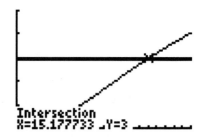

Intersection
X=15.177733  Y=3

(b) The biomass $B$ is obtained by multiplying the population size by the weight of a fish:

$$B = N \times w$$
$$B = 1000e^{-0.1t} \times 6.32(1 - 0.93e^{-0.095t})^3.$$

To make the graph, we use a horizontal span of 0 to 20 years as suggested. From the table below, we chose a vertical span of 0 to 700 pounds.

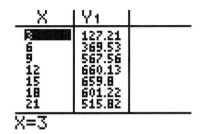 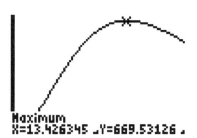

The horizontal axis is age, and the vertical axis is biomass.

(c) The maximum biomass occurs at the peak of the graph above, and we see from there that it occurs at $t = 13.43$ years.

(d)    i. From Part (a), we know that a 3-pound plaice is 15.18 years old, so we can only harvest plaice 15.18 years old or older. In the graph below, we have placed the cursor at $t = 15.18$, and we are looking for the maximum considering the graph only from that point on. But biomass is decreasing beyond 15.18, so 15.18 years is the age giving the largest biomass.

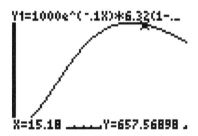

ii. As in Part (a), we can find the age of a 2-pound plaice as the solution to $6.32(1 - 0.93e^{-0.095t})^3 = 2$. We have done this by the crossing graphs method in the left-hand picture below and find that a 2-pound plaice is 11.28 years old. We can harvest plaice 11.28 years old or older. We have marked the graph in the right-hand picture below at this point, and we see that the part of the graph from there on includes the maximum at $t = 13.43$. Thus, if we are able to catch 2-pound plaice, we harvest at the maximum biomass which occurs at 13.43 years.

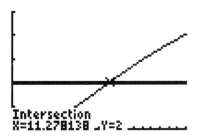 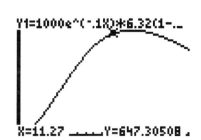

An alternative solution is to find the weight of a plaice at the age $t = 13.43$, found in Part (c), that gives the maximum biomass. This is $w(13.43)$, about 2.56 pounds. Since the weight increases with age, a 2-pound plaice is younger than 13.43 years, and so (as marked in the right-hand picture above) the relevant portion of the graph of biomass will include the maximum at $t = 13.43$ years.

21. **Rate of growth**:

   (a) Since the maximum size of a North Sea cod is about 40 pounds, we use a horizontal span of 0 to 40 pounds. From the table of values below, we choose a vertical span of 0 to 5 pounds per year. The graph shows weight on the horizontal axis and rate of growth on the vertical axis.

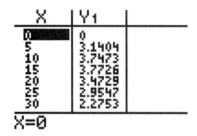
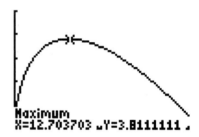

   (b) The peak of the graph is at $w = 12.70$ pounds and $G = 3.81$ pounds per year. For this part of the exercise, we are interested in cod weighing 5 pounds or more. Thus we should be looking at the graph from $w = 5$ on. This includes the peak of the graph, and so the maximum growth rate is $G = 3.81$ pounds per year.

   (c) For cod weighing at least 25 pounds, we should look at the graph of $G$ against $w$ from $w = 25$ on. This is beyond the peak of the graph, and from this point on the graph is decreasing. Thus the maximum value of $G$ occurs when the weight is $w = 25$ pounds. That maximum is 2.95 pounds per year.

23. **Size of high schools**:

   (a) The study was for schools with a maximum enrollment of 2400 students, and so we use a horizontal span of 0 to 2400. The table below suggests a vertical span of 0 to 750 dollars per pupil. The horizontal axis on the graph corresponds to enrollment, and the vertical axis corresponds to cost per pupil.

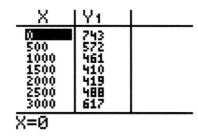 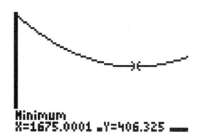

(b) The enrollment giving the minimum per-pupil cost occurs at the bottom of the graph, and from the picture above, this is at $n = 1675$ students (with a cost of $C = \$406.33$ per pupil).

(c) If a high school had an enrollment of 1200, then its cost would be $C(1200) = \$433.40$ per pupil. Increasing the school size to the optimum size of 1675 would lower the cost to $\$406.33$ per pupil, a savings of $\$27.07$ per pupil.

25. **Water flea**:

(a) The exercise suggests a horizontal span of 0 to 350, and a table of values led us to choose a vertical span of $-15$ to 15. The graph is on the left below.

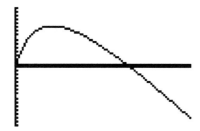 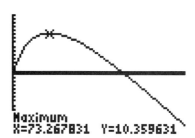

(b) To find where the greatest rate of growth occurs, we locate the peak of the graph. In the figure on the right above, we see that the peak occurs when $N = 73.27$ and $G = 10.36$. Thus the greatest rate of growth occurs at a population level of 73.27.

(c) It is evident from the graph or a table of values that $G = 0$ when $N = 0$. We find the other solution to the equation $G(N) = 0$ using the single-graph method. As the graph below indicates, that solution is $N = 228$. Thus the desired values are $N = 0$ and $N = 228$. (Another way to do this part is to notice that in the formula for $G$ the numerator is factored, and this makes it easy to find the desired solutions.) At the two population levels of 0 and 228, the growth rate is 0, that is, the size of the population is not changing.

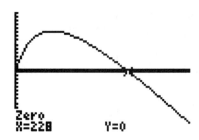

(d) From the graph or a table of values we see that if the population size is 300 then the rate of population growth is $-7.51$ per day. Because the growth rate is negative, at this level the population is decreasing in size (at a rate of 7.51 per day).

27. **Research project**: Answers will vary.

## Skill Building Exercises

S-1. **Maximum and minimum values**: We calculate $f(1) = 1^3 - 3 \times 1 + 1 = -1$.

S-3. **Maximum and minimum values**: We calculate $f(1) = 1 + 1^{-1} = 2$.

S-5. **Finding maxima and minima**: To find the maximum value of $f(x) = 5x + 4 - x^2$ on the horizontal span of 0 to 5, we graph the function. A table of values leads us to choose a vertical span of 0 to 15. The graph, below, shows a maximum value of 10.25 at $x = 2.5$.

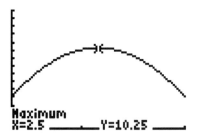

S-7. **Finding maxima and minima**: To find the minimum value of $f(x) = x + \dfrac{x + 5}{x^2 + 1}$ on the horizontal span of 0 to 5, we graph the function. A table of values leads us to choose a vertical span of 0 to 7. The graph, below, shows a minimum value of 3.40 at $x = 1.92$.

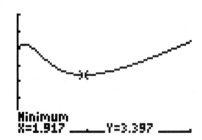

S-9. **Finding maxima and minima**: To find the maximum value of $f(x) = x^{1/x}$ on the horizontal span of 0 to 10, we graph the function. A table of values leads us to choose a vertical span of 0 to 3. The graph, below, shows a maximum value of 1.44 at $x = 2.72$.

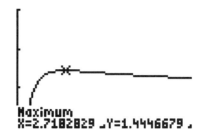

S-11. **Finding maxima and minima**: To find maxima and minima of $f(x) = x^3 - 6x + 1$ with a horizontal span from $-2$ to 2 and a vertical span from $-10$ to 10, we graph the function. The graph on the left below shows the maximum is at $x = -1.41$, $y = 6.66$, while the graph on the right below shows the minimum is at $x = 1.41$, $y = -4.66$.

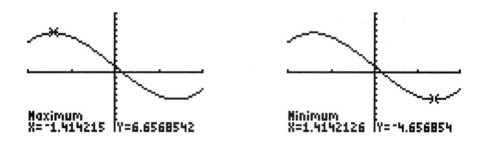

S-13. **Finding maxima and minima**: To find maxima and minima of $f(x) = x^3 - 2^x$ with a horizontal span from 0 to 10, we graph the function. A table of values leads us initially to choose a vertical span of $-30$ to 300. The graph on the left below shows the maximum is at $x = 8.18$, $y = 257.33$, while the graph on the right below uses a horizontal span of 0 to 2 and a vertical span of $-3$ to 1 in order to show the minimum is at $x = 0.59$, $y = -1.30$.

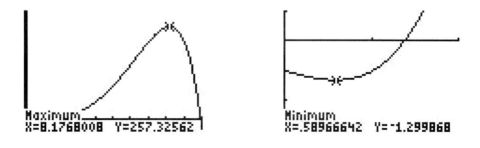

S-15. **Finding maxima and minima**: To find the minimum value of $f(x) = e^x - 2^x - x$ on the horizontal span of $-2$ to $2$, we graph the function. A table of values leads us to choose a vertical span of $-1$ to $3$. The graph, below, shows the minimum is at $x = 0.79, y = -0.32$.

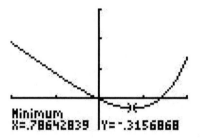

S-17. **Finding maxima and minima**: To find the maximum value of $f(x) = 50/(1 + 1.2^{-x}) - x$ on the horizontal span of $0$ to $20$, we graph the function. A table of values leads us to choose a vertical span of $15$ to $40$. The graph, below, shows the maximum is at $x = 10.65$, $y = 33.08$.

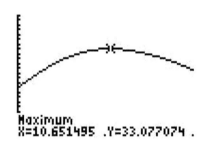

S-19. **Finding maxima and minima**: To find the maximum value of $f(x) = 1 - 1/(\sqrt{x}(5 - \sqrt{x}))$ on the horizontal span of $1$ to $10$, we graph the function. A table of values leads us to choose a vertical span of $0.5$ to $1$. The graph, below, shows the maximum is at $x = 6.25$, $y = 0.84$.

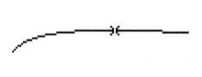

S-21. **Finding maxima and minima**: To find maximum and minimum of $f(x) = (x-3)^2 e^{-1/x^2}$ with a horizontal span from 1 to 4, we graph the function. A table of values leads us initially to choose a vertical span of 0 to 2, but instead we use a vertical span of $-1$ to 2 to improve the display. The graph on the left below shows the maximum is at $x = 1.21$, $y = 1.62$, while the graph on the right shows the minimum is at $x = 3, y = 0$.

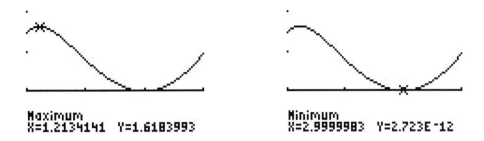

S-23. **Finding maxima and minima**: To find the maximum value of $f(x) = x^3 e^{-x}$ on the horizontal span of 0 to 5, we graph the function. A table of values leads us to choose a vertical span of $-1$ to 2. The graph, below, shows the maximum is at $x = 3.00, y = 1.34$.

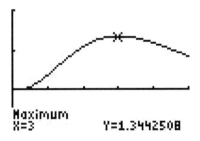

S-25. **Endpoint minimum**: To find the minimum value of $f(x) = x^3 + x$ on the horizontal span of 0 to 5, we graph the function. A table of values leads us to choose a vertical span of $-20$ to 130. The graph, below, is increasing, so the minimum value of 0 occurs at the endpoint $x = 0$.

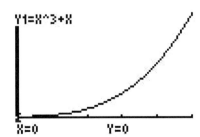

S-27. **Endpoint maximum**: To find the maximum value of $f(x) = 200 - x^3$ on the horizontal span of 0 to 5, we graph the function. A table of values leads us to choose a vertical span of 50 to 250. The graph, below, is decreasing, so the maximum value of 200 occurs at the endpoint $x = 0$.

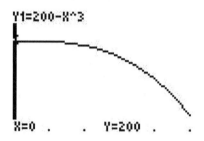

## Chapter 2 Review Exercises

1. **Finding a minimum**: To find the minimum value of $f$, we 288make a table. We enter the function as Y1 $= X^3 - 9X^2/2 + 6X + 1$ and use a table starting value of 1 and a table increment value of 1, resulting in the following table.

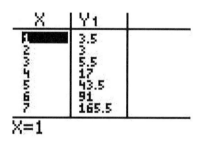

The table shows that $f$ has a minimum value of 3 at $x = 2$.

2. **A population of foxes**:

    (a) We show the graph with a horizontal span of 0 to 20 and a vertical span of 0 to 150.

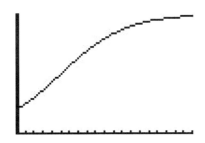

(b) From a table of values or the graph we find that the value of the function at $t = 9$ is 112.08, which we round to 112. Thus there are 112 foxes 9 years after introduction.

(c) Tracing the graph shows that it is concave up from $t = 0$ to about $t = 5$. It is concave down afterwards.

(d) Scrolling down a table on the calculator (or tracing the graph), we see that the limiting value of the function is 150.

3. **Linear equations**: To solve $11 - x = 3 + x$ for $x$, we follow the usual procedure:

$$\begin{aligned}
11 - x &= 3 + x \\
-x - x &= 3 - 11 \\
-2x &= -8 \\
x &= \frac{-8}{-2} = 4.
\end{aligned}$$

4. **Water jug**:

(a) On the left below we show the graph with a horizontal span of 0 to 7.5 and a vertical span of 0 to 16.

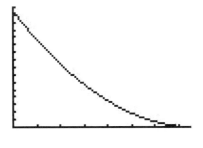
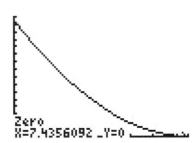

(b) From a table of values or the graph we find that $D(3)$, the value of the function at $t = 3$, is 5.72. Thus the depth of the water is 5.72 inches above the spigot after the spigot has been open 3 minutes.

(c) The time when the jug will be completely drained corresponds to the point when the graph crosses the horizontal axis. In the graph on the right above we find using the single-graph method that this occurs at $t = 7.44$. Thus the jug will be completely drained after 7.44 minutes.

(d) The graph is steeper at the beginning than it is toward the end, so the water drains faster near the beginning.

5. **Maxima and minima**: To find the maximum and minimum values of $x^2 + 100/x$ with a horizontal span of 1 to 5, we graph the function. A table of values leads us to choose a vertical span of 0 to 115. The graph on the left below shows that the maximum is at the left endpoint $x = 1$, where the value is $y = 101$. The graph on the right shows that the minimum is at $x = 3.68$, where the value is $y = 40.72$.

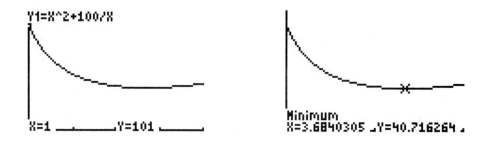

6. **George Reserve population**: Below is a table of values for the function $N$.

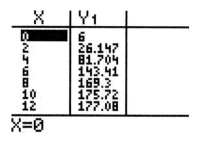

(a) We evaluate the function $N$ at $t = 0$. According to the table that value is 6. Hence there were 6 deer introduced into the deer reserve.

(b) According to the table $N(4)$ is 81.70, or about 82. This means that there were 82 deer in the reserve 4 years after introduction.

(c) Scrolling down the table on the calculator, we see that the limiting value of the function is 177.43, or about 177, and this is the carrying capacity.

(d) The average rate of increase from $t = 0$ to $t = 2$ is

$$\frac{N(2) - N(0)}{2 - 0} = \frac{26.147 - 6}{2} = 10.07 \text{ deer per year.}$$

Continuing in this way, we get the following table, where the rate of change is measured in deer per year.

| Interval | 0 to 2 | 2 to 4 | 4 to 6 | 6 to 8 |
|---|---|---|---|---|
| Rate of change | 10.07 | 27.78 | 30.85 | 12.94 |

The population increases more and more rapidly, and then the rate of growth decreases.

7. **Making a graph**: To find an appropriate window that will show a good graph of the function, first we enter the function as $Y1 = 315/(1 + 14e^{(-0.23X)})$, and then we make a table of values. The table shows that a vertical span from 0 to 330 will display the graph, and the graph is shown below.

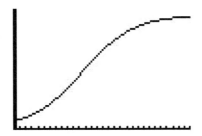

8. **Forming a pen**:

   (a) We use two sides that are $W$ feet long and two that are $L$ feet long, so the total amount that we need is $F = 2W + 2L$ feet of fence.

   (b) The area of a rectangle is the width times the length, and because that area is to be 144 square feet we have $144 = W \times L$.

   (c) To solve $144 = W \times L$ for $W$ we divide both sides by $L$. The result is $W = \dfrac{144}{L}$.

   (d) By Part (a) the total amount of fence needed is $F = 2W + 2L$ feet, and by the equation in Part (c) this gives $F = 2 \times \dfrac{144}{L} + 2L$ or $F = \dfrac{288}{L} + 2L$.

   (e) After examining a table of values, we used a horizontal span of 0 to 24 and a vertical span of 0 to 100. The graph is on the left below.

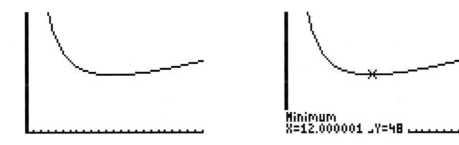

(f) In the graph on the right above we have located the minimum point, and we see
that it occurs where $L = 12$ feet. From the equation in Part (c) we find that the
corresponding width is $W = \dfrac{144}{12} = 12$ feet. Thus a square (12 by 12) pen requires
a minimum amount of fence, which makes sense. The graph illustrates that, as the
length increases, first the area decreases to a minimum, and then it increases.

9. **The crossing-graphs method**: To solve the equation using the crossing-graphs method,
we enter $6 + 69 \times 0.96^X$ as Y1 and 32 as Y2. Using a horizontal span of 0 to 30 and a
vertical span of 0 to 80, we obtain the graph below, which shows that the solution is
$t = 23.91$.

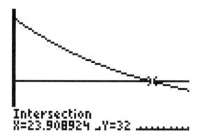

10. **Gliding pigeons**:

(a) We show the graph on the left below with a horizontal span of 0 to 20 and a vertical
span of 0 to 10.

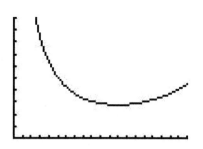 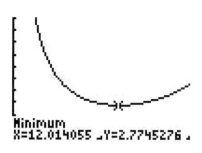

(b) In the graph on the right above we have located the minimum point, and we see
that it occurs where $u = 12.01$ meters per second.

(c) The graph shows that if $u$ is very small then $s$ is very large. Hence if the airspeed
is very slow the pigeon sinks quickly.

11. **Finding a maximum**: To locate the maximum of $f$, we make a table. We enter the
function as Y1 $= 12X - X^2 - 25$ and use a table starting value of 3 and a table increment
value of 1, resulting in the following table.

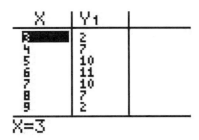

The table shows that $f$ has a maximum value of 11 at $x = 6$.

12. **Thrown ball**:

(a) On the left below we show the table only from $t = 0$ to $t = 1.5$. Here we use a table starting value of 0 and a table increment value of 0.25.

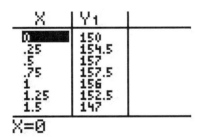   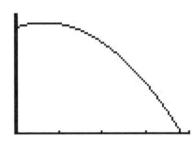

(b) Using a horizontal span of 0 to 4 and a vertical span of 0 to 170, we obtain the graph on the right above.

(c) In the graph below we have located the maximum point, and we see that it occurs at $t = 0.69$ second.

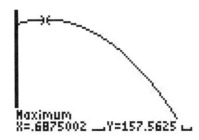

(d) The graph above shows that the maximum value is 157.56 feet. This means that the ball is 157.56 feet high at the peak.

(e) We use the crossing-graphs method. We enter $Y2 = 155$ and, to get a better picture of the intersection points, we change the vertical span to go from 150 to 160. The graphs below show that the two times are $t = 0.29$ second and $t = 1.09$ seconds.

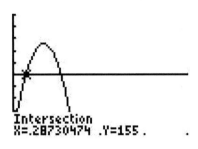

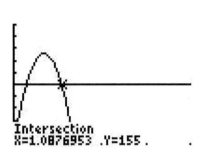

13. **Linear equations**: To solve $L = 98.42 + 1.08W - 4.14A$ for $W$, we follow the usual procedure:

$$
\begin{aligned}
L &= 98.42 + 1.08W - 4.14A \\
L - 98.42 + 4.14A &= 1.08W \\
W &= \frac{L - 98.42 + 4.14A}{1.08}.
\end{aligned}
$$

14. **Growth of North Sea sole**:

(a) We show the graph on the left below with a horizontal span of 1 to 10 and a vertical span of 0 to 15.

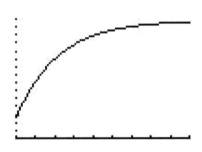

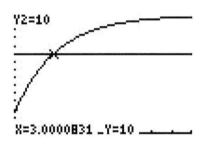

(b) In functional notation the length of a 4-year-old sole is $L(4)$. From a table of values or the graph we find that the value is 11.77 inches.

(c) To use the crossing-graphs method we enter Y2 = 10. The graph on the right above shows that the age is $t = 3$ years.

(d) Scrolling down a table on the calculator (or tracing the graph), we see that the limiting value of the function is 14.8. This means that the limiting length for a North Sea sole is 14.8 inches.

15. **Minimum**: To find the minimum value of $x^2 + 20/(x + 1)$ with a horizontal span of 0 to 10, we graph the function. A table of values leads us to choose a vertical span of 0

to 105. The graph below shows that the minimum is at $x = 1.54$, where the value is $y = 10.25$.

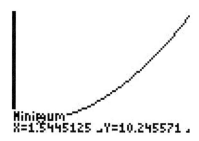

16. **Profit**:

(a) In functional notation the profit at a production level of 7 million items is $P(7)$.

(b) When we put $N = 0$ into the formula we find $P = -6.34$. This means that the loss at a production level of $N = 0$ million items is 6.34 million dollars.

(c) The break-even points occur where the graph crosses the horizontal axis. To make the graph we use a horizontal span of 0 to 10 and a vertical span of $-10$ to 20. We find by means of the single-graph method that the graph crosses the horizontal axis at $N = 0.68$ and at $N = 9.32$, as shown in the graphs below. Hence the two break-even points are at 0.68 million items and at 9.32 million items.

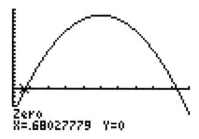
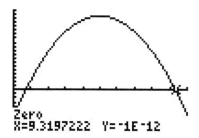

(d) In the graph below we have located the maximum point, and we see that it occurs where $N = 5$ and $P = 18.66$. Thus the production level that gives maximum profit is 5 million items, and the amount of the maximum profit is 18.66 million dollars.

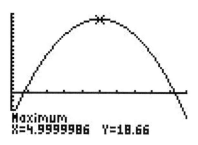

17. **Making a graph**: To find an appropriate window setup which will show a good graph of the function, first we enter the function as $Y1 = 0.5X/(6380 + X)$, and then we make a table of values. The table shows that a vertical span from 0 to 0.5 will display the graph, and the graph is shown below.

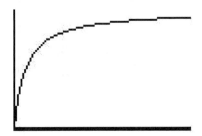

18. **Temperature conversions**:

(a) To solve these equations for $C$, we follow the usual procedure. For the first, we have

$$
\begin{aligned}
F &= 1.8C + 32 \\
F - 32 &= 1.8C \\
C &= \frac{F - 32}{1.8}.
\end{aligned}
$$

For the second, we have

$$
\begin{aligned}
K &= C + 273.15 \\
C &= K - 273.15.
\end{aligned}
$$

(b) We put the formula for $C$ in terms of $K$ from Part (a) into the original formula for $F$ in terms of $C$. The result is

$$
F = 1.8(K - 273.15) + 32.
$$

(c) We put the formula for $C$ in terms of $F$ from Part (a) into the original formula for $K$ in terms of $C$. The result is

$$
K = \frac{F - 32}{1.8} + 273.15.
$$

This can also be done by solving the formula from Part (b) to get $K$ in terms of $F$.

(d) We put $C = 0$ into the two original formulas. The results are $F = 32$ and $K = 273.15$. Thus 0 degrees Celsius corresponds to 32 degrees Fahrenheit and 273.15 kelvins.

(e) First we put $F = 72$ into the formula $C = \dfrac{F - 32}{1.8}$ from Part (a). The result is $C = 22.22$. Thus 72 degrees Fahrenheit corresponds to 22.22 degrees Celsius. Next we put this result, $C = 22.22$, into the original formula $K = C + 273.15$. The result is $K = 295.37$. (This can also be done directly by using the formula from Part (c).) Thus 72 degrees Fahrenheit corresponds to 295.37 kelvins.

19. **Lidocaine:**

(a) We show the graph on the left below with a horizontal span of 0 to 240 and a vertical span of 0 to 30.

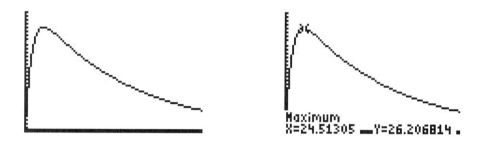

(b) In the graph on the right above we have located the maximum point, and we see that it occurs at $t = 24.51$. Thus the drug reaches its maximum level after 24.51 minutes. For future reference, note from this graph that the maximum level is 26.21 mg.

(c) We use the crossing-graphs method and enter Y2 $= 7.5$. The graphs below show that the two intersection points are at $t = 2.277$ minutes and at $t = 198.295$ minutes. The difference is 196.02 minutes, and that is how long the drug is at or above the level of 7.5 mg.

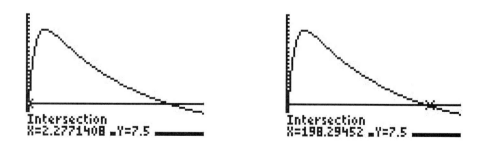

(d) As we noted in Part (b), the maximum level is 26.21 mg. Thus the amount never exceeds 30 mg, so this dose is not lethal for a person of typical size.

(e) We use the crossing-graphs method and enter Y2 = 15. The graph below shows that the first intersection point is at $t = 5.54$ minutes. Thus this dose is lethal for a small person, and it will be lethal after 5.54 minutes.

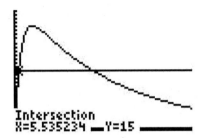

20. **The single-graph method**: To solve the equation using the single-graph method, we enter the function on the left as Y1. Using a horizontal span of $-2$ to 2 and a vertical span of $-1$ to 1, we obtain the graphs below, which show that the solutions are $x = -1.22$, $x = 0$, and $x = 1.22$.

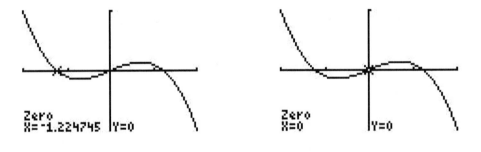

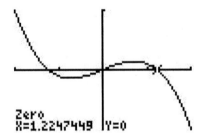

21. **Inequality**: We want to solve the inequality $x \geq 5e^{-x}$. To do this we graph $x$ and $5e^{-x}$ (thick graph) using a horizontal span of 0 to 2 and a vertical span of 0 to 5. The graph is below.

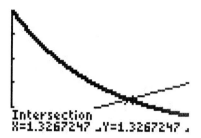

Intersection
X=1.3267247   Y=1.3267247

The inequality is satisfied when the graph of $x$ is at or above the graph of $5e^{-x}$. To find the intersection point we use the crossing-graphs method. We see from the figure that the crossing point occurs when $x = 1.33$. Therefore, the inequality $x \geq 5e^{-x}$ holds when $x \geq 1.33$.

## A FURTHER LOOK: LIMITS

1. **Verifying the basic exponential limit**: We show tables for two values of $a$. The table below on the left corresponds to $a = 0.5$, so the function is $0.5^t$. The table on the right corresponds to $a = 0.3$, so the function is $0.3^t$. In both cases the function values are approaching 0, in support of the statement of the basic exponential limit.

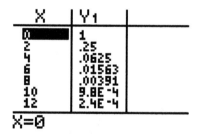

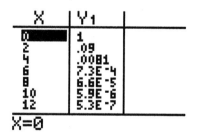

2. **Verifying the basic power limit**: We show tables for two values of $n$. The table below on the left corresponds to $n = 2$, so the function is $1/t^2$. The table on the right corresponds to $n = 3$, so the function is $1/t^3$. In both cases the function values are approaching 0, in support of the statement of the basic power limit.

| X | Y1 | |
|---|---|---|
| **1** | 1 | |
| 3 | .11111 | |
| 5 | .04 | |
| 7 | .02041 | |
| 9 | .01235 | |
| 11 | .00826 | |
| 13 | .00592 | |

X=1

| X | Y1 | |
|---|---|---|
| **1** | 1 | |
| 3 | .03704 | |
| 5 | .008 | |
| 7 | .00292 | |
| 9 | .00137 | |
| 11 | 7.5E-4 | |
| 13 | 4.6E-4 | |

X=1

3. **Calculating with the basic exponential limit**: Because $\lim\limits_{t\to\infty} 0.5^t = 0$ by the basic exponential limit, we have

$$\lim_{t\to\infty} 15(1 - 3 \times 0.5^t) = 15(1 - 3 \times 0) = 15.$$

4. **Calculating with the basic exponential limit**: Because $\lim\limits_{t\to\infty} 0.5^t = 0$, we have

$$\lim_{t\to\infty} \sqrt{7 + 0.5^t} = \sqrt{7 + 0} = \sqrt{7}.$$

5. **Calculating with the basic exponential limit**: Because $\lim\limits_{x\to\infty} 0.5^x = 0$ and $\lim\limits_{x\to\infty} 0.25^x = 0$, we have

$$\lim_{x\to\infty} \frac{\sqrt{4 + 0.5^x}}{\sqrt{9 + 0.25^x}} = \frac{\sqrt{4 + 0}}{\sqrt{9 + 0}} = \frac{2}{3}.$$

6. **Calculating with the basic power limit**: Because $\lim\limits_{t\to\infty} 1/t = 0$ and $\lim\limits_{t\to\infty} 1/t^2 = 0$, we have

$$\lim_{t\to\infty} \frac{2 + \frac{1}{t}}{3 + \frac{4}{t^2}} = \frac{2 + 0}{3 + 4 \times 0} = \frac{2}{3}.$$

7. **Calculating with the basic power limit**: Recall from the basic properties of exponents that $t^{-1/3} = \dfrac{1}{t^{1/3}}$. Thus

$$\lim_{t\to\infty} \left(5 + t^{-1/3}\right) = \lim_{t\to\infty} \left(5 + \frac{1}{t^{1/3}}\right) = 5 + 0 = 5;$$

here we have applied the basic power limit with $n = \dfrac{1}{3}$.

8. **Calculating with the basic power limit**: Because $\lim\limits_{x\to\infty} 1/x = 0$, we have

$$\lim_{x\to\infty} \frac{\sqrt{1 + \frac{1}{x}}}{\sqrt{1 - \frac{1}{x}}} = \frac{\sqrt{1 + 0}}{\sqrt{1 - 0}} = 1.$$

9. **Calculating with quotients of polynomials**: Calculating the limit using the leading terms, we have
$$\lim_{t \to \infty} \frac{6t^3 + 5}{2t^3 + 3t + 1} = \lim_{t \to \infty} \frac{6t^3}{2t^3} = \lim_{t \to \infty} \frac{6}{2} = 3.$$

10. **Calculating with quotients of polynomials**: Calculating the limit using the leading terms, we have
$$\lim_{x \to \infty} \frac{x^2 + 1}{x + 1} = \lim_{x \to \infty} \frac{x^2}{x} = \lim_{x \to \infty} x.$$
Since $x$ increases without bound, the limit does not exist.

11. **Calculating with quotients of polynomials**: Calculating the limit using the leading terms, we have
$$\lim_{x \to \infty} \frac{7x^2 + 9x + 1}{2x^4 + x + 2} = \lim_{x \to \infty} \frac{7x^2}{2x^4} = \lim_{x \to \infty} \frac{7}{2x^2} = 0.$$
Here we have applied the basic power limit to $\frac{1}{x^2}$.

12. **Calculating with quotients of polynomials**: Calculating the limit using the leading terms, we have
$$\lim_{x \to \infty} \frac{5x^3 + x + 4}{10x^3 + 9} = \lim_{x \to \infty} \frac{5x^3}{10x^3} = \lim_{x \to \infty} \frac{5}{10} = \frac{1}{2},$$
or $0.5$.

13. **Carrying capacity for the general logistic function**: The carrying capacity is the limiting value of the function. Now the basic exponential limit says that $\lim_{t \to \infty} a^t = 0$ (because $a < 1$), so the limiting value is
$$\lim_{t \to \infty} \frac{k}{1 + pa^t} = \frac{k}{1 + p \times 0} = k.$$
Thus the carrying capacity is $k$.

14. **Water in a tank**: The amount of water in the tank after a long period of time is given by the following limit:
$$\lim_{t \to \infty} (a - bc^t) = a - b \times 0 = a.$$
Here we have used the basic exponential limit for $c^t$ because $c < 1$. Thus the amount of water in the tank after a long period of time is $a$ cubic feet.

15. **Weight of a fish**: The maximum weight for this fish is given by the following limit:
$$\lim_{t \to \infty} 25 - 12 \times 0.7^t = 25 - 12 \times 0 = 25.$$
Here we have used the basic exponential limit for $0.7^t$. Thus the maximum weight for this fish is 25 pounds.

## A FURTHER LOOK: SHIFTING AND STRETCHING

1. **Shifting**: We shift the original graph 2 units to the left:

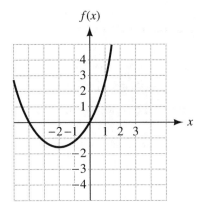

2. **Shifting**: We shift the original graph 2 units to the right:

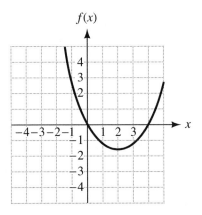

3. **Shifting**: We shift the original graph 2 units to the left and then up 2 units:

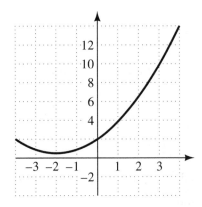

4. **Shifting**: We shift the original graph 2 units to the right and then up 2 units:

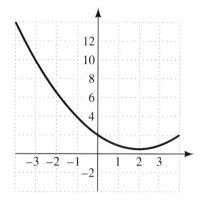

5. **Stretching**: We stretch the graph vertically by a factor of 2:

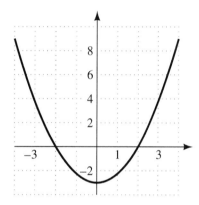

6. **Stretching**: We compress the original graph horizontally by a factor of 2:

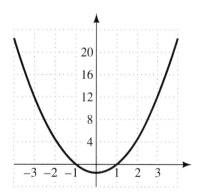

7. **Stretching**: We stretch the original graph horizontally by a factor of 2 and then we stretch it vertically by a factor of 2:

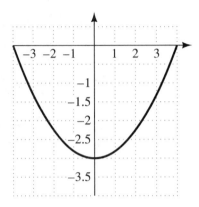

8. **Stretching**: We compress the original graph horizontally by a factor of 2 and then we stretch it vertically by a factor of 2:

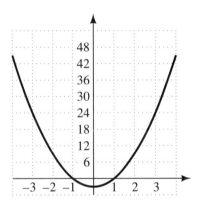

9. **Shifting and stretching**: We compress the original graph horizontally by a factor of 2 and then we shift it up 1 unit:

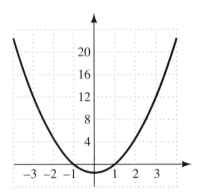

10. **Shifting and stretching**: We stretch the original graph horizontally by a factor of 2 and then we shift it up 1 unit:

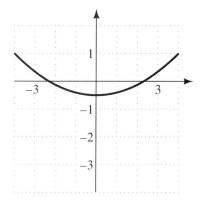

11. **Shifting and stretching**: We stretch the original graph horizontally by a factor of 2, then we stretch it vertically by a factor of 2, and then we shift it up 1 unit:

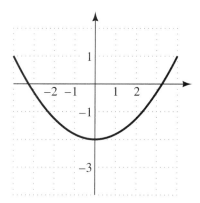

12. **Shifting and stretching**: We shift the original graph 1 unit to the right, then we stretch it vertically by a factor of 2, and then we shift it up 1 unit:

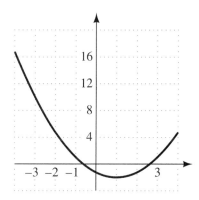

13. **Shifting and stretching**: We shift the original graph 1 unit to the right, then we compress it vertically by a factor of 2, and then we shift it down 1 unit:

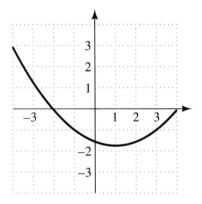

14. **Reflections**: Here is the graph:

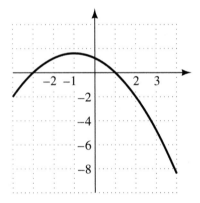

15. **Reflections**: The graph of $g$ is reflected first through the horizontal axis and then through the vertical axis. Here is the result:

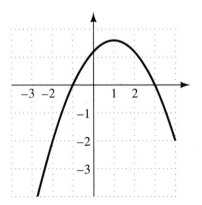

## A FURTHER LOOK: OPTIMIZING WITH PARABOLAS

1. **Locating the vertex of a parabola**: Since the vertex of the graph of $ax^2 + bx + c$ occurs at $x = -\dfrac{b}{2a}$, so the vertex of $x^2 + 6x - 4$ occurs at $x = -\dfrac{6}{2 \times 1} = -3$. The vertical coordinate of the vertex is found by getting the function value at $x = -3$:

$$\text{Vertical coordinate of vertex} = (-3)^2 + 6 \times (-3) - 4 = -13.$$

   Thus the vertex of the parabola is $(-3, -13)$. Because the leading coefficient 1 is positive, this is a minimum.

2. **Locating the vertex of a parabola**: Since the vertex of the graph of $ax^2 + bx + c$ occurs at $x = -\dfrac{b}{2a}$, the vertex of $3x^2 - 30x + 1$ occurs at $x = -\dfrac{-30}{2 \times 3} = 5$. The vertical coordinate of the vertex is found by getting the function value at $x = 5$, which is $3 \times 5^2 - 30 \times 5 + 1 = -74$. Thus the vertex of the parabola is $(5, -74)$. Because the leading coefficient 3 is positive, this is a minimum.

3. **Locating the vertex of a parabola**: Since the vertex of the graph of $ax^2 + bx + c$ occurs at $x = -\dfrac{b}{2a}$, the vertex of $-2x^2 + 12x - 3$ occurs at $x = -\dfrac{12}{2 \times -2} = 3$. The vertical coordinate of the vertex is found by getting the function value at $x = 3$, which is $-2 \times 3^2 + 12 \times 3 - 3 = 15$. Thus the vertex of the parabola is $(3, 15)$. Because the leading coefficient $-2$ is negative, this is a maximum.

4. **Locating the vertex of a parabola**: Since the vertex of the graph of $Ax^2 + Bx + C$ occurs at $x = -\dfrac{B}{2A}$ (here we used capital letters to avoid confusion with the $a$ of this function), the vertex of $a^2x^2 + 4$ occurs at $x = -\dfrac{0}{2 \times a^2} = 0$. The vertical coordinate of the vertex is found by getting the function value at $x = 0$, which is $a^2 0^2 + 4 = 4$. Thus the vertex of the parabola is $(0, 4)$. Because the leading coefficient $a^2$ is positive, this is a minimum.

5. **Finding the least area of a triangle**:

   (a) Assume that $x$ is measured in yards, and let $y$ be the base of the triangle (in yards). Since we are to use 100 yards of fence, we have $x + y = 100$. Solving for $y$ gives $y = 100 - x$. Thus the base of the triangle is $100 - x$ yards.

   (b) Because the area of a triangle is half of the product of the base with the height, from Part (a) we see that the area is $\dfrac{1}{2}(100 - x)x$ square yards.

   (c) By Part (b) the area is given by the quadratic function $\dfrac{1}{2}(100 - x)x$. We write this in standard form as $-\dfrac{1}{2}x^2 + 50x$. The vertex occurs at $x = -\dfrac{50}{2 \times -\frac{1}{2}} = 50$. Because the leading coefficient $-\dfrac{1}{2}$ is negative, this is a maximum. Thus $x = 50$ yards gives a maximum area.

(d) To find the maximum area, we calculate the value of the area function $\dfrac{1}{2}(100 - x)x$ at $x = 50$:

$$\text{Greatest area } = \frac{1}{2}(100 - 50)(50) = 1250 \text{ square yards.}$$

The maximum area is 1250 square yards.

6. **Distance from a line**: To find the $x$ value giving the point on the line nearest the origin, we find where the minimum of $D$, the square of the distance function, occurs. To analyze $D$ we expand the expression and collect terms:

$$D = x^2 + (x + 1)^2 = x^2 + (x^2 + 2x + 1) = 2x^2 + 2x + 1.$$

Now $2x^2 + 2x + 1$ is a quadratic function in standard form, and the vertex occurs at $x = -\dfrac{2}{2 \times 2} = -\dfrac{1}{2}$. Because the leading coefficient 2 is positive, this is a minimum. Thus the desired $x$ value is $x = -\dfrac{1}{2}$.

7. **Size of high schools**: We want to minimize the function $C = 65{,}000 - 500n + n^2$, where $n$ is the number of pupils enrolled and $C$ is the cost (in dollars) per pupil. This is a quadratic function, and the vertex occurs at $n = -\dfrac{-500}{2 \times 1} = 250$. Because the leading coefficient 1 is positive, this is a minimum. Thus an enrollment size of 250 gives a minimum cost per pupil. The minimum cost is $C(250) = 65{,}000 - 500 \times 250 + 250^2 = 2500$ dollars per pupil.

8. **Bending metal**: We want to minimize the function $T = 4t^2 - 16t + 130$, where $t$ is the time in minutes spent cooling and $T$ the temperature in degrees Fahrenheit. This is a quadratic function, and the vertex occurs at $t = -\dfrac{-16}{2 \times 4} = 2$. Since the leading coefficient 4 is positive, the vertex is a minimum. Thus the minimum temperature occurs at $t = 2$ minutes, at which time the temperature is $T(2) = 4 \times 2^2 - 16 \times 2 + 130 = 114$ degrees Fahrenheit.

9. **Concentration of medicine**: We want to maximize the function $C = 6 + 6t - t^2$, where $t$ is the time in hours after ingestion of the medicine and $C$ the concentration of the drug in milligrams per liter. This is a quadratic function, and the vertex occurs at $t = -\dfrac{6}{2 \times -1} = 3$. Since the leading coefficient $-1$ is negative, the vertex is a maximum. Thus the maximum concentration occurs at $t = 3$ hours, at which time the concentration is $C(3) = 6 + 6 \times 3 - 3^2 = 15$ milligrams per liter.

10. **A rock**: We want to maximize the function $S = -16t^2 + 20t + 5$, where $t$ is the time in seconds after the rock is tossed and $S$ the height of the rock in feet. This is a quadratic function, and the vertex occurs at $t = -\dfrac{20}{2 \times -16} = \dfrac{5}{8}$, or 0.63 second. Since the leading coefficient $-16$ is negative, the vertex is a maximum. Thus the maximum height occurs at $t = \dfrac{5}{8}$, or 0.63 second, at which time the height is $S(5/8) = -16(5/8)^2 + 20(5/8) + 5 = 45/4 = 11.25$ feet.

11. **A fish farm**: We want to maximize the function $G = n(1-n)/2$, where $n$ is the population size in tons and $G$ is the population growth rate in tons per year. Multiplying out the expression, we find $G = n(1-n)/2 = (n - n^2)/2 = \dfrac{n}{2} - \dfrac{n^2}{2}$. We see that this is a quadratic function, and the vertex occurs at $n = -\dfrac{1/2}{2 \times -1/2} = 1/2$, or 0.50 ton. Since the leading coefficient $-1/2$ is negative, the vertex is a maximum. Thus the maximum growth rate occurs at $n = 0.50$ ton, for which the growth rate is $G(0.5) = 0.5(1 - 0.5)/2 = 0.125$, or 0.13 ton per year.

12. **Net profit**: We want to maximize the function $N = 300 + 1000d - d^2$, where $d$ is the amount of advertising dollars spent and $N$ is the net profit from sales. This is a quadratic function, and the vertex occurs at $d = -\dfrac{1000}{2 \times -1} = 500$. Since the leading coefficient $-1$ is negative, the vertex is a maximum. Thus the maximum profit occurs when $d = 500$ advertising dollars, for which the net profit is $N(500) = 300 + 1000 \times 500 - 500^2 = 250{,}300$ dollars net profit.

13. **A window**:

   (a) The perimeter of the opening consists of the circumference of two half-circles each of radius $d/2$ (so, in all, the circumference of a full circle of radius $d/2$) plus the two sides of the rectangle, each of length $h$. So the perimeter is $2\pi(d/2) + 2h = \pi d + 2h$.

   (b) The area of the opening equals the areas of two half-circles of radius $d/2$ (so in all the area of a full circle of radius $d/2$) plus the area of the rectangle of height $h$ and width $d$, so the combined area is $A = dh + \pi(d/2)^2 = dh + \pi(d^2/4)$.

   (c) If the perimeter is 10 feet, then by Part (a), $\pi d + 2h = 10$. Solving for $h$, we have $2h = 10 - \pi d$ and so $h = \dfrac{10 - \pi d}{2}$.

   (d) To find the area, we have a formula for $A$ in terms of $d$ and $h$ in Part (b) and a formula for $h$ in terms of $d$ in Part (c) (since the perimeter is 10 feet), so we can substitute the second into the first to get a formula for $A$ in terms of $d$ only:

$$
\begin{aligned}
A &= dh + \pi\frac{d^2}{4} \\
&= d\left(\frac{10 - \pi d}{2}\right) + \pi\frac{d^2}{4} \\
&= 5d - \frac{\pi}{2}d^2 + \frac{\pi}{4}d^2 \\
&= 5d - \frac{\pi}{4}d^2
\end{aligned}
$$

This is a quadratic function, and the vertex occurs at $d = -\dfrac{5}{2 \times (-\pi/4)} = \dfrac{10}{\pi} = 3.18$ feet. Since the leading coefficient $-\pi/4$ is negative, the vertex is a maximum. Thus the maximum area occurs when $d = \dfrac{10}{\pi} = 3.18$ feet and, using Part (c), $h = \dfrac{10 - \pi(10/\pi)}{2} = 0$ feet (!), i.e. the maximum area is achieved when there is no rectangle and the window is simply a circle.

# Solution Guide for Chapter 3: Straight Lines and Linear Functions

## 3.1 THE GEOMETRY OF LINES

1. **A line with given intercepts**: The picture is drawn below. Since the line is falling from left to right, we expect the slope of the line to be negative. The slope is

$$m = \frac{\text{Vertical change}}{\text{Horizontal change}} = \frac{0-4}{3-0} = -\frac{4}{3}.$$

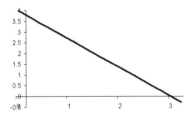

3. **Another line with given vertical intercept and slope**: Because the graph falls by 2 units for each unit of run, we need to move $\frac{8}{2} = 4$ units to the right of the vertical axis to get to the horizontal intercept. Thus the horizontal intercept is 4. More formally, if we move from the location of the horizontal intercept to that of the vertical, we have

$$\text{Slope} = \frac{\text{Vertical change}}{\text{Horizontal change}} = \frac{\text{Vertical intercept} - 0}{0 - \text{Horizontal intercept}},$$

so

$$-2 = \frac{8}{-\text{Horizontal intercept}},$$

and solving gives the value of 4 for the horizontal intercept.

5. **Lines with the same slope**: The first line should have a vertical intercept at 3 and rise 2 units for each unit of run. The second should have a vertical intercept of 1 and rise 2 units for each unit of run. Our picture is shown below. The lines do not cross. Different lines with the same slope are parallel.

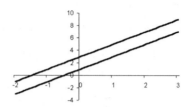

7. **Slides**:

   (a) We focus on the line formed by the slide, where a change of 7 feet in the horizontal direction corresponds to a change of 5 feet in the vertical direction. The slope is thus

   $$\text{Slope} = \frac{\text{Vertical change}}{\text{Horizontal change}} = \frac{5}{7} = 0.71 \text{ foot per foot.}$$

   (b) A faster ride corresponds to a steeper line, so the slope should be increased.

9. **A wrap skirt**: We focus on the line at the right of the figure, where a change of 3 inches in the horizontal direction corresponds to a change of 20 inches in the vertical direction, if we take the positive direction to be downward. The slope of the line is thus

   $$\text{Slope} = \frac{\text{Vertical change}}{\text{Horizontal change}} = \frac{20}{3}.$$

   An altered skirt that is 24 inches long corresponds to a change of $24 - 20 = 4$ inches in the vertical direction from the bottom hem of the 20-inch skirt. Now

   $$
   \begin{aligned}
   \text{Vertical change} &= \text{Horizontal change} \times \text{Slope} \\
   4 &= \text{Horizontal change} \times \frac{20}{3} \\
   \frac{4}{20/3} &= \text{Horizontal change} \qquad \textbf{Divide both sides by 20/3.} \\
   0.6 &= \text{Horizontal change.}
   \end{aligned}
   $$

   Therefore, for the altered skirt there is an increase of 0.6 inch to the right in the length of the hem. The increase to the left is the same, so the total increase is 1.2 inches. Therefore, the bottom hem of the altered skirt is $63 + 1.2 = 64.2$ inches long.

11. **A topographical map**: We focus on the line from the eye to the peak. A change of 10 feet in the horizontal direction from the eye to the pole corresponds to a change of $8 - 6 = 2$ feet in the vertical direction. The slope of the line is thus

$$\text{Slope} = \frac{\text{Vertical change}}{\text{Horizontal change}} = \frac{2}{10} = 0.2.$$

Moving from the eye to the peak corresponds to a change of 2 miles or $2 \times 5280 = 10{,}560$ feet in the horizontal direction. This move gives a vertical change of

$$\text{Vertical change} = \text{Slope} \times \text{Horizontal change} = 0.2 \times 10{,}560 = 2112 \text{ feet.}$$

This is the vertical change from the 6-foot-high eye, so the height of the peak is $6 + 2112 = 2118$ feet.

13. **A ramp to a building**:

   (a) The slope of the graph is 0.4, so one foot of run results in 0.4 foot of rise. Thus, one foot from the base of the ramp, it is 0.4 foot high.

   (b) To get to the steps we have moved horizontal 15 feet from the base of the ramp. Since the slope is 0.4, the ramp rises

$$\text{Rise} = \text{Run} \times \text{Slope} = 15 \times 0.4 = 6 \text{ feet.}$$

15. **A cathedral ceiling**:

   (a) When we move 3 feet east from the west wall, the ceiling rises $10.5 - 8 = 2.5$. So the slope is

$$\text{Slope} = \frac{\text{Rise}}{\text{Run}} = \frac{2.5}{3} = 0.83 \text{ foot per foot.}$$

   (b) If we move 17 feet to the right of the west wall then the vertical change is

$$\text{Rise} = \text{Slope} \times \text{Run} = 0.83 \times 17 = 14.11 \text{ feet.}$$

So the height of the ceiling at that point is $8 + 14.11 = 22.11$ feet.

   (c) Since we start at 8 feet high on the west wall and we want to reach a 12-foot height, then the vertical change is 4 feet. Now

$$
\begin{aligned}
\text{Vertical change} \ &= \ \text{Horizontal change} \times \text{Slope} \\
4 \ &= \ \text{Horizontal change} \times 0.83 \\
\frac{4}{0.83} \ &= \ \text{Horizontal change} \qquad \textbf{Divide both sides by 0.83.} \\
4.82 \ &= \ \text{Horizontal change.}
\end{aligned}
$$

So if we place the light 4.82 feet from the west wall, then we will still be able to change the bulb.

17. **Cutting plywood siding**: The shape of the first piece shows that for a horizontal change (to the left) of 4 feet the vertical change along the roof line is $2.5 - 1 = 1.5$ feet. Now each piece has width 4 feet, so we observe that for each piece the longer side and the shorter side differ in length by 1.5 feet.

   (a) By the observation above, the length $h$ is $2.5 + 1.5 = 4$ feet.

   (b) By the observation above, the length $k$ is $h + 1.5 = 4 + 1.5 = 5.5$ feet, or 5 feet 6 inches.

   *Alternative approach*: In both parts we focus on the part of the roof line sloping upward from right to left, and we take movement toward the center as going in the positive direction. Moving along this line from the outer wall to the far corner of the first piece of siding, we have a horizontal change of 4 feet and a vertical change of $2.5 - 1 = 1.5$ feet. Thus the slope of this line is $\dfrac{1.5}{4} = 0.375$ foot per foot. For Part(a): Moving along upper edge of the second piece of siding toward the peak, we have a horizontal change of 4 feet. Because the slope is 0.375, from

   $$\text{Vertical change} = \text{Slope} \times \text{Horizontal change},$$

   we find

   $$\text{Vertical change} = 0.375 \times 4 = 1.5 \text{ feet.}$$

   To find $h$ we add this vertical change to the length 2.5 feet of the short side of the second piece. Thus the length $h$ is $2.5 + 1.5 = 4$ feet.

   For Part(b): Moving along upper edge of the third piece of siding toward the peak, we have a horizontal change of 4 feet. Because the slope is 0.375, from

   $$\text{Vertical change} = \text{Slope} \times \text{Horizontal change},$$

   we again find

   $$\text{Vertical change} = 0.375 \times 4 = 1.5 \text{ feet.}$$

   To find $k$ we add this vertical change to the length 4 feet of the short side of the third piece. Thus the length $h$ is $4 + 1.5 = 5.5$ feet.

19. **Looking over a wall**: Assuming that the man can just see the top of the building over the wall, we focus on the line of sight from the man to the top of the building. Taking south as the positive direction, we compute the slope of the line by considering the change from the top of the wall to the top of the building. We find that the slope of the line is

$$\text{Slope} = \frac{\text{Vertical change}}{\text{Horizontal change}} = \frac{50-35}{20} = \frac{15}{20} = 0.75 \text{ foot per foot.}$$

Now we find the horizontal distance from the wall to the man by considering the change along the line from the man to the top of the wall. We use

$$\text{Vertical change} = \text{Slope} \times \text{Horizontal change,}$$

which says

$$35 - 6 = 0.75 \times \text{Horizontal distance.}$$

Dividing both sides by 0.75 shows that the horizontal distance is $\dfrac{35-6}{0.75} = 38.67$ feet. Thus the man must be at least 38.67 feet north of the wall.

21. **A road up a mountain**:

   (a) The vertical change is $4960 - 4130 = 830$ feet. The horizontal change is 3 miles. So the slope is
   $$\frac{\text{Vertical change}}{\text{Horizontal change}} = \frac{830}{3} = 276.67 \text{ feet per mile.}$$

   (b) The vertical change in moving 5 miles from the first sign is
   $$\text{Rise} = \text{Run} \times \text{Slope} = 5 \times 276.67 = 1383.35 \text{ feet.}$$

   Since we started at 4130 feet at the first sign, the elevation is $4130 + 1383.35 = 5513.35$ feet when we are five miles from the first sign.

   (c) The vertical change from the first sign to the peak is $10,300 - 4130 = 6170$ feet.

   $$
   \begin{aligned}
   \text{Vertical change} &= \text{Horizontal change} \times \text{Slope} \\
   6170 &= \text{Horizontal change} \times 276.67 \\
   \frac{6170}{276.67} &= \text{Horizontal change} && \textbf{Divide both sides by 276.67.} \\
   22.30 &= \text{Horizontal change.}
   \end{aligned}
   $$

   Thus the peak is 22.30 horizontal miles away.

23. **Earth's umbra**: We take the positive horizontal axis to pass from the point where the umbra ends through the center of Earth. We find the slope of the upper line shown in the figure by considering the change from the point where the umbra ends to the surface of Earth. We find that the slope is

$$\text{Slope} = \frac{\text{Vertical change}}{\text{Horizontal change}} = \frac{\text{Radius of Earth}}{\text{Distance from Earth to end of umbra}} = \frac{3960}{860{,}000} \text{ mile per mile.}$$

Now we find the radius of the umbra at the indicated point by considering the change along the line from the point where the umbra ends to the radius at the indicated point. We use

$$\text{Vertical change} = \text{Slope} \times \text{Horizontal change.}$$

Now the vertical change is the radius of the umbra we are asked to find, and the horizontal change is $860{,}000 - 239{,}000 = 621{,}000$. Using the slope calculated above, we find

$$\text{Radius of the umbra} = \frac{3960}{860{,}000} \times 621{,}000 = 2859.49 \text{ miles.}$$

Thus the radius of the umbra at the indicated point is about 2859 miles. Since the radius of the moon is smaller than this, the moon can fit inside Earth's umbra. When this happens there is a total lunar eclipse.

25. **The umbra of the moon**: We take the positive horizontal axis to pass from the apex (at the center of Earth) to the center of the sun. We find the slope of the line from the apex to the surface of the sun by considering the change from the apex to the surface of the moon. (This can be visualized by replacing Earth by the moon in Figure 3.23.) We find that the slope is

$$\text{Slope} = \frac{\text{Vertical change}}{\text{Horizontal change}} = \frac{\text{Radius of moon}}{\text{Distance from Earth to moon}} = \frac{1100}{239{,}000} \text{ mile per mile.}$$

Now we find the radius of the sun by considering the change from the apex to the surface of the sun. We use

$$\text{Vertical change} = \text{Slope} \times \text{Horizontal change.}$$

Now the vertical change is the desired radius of the sun, and the horizontal change is the distance from Earth to the sun. Using the slope calculated above, we find

$$\text{Radius of the sun} = \frac{1100}{239{,}000} \times 93{,}498{,}600 = 430{,}328.28 \text{ miles.}$$

Thus the radius of the sun is about 430,328 miles.

## Skill Building Exercises

S-1. **A ramp**: We focus on the line formed by the ramp, where a change of 12 feet in the horizontal direction corresponds to a change of 3 feet in the vertical direction. The slope is thus

$$\text{Slope} = \frac{\text{Vertical change}}{\text{Horizontal change}} = \frac{3}{12} = \frac{1}{4} = 0.25 \text{ foot per foot.}$$

S-3. **Slope from rise and run**: We have $\text{Slope} = \dfrac{\text{Rise}}{\text{Run}}$. In this case, the rise is 8 feet, the height of the wall, and the run is 2 feet, since that is the horizontal distance from the wall. Thus the slope is $\dfrac{8}{2} = 4$ feet per foot.

S-5. **Height from slope and horizontal distance**: We have

$$\text{Vertical change} = \text{Slope} \times \text{Horizontal change.}$$

In this case, the slope of the ladder is 2.5, while the horizontal distance is 3 feet, so the vertical height is $2.5 \times 3 = 7.5$ feet.

S-7. **Horizontal distance from height and slope**: We have

$$\text{Vertical change} = \text{Slope} \times \text{Horizontal change.}$$

In this case, the slope of the ladder is 1.75, while the vertical distance is 9 feet, so we have $9 = 1.75 \times \text{Horizontal change}$. Thus the horizontal distance is $\dfrac{9}{1.75} = 5.14$ feet.

S-9. **Slope from two points**: We have $\text{Slope} = \dfrac{\text{Vertical change}}{\text{Horizontal change}}$. In this case, the vertical change is $-2$ feet (12 feet dropped to 10 feet) while the horizontal change is 3 feet (since west is the positive direction). Thus the slope is $\dfrac{-2}{3} = -\dfrac{2}{3}$ foot per foot.

S-11. **A circus tent**: We have Vertical change $=$ Slope $\times$ Horizontal change. In this case, the slope is $-0.8$, while the horizontal change is 7 feet, so the vertical change is $-0.8 \times 7 = -5.6$ feet. Thus the height of the tent if you walk 7 feet west is $22 - 5.6 = 16.4$ feet.

S-13. **Slope**: We have $\text{Slope} = \dfrac{\text{Rise}}{\text{Run}}$. In this case, the rise is 100 feet, the height of the building, and the run is 70 feet, since that is the horizontal distance to the building. Thus the slope is $\dfrac{100}{70} = 1.43$ feet per foot.

S-15. **Slopes of lines in the coordinate plane**: The slope of the line through the points $(2, 2)$ and $(4, 1)$ is

$$\text{Slope} = \frac{b_2 - b_1}{a_2 - a_1} = \frac{1 - 2}{4 - 2} = \frac{-1}{2},$$

or $-0.5$.

**S-17. Slopes of lines in the coordinate plane:** The slope of the line through the points $(-3.6, 2.5)$ and $(2.4, -1.6)$ is

$$\text{Slope} = \frac{b_2 - b_1}{a_2 - a_1} = \frac{(-1.6) - 2.5}{2.4 - (-3.6)} = \frac{-4.1}{6.0} = -0.68.$$

**S-19. Slopes of lines in the coordinate plane:** The line through $(1.2, 3.1)$ must fall by 3.1 vertical units to reach the horizontal axis. Since the slope is $-0.8$, a horizontal change of $\frac{-3.1}{-0.8} = 3.88$ units will result in a vertical change of $3.88 \times -0.8 = -3.1$ units, so the line meets the horizontal axis at the point $(1.2 + 3.88, 3.1 - 3.1) = (5.08, 0)$, or $x = 5.08$.

**S-21. Slopes of lines in the coordinate plane:** The line through $(-4, 3)$ must increase by 4 horizontal units to reach the vertical axis. Since the slope is 3, a horizontal change of 4 units will result in a vertical change of $4 \times 3 = 12$ units, so the line meets the vertical axis at the point $((-4) + 4, 3 + 12) = (0, 15)$, or $y = 15$.

**S-23. Slopes of lines in the coordinate plane:** The line through $(3.2, -1.5)$ must decrease by 3.2 horizontal units to reach the vertical axis. Since the slope is $-2.3$, a horizontal change of $-3.2$ units will result in a vertical change of $-2.3 \times -3.2 = 7.36$ units, so the line meets the vertical axis at the point $(3.2 - 3.2, (-1.5) + 7.36) = (0, 5.86)$, or $y = 5.86$.

**S-25. Slopes of lines in the coordinate plane:** The line through $(1, 5)$ must increase by 5.6 horizontal units to reach the point on it with horizontal coordinate 6.6. Since the slope is $-1.1$, a horizontal change of 5.6 units will result in a vertical change of $-1.1 \times 5.6 = -6.16$ units, so the point $(1 + 5.6, 5 + (-6.16)) = (6.6, -1.16)$ lies on the line.

## 3.2  LINEAR FUNCTIONS

1. **A dairy:**

   (a) Now

$$
\begin{aligned}
\text{Total yearly expense} \;&=\; \text{Cost for caring and feeding cows } + \text{ Cost of maintenance} \\
E \;&=\; 2000 \text{ dollars per cow } \times \text{ Number of cows } + 25{,}000 \text{ dollars} \\
E \;&=\; 2000C + 25{,}000 \text{ dollars.}
\end{aligned}
$$

Here is an alternative approach that uses the language of linear functions: The information given tells us that the rate of change of $E$ is constant, namely 2000 dollars per cow, so $E$ is a linear function of $C$ with slope 2000 dollars per cow. The initial value is the maintenance cost, which is \$25,000. Therefore, the formula for $E$ is $E = 2000C + 25{,}000$.

(b) Expressed in functional notation, the total expense if the dairy has 30 cows is $E(30)$.

(c) By the formula from part a, $E(30) = 2000 \times 30 + 25{,}000 = 85{,}000$ dollars. Thus, the total expense if the dairy has 30 cows is \$85,000.

3. **Gasoline prices**:

(a) The information given tells us that the rate of change of $G$ is constant, namely 2.5 cents per gallon per year, so $G$ is a linear function of $t$ with slope 2.5 cents per gallon per year. The initial value is the price in 1960, which is 31 cents per gallon. Therefore, the formula for $G$ is $G = 2.5t + 31$.

(b) Because 1990 is 30 years after 1960 and $G(30) = 2.5 \times 30 + 31 = 106$, the model yields the price 106 cents per gallon or \$1.06 per gallon.

(c) Answers will vary.

5. **Walking**:

(a) Let $C$ denote the total number of calories burned. Now

$$
\begin{aligned}
\text{Calories burned} \;&=\; \text{Calories burned during walk } + \text{ Calories burned during workout} \\
C \;&=\; 258 \text{ calories per hour } \times \text{ Number of hours } + 700 \text{ calories} \\
C \;&=\; 258h + 700 \text{ calories.}
\end{aligned}
$$

Here is an alternative approach that uses the language of linear functions: The information given tells us that the rate of change of $C$ is constant, namely 258 calories per hour, so $C$ is a linear function of $h$ with slope 258 calories per hour. The initial value is the total number of calories burned if we don't take the walk, which is 700. Therefore, the formula for $C$ is $C = 258h + 700$.

(b) We want to find the value of $h$ for which $C = 1100$, so we need to solve the equation $258h + 700 = 1100$. We solve it as follows:

$$
\begin{aligned}
258h + 700 \;&=\; 1100 \\
258h \;&=\; 1100 - 700 = 400 \qquad \textbf{Subtract 700 from both sides.} \\
h \;&=\; \frac{400}{258} \qquad \textbf{Divide both sides by 258.} \\
h \;&=\; 1.55.
\end{aligned}
$$

Therefore, you need to walk 1.55 hours in order to burn a total of 1100 calories.

7. **Nail growth**: As in the solution to Exercise 1, there are a couple of ways to find the formula. Here we use the language of linear functions: The information given tells us that the rate of change of $L$ is constant, namely 3 millimeters per month, so $L$ is a linear function of $t$ with slope 3 millimeters per month. The initial value is 12 millimeters. Therefore, the formula for $L$ is $L = 3t + 12$.

9. **Getting Celsius from Fahrenheit**:

   (a) Water freezes at 0 degrees Celsius, which is 32 degrees Fahrenheit. So $C = 0$ when $F = 32$. Water boils at 100 degrees Celsius, which is 212 degrees Fahrenheit. So $C = 100$ when $F = 212$. These two bits of information allow us to get the slope of the function:

   $$\text{Slope} = \frac{\text{Change in } C}{\text{Change in } F} = \frac{100 - 0}{212 - 32} = 0.56.$$

   Thus $C = 0.56F + b$, and we need to find $b$. When $C = 0$, then $F = 32$. This gives $0 = 0.56 \times 32 + b$. Solving for $b$, we get $b = -17.92$. Thus $C = 0.56F - 17.92$.

   There are some variations that yield slightly different formulas. The differences are due to the rounding of $\frac{5}{9}$ as 0.56. If the fraction is left as a fraction, then the formula will be $C = \frac{5}{9}F - \frac{160}{9}$, which, to two decimal places, is $C = 0.56F - 17.78$. If the second data point, $C = 100$ when $F = 212$, is used to find $b$ from $C = 0.56F + b$, the resulting calculation will yield $b = -18.72$, giving the formula $C = 0.56F - 18.72$.

   (b) The slope we found in Part (a) was 0.56. This means that for every one degree increase in the Fahrenheit temperature, the Celsius temperature will increase by 0.56 degree.

   (c) We need to solve $F = 1.8C + 32$ for $C$. We have

   $$
   \begin{aligned}
   F &= 1.8C + 32 \\
   F - 32 &= 1.8C & &\text{Subtract 32 from each side.} \\
   \frac{1}{1.8} \times (F - 32) &= C & &\text{Divide both sides by 1.8.} \\
   \frac{1}{1.8}F - \frac{1}{1.8} \times 32 &= C \\
   0.56F - 17.78 &= C
   \end{aligned}
   $$

   This is the same as the formula we found in Part (a).

11. **Digitized pictures on a disk drive:**

(a) Each additional picture stored increases the total storage space used by 2 megabytes. This means that the change in total storage space used is always the same, 2 megabytes, for a change of 1 in the number of pictures that are stored. Thus the storage space used $S$ is a linear function of the number of pictures stored $n$.

(b) From Part (a), $S$ is a linear function of $n$ with slope 2, since the slope represents the additional storage space used with the addition of 1 picture. The initial value of $S$ is the amount of storage space used if $n = 0$, that is, if no pictures are stored. Since the formatting information, operating system, and applications software use 6000 megabytes, that is the initial value. Thus a formula for $S(n)$ is $S = 2n + 6000$.

(c) The total amount of storage space used on the disk drive if there are 350 pictures stored on the drive is expressed in functional notation as $S(350)$. (Of course, this depends on the choice of function name we made in Part (b).) Its value is $S(350) = 2 \times 350 + 6000 = 6700$ megabytes.

(d) If there are 769,000 megabytes free and the drive holds 800,000 megabytes, then there are $800,000 - 769,000 = 31,000$ megabytes used. To find out how many pictures are stored, we need to solve this equation for $n$:

$$
\begin{aligned}
2n + 6000 &= 31,000 \\
2n &= 31,000 - 6000 = 25,000 \qquad \textbf{Subtract 6000 from both sides.} \\
n &= \frac{25,000}{2} = 12,500 \text{ pictures} \qquad \textbf{Divide both sides by 2.}
\end{aligned}
$$

Thus there are 12,500 digitized pictures stored. There are 769,000 megabytes of room left, and that space can hold $\dfrac{769,000}{2} = 384,500$ more pictures before the disk drive is full.

13. **Total cost:**

(a) Because the variable cost is a constant $20 per widget, for each additional widget produced per month the monthly cost increases by the same amount, $20. This means that $C$ always increases by 20 when $N$ increases by 1. Thus $C$ has a constant rate of change and so is a linear function of $N$. The slope is the constant rate of change, namely 20 dollars per widget. The initial value is the monthly cost when no widgets are manufactured, and this is the amount of the fixed costs, namely $1500. Since the slope is 20 and the initial value is 1500, the formula is $C = 20N + 1500$.

(b) Let $A$ represent the monthly costs (in dollars) for this other manufacturer and $N$ the number of widgets produced in a month. As in Part (a), the slope of this linear function is given by the variable cost, which in this case is 12 dollars per widget. Thus $A = 12N + b$ for some constant $b$. In fact, the constant $b$ is the initial value of the function, and, as in Part (a), this represents the fixed costs. We find $b$ using the fact that $A = 3100$ when $N = 150$: We have $3100 = 12 \times 150 + b$, so $b = 3100 - 12 \times 150 = 1300$. Thus the amount of fixed costs is $1300.

Another way of finding this is to start with the total cost of $3100 at a production level of $N = 150$ and subtract $12 for each widget produced to account for the variable cost; the resulting amount of $3100 - 12 \times 150 = 1300$ dollars is the amount of fixed costs.

(c) Let $B$ represent the monthly costs (in dollars) for this manufacturer and $N$ the number of widgets produced in a month. As in Parts (a) and (b), the slope of this linear function is the variable cost, and the initial value of the function represents the fixed costs. We know that $B = 2700$ when $N = 100$ and that $B = 3500$ when $N = 150$. Thus the slope of $B$ is given by

$$\text{Slope} = \frac{\text{Change in } B}{\text{Change in } N} = \frac{3500 - 2700}{150 - 100} = \frac{800}{50} = 16 \text{ dollars per widget.}$$

Hence the variable cost is $16 per widget. We now know that $B = 16N + b$, where $b$ is the initial value (representing the fixed costs). We find $b$ using the fact that $B = 2700$ when $N = 100$: We have $2700 = 16 \times 100 + b$, so $b = 2700 - 16 \times 100 = 1100$. Thus the amount of fixed costs is $1100.

15. **Slowing down in a curve**:

(a) Because $S$ decreases by 0.746 when $D$ increases by 1, the slope of $S$ is $-0.746$ mile per hour per degree. The initial value of $S$ is 46.26 miles per hour because that is the speed on a straight road. Thus the formula is $S = -0.746D + 46.26$. Of course, this can also be written as $S = 46.26 - 0.746D$.

(b) The speed for a road with a curvature of 10 degrees is expressed as $S(10)$ in functional notation. The value is $S(10) = -0.746 \times 10 + 46.26 = 38.8$ miles per hour.

17. **Currency conversion**:

(a) We know we can get $P = 46$ pounds for $D = 70$ dollars, and we can get $P = 0$ pounds for $D = 0$ dollars. Thus

$$\text{Slope} = \frac{\text{Change in pounds}}{\text{Change in dollars}} = \frac{46}{70} = 0.66 \text{ pound per dollar.}$$

This means each dollar is worth 0.66 pound.

(b) Since we get 0.66 pound per dollar, for 130 dollars we get $0.66 \times 130 = 85.80$ pounds. So the tourist received 85.80 pounds for the 130 dollars.

(c) We want to know how many dollars we get for 12.32 pounds. That is, we want to solve $0.66D = 12.32$ for $D$. We need only divide each side by 0.66 to get $D = 18.67$ dollars. The tourist received \$18.67.

19. **Adult male height and weight**:

(a) According to this rule of thumb, if a man is 1 inch taller than another, then we expect him to be heavier by 5 pounds. This says that the rate of change in adult male weight is constant, namely 5 pounds per inch, and thus weight is a linear function of height. The slope is the rate of change, 5 pounds per inch.

(b) Let $h$ denote the height in inches and $w$ the weight in pounds. Part (a) tells us that $w$ is a linear function of $h$ with slope 5. Thus $w = 5h + b$. To get the value of $b$, we note that when the weight is $w = 170$ pounds, the height is $h = 70$ inches. Thus $170 = 5 \times 70 + b$, which gives $b = 170 - 5 \times 70 = -180$. We get $w = 5h - 180$.

(c) If the weight is $w = 152$ pounds, then $5h - 180 = 152$. We want to solve this for $h$ to find the height:

$$
\begin{aligned}
5h - 180 &= 152 \\
5h &= 152 + 180 \qquad \textbf{Add } 180 \textbf{ to each side.} \\
h &= \frac{152 + 180}{5} \qquad \textbf{Divide both sides by } 5. \\
h &= 66.4.
\end{aligned}
$$

We would expect the man to be 66.4 inches tall.

(d) According to the rule of thumb, if a man is 75 inches tall then his weight is $w = 5 \times 75 - 180 = 195$ pounds. The atypical man with a height of 75 inches and a weight of 190 pounds would therefore be light for his height.

21. **Lean body weight in females**: We noted in Exercise 20 that lean body weight in young adult males increases by 1.08 pounds for every pound of increase in total weight, assuming that the abdominal circumference remains the same. For young adult females, the formula shows that the slope of lean body weight as a function of total weight alone (for fixed values of $R$, $A$, $H$, and $F$) is 0.73 pound per pound. This means that their lean body weight increases by only 0.73 pound for every pound of increase in their total weight, assuming that all other factors remain the same.

23. **Vertical reach of fire hoses**:

(a) Because the vertical factor $V$ always increases by 5 when $d$ increases by $\frac{1}{8}$ inch, the vertical factor has a constant rate of change and hence is linear.

(b) First we find the slope of $V$. Because $V$ always increases by 5 when $d$ increases by $\frac{1}{8}$, the slope is given by

$$\text{Slope} = \frac{\text{Change in } V}{\text{Change in } d} = \frac{5}{\frac{1}{8}} = 40.$$

We now know that $V = 40d + b$, where $b$ is a constant. We find $b$ using the fact that $V = 85$ when $d = 0.5$: We have $85 = 40 \times 0.5 + b$, so $b = 85 - 40 \times 0.5 = 65$. Thus the formula is $V = 40d + 65$.

(c) We are given that $p = 50$ and $d = 1.75$. Using the formula for $V$ from Part (b), we have $V = 40 \times 1.75 + 65$, and thus

$$S = \sqrt{Vp} = \sqrt{(40 \times 1.75 + 65)50} = 82.16 \text{ feet}.$$

Hence the vertical stream will travel 82.16 feet high.

(d) We are given that $d = 1.25$ and $p = 70$. Using the formula for $V$ from Part (b), we have $V = 40 \times 1.25 + 65$, and thus

$$S = \sqrt{Vp} = \sqrt{(40 \times 1.25 + 65)70} = 89.72 \text{ feet}.$$

Hence the horizontal stream will travel 89.72 feet high. This is greater than the height of 60 feet, so the stream can reach the fire.

25. **More on budget constraints**:

(a) If you buy $a$ pounds of apples and each pound costs \$1, you will have spent $1 \times a = a$ dollars on apples.

(b) If you buy $g$ pounds of grapes and each pound costs \$2, you will have spent $2 \times g = 2g$ dollars on grapes.

(c) Since you will spend a total of \$5 on some combination of grapes and apples, then

$$\text{Money spent on apples } + \text{ Money spent on grapes} = \$5,$$

so $a + 2g = 5$.

(d) To solve $a + 2g = 5$ for $g$, subtract $a$ from each side, and divide by 2, to get $g = (-a + 5)/2 = -0.50a + 2.5$, which is the same as the answer for Part (d) from Exercise 24.

27. **Sleeping longer**: If we let $M$ be the number of hours the man sleeps and $t$ the number of days since his observation, then $M$ is a linear function of $t$, with slope $\frac{1}{4}$ hour per day (since 15 minutes is $\frac{1}{4}$ hour) and initial value 8 hours. Thus, $M = \frac{1}{4}t + 8$. Since we want to know when the man sleeps 24 hours, we want to find $t$ so that $M(t) = 24$. Thus we want to solve the equation $\frac{1}{4}t + 8 = 24$. We find that $\frac{1}{4}t = 16$, so $t = 64$. Thus, 64 days after his observation the man will sleep 24 hours.

To answer this without using the language of linear functions, note that the exercise is asking how many $\frac{1}{4}$-hour segments we need to add to 8 to get 24, i.e., how many $\frac{1}{4}$-hour segments there are in 16 hours. Clearly there are 64.

29. **Life on other planets**:

(a) To calculate the most pessimistic value for $N$ we take the pessimistic value for each of the variables other than $S$ (because $N$ is an increasing function of each variable). Thus the most pessimistic value for $N$ is

$$N = (2 \times 10^{11}) \times 0.01 \times 0.01 \times 0.01 \times 0.01 \times 10^{-8} = 2 \times 10^{-5}.$$

Hence the most pessimistic value for $N$ is $2 \times 10^{-5}$, or 0.00002. To calculate the most optimistic value for $N$ we take the optimistic value for each of the variables other than $S$. Thus the most optimistic value for $N$ is

$$N = (2 \times 10^{11}) \times 0.5 \times 1 \times 1 \times 1 \times 10^{-4} = 10{,}000{,}000.$$

Hence the most optimistic value for $N$ is 10,000,000.

(b) Keeping $L$ as a variable and using pessimistic values for each of the other variables (except for $S$) gives

$$N = (2 \times 10^{11}) \times 0.01 \times 0.01 \times 0.01 \times 0.01 \times L = 2000L.$$

Now we want to find what value of the variable $L$ gives the function value $N = 1$. Thus we need to solve the equation $2000L = 1$. The result is $L = \frac{1}{2000} = 5 \times 10^{-4}$. Thus a value of $L = 5 \times 10^{-4}$, or $L = 0.0005$, will give 1 communicating civilization per galaxy with pessimistic values for the other variables.

(c) Because there is such a difference, even in orders of magnitude, between the pessimistic and optimistic values for the variables and for the function $N$, the formula does not seem helpful in determining if there is life on other planets.

## Skill Building Exercises

**S-1. The chess club**: The slope is 3 people per week, which means that, since the first of the year, membership in the chess club has increased by 3 people each week.

**S-3. Money in the cookie jar**: The slope is 8 dollars per week, which means that, since the birthday, the amount of money in the jar increased by 8 dollars each week.

**S-5. Slope from two values**: We have that the slope $m$ is:

$$m = \frac{\text{Change in function}}{\text{Change in variable}} = \frac{19 - 7}{5 - 2} = \frac{12}{3} = 4.$$

**S-7. Function value from slope and run**: We have that the slope is $m = 2.7$:

$$2.7 = \frac{\text{Change in function}}{\text{Change in variable}} = \frac{f(5) - f(3)}{5 - 3} = \frac{f(5) - 7}{5 - 3} = \frac{f(5) - 7}{2}.$$

Thus $f(5) - 7 = 2 \times 2.7 = 5.4$, and so $f(5) = 5.4 + 7 = 12.4$.

**S-9. Run from slope and rise**: We have that the slope is $m = -3.4$:

$$-3.4 = \frac{\text{Change in function}}{\text{Change in variable}} = \frac{f(x) - f(1)}{x - 1} = \frac{0 - 6}{x - 1} = \frac{-6}{x - 1}.$$

Thus $-6 = -3.4 \times (x - 1)$, so $x - 1 = \frac{-6}{-3.4} = 1.76$, and therefore $x = 1.76 + 1 = 2.76$.

**S-11. Linear equation from slope and point**: Since $f$ is a linear function with slope 4, $f = 4x + b$ for some $b$. Now $f(3) = 5$, so $5 = 4 \times 3 + b$, and therefore $b = 5 - 4 \times 3 = -7$. Thus $f = 4x - 7$.

**S-13. Linear equation from two points**: We first compute the slope:

$$m = \frac{\text{Change in function}}{\text{Change in variable}} = \frac{f(9) - f(4)}{9 - 4} = \frac{2 - 8}{9 - 4} = \frac{-6}{5} = -1.2.$$

Since $f$ is a linear function with slope $-1.2$, $f = -1.2x + b$ for some $b$. Now $f(4) = 8$, so $8 = -1.2 \times 4 + b$, and therefore $b = 8 + 1.2 \times 4 = 12.8$. Thus $f = -1.2x + 12.8$.

**S-15. Properties of linear functions**: The change in $y$ is the slope times the change in $x$. Since the increase of 5 units is a change in $x$ of 5, the change in $y$ is $3.3 \times 5 = 16.5$ units, so an increase of 16.5 units.

**S-17. Properties of linear functions**: The change in $y$ is the slope times the change in $x$. Since the increase of 6.3 units is a change in $x$ of 6.3, the change in $y$ is $-2.6 \times 6.3 = -16.38$ units, so a decrease of 16.38 units.

S-19. **Properties of linear functions**: The change in $x$ is the change in $y$ divided by the slope. Since the increase of 3 units is a change in $y$ of 3, the change in $x$ is $\dfrac{3}{3.1} = 0.97$ unit, so an increase of 0.97 unit.

S-21. **Properties of linear functions**: The change in $x$ is the change in $y$ divided by the slope. Since the increase of 4.6 units is a change in $y$ of 4.6, the change in $x$ is $\dfrac{4.6}{-3.2} = -1.44$ units, so a decrease of 1.44 units.

S-23. **Properties of linear functions**: The slope is the change in $y$ divided by the change in $x$. Since the decrease of 1.1 units is a change in $y$ of $-1.1$ and the increase of 3.3 units is a change in $x$ of 3.3, then the slope is $\dfrac{-1.1}{3.3} = -0.33$ unit per unit.

S-25. **Properties of linear functions**: The rate of change is the change in $y$ divided by the change in $x$. In this case, the change in $y$ is 9 and then 8 for a change in $x$ of 7, so the rate of change is not constant; $y$ is not a linear function of $x$.

S-27. **Properties of linear functions**: The slope is the change in $y$ divided by the change in $x$, so the slope is $\dfrac{-3.6}{4.4}$. The change in $y$ is the slope times the change in $x$. Since the decrease of 3.3 units is a change in $x$ of $-3.3$, the change in $y$ is $\dfrac{-3.6}{4.4} \times -3.3 = 2.70$ units, so an increase of 2.70 units in $y$.

## 3.3  MODELING DATA WITH LINEAR FUNCTIONS

1. **Employee turnover**:

   (a) The table below shows the differences. For example, the first difference in $C$ is $400 - 250 = 150$.

   | Change in employee turnover | 10 to 20 | 20 to 30 | 30 to 40 |
   |---|---|---|---|
   | Change in cost | 150 | 150 | 150 |

   Because the change in $C$ is 150 for each change of 10 in $E$, the data are linear.

   (b) The difference in the $E$ values is always 10, and the difference in the $C$ values is always 150. This shows that the linear function $C$ has slope $\dfrac{150}{10} = 15$ million dollars per percentage point.

   (c) Because the slope is 15, we have $C = 15E + b$. To find $b$, we use the first point given, $C = 250$ million dollars when $E = 10$ percent. Thus $250 = 15 \times 10 + b$, so $b = 250 - 150 = 100$. Thus, the model is $C = 15E + 100$.

(d) We want to find $C$ when $E = 33$ percent. By the formula from part c, $C(33) = 15 \times 33 + 100 = 595$ million dollars. Thus, if employee turnover is 33% then the cost is 595 million dollars.

3. **Making ice**:

(a) The table below shows the differences over successive time periods. For example, the first difference is $273 - 200 = 73$.

| Change in time | 12:00 to 1:00 | 1:00 to 2:00 | 2:00 to 3:00 |
|---|---|---|---|
| Change in ice (pounds) | 73 | 73 | 73 |

The table of differences shows a constant change of 73 pounds over each 1-hour period. This shows that the data are linear.

(b) The difference in the $t$ values is always 1, and the difference in the $I$ values is always 73. This shows that the linear function $I$ has slope 73 pounds per hour. Now $t = 0$ corresponds to 12:00, so (from the data table) the initial value of $I$ is 200, and thus the equation is $I = 73t + 200$.

(c) Now game time corresponds to $t = 7$, so we plug $t = 7$ into the formula from Part (b): $I = 73 \times 7 + 200 = 711$. Thus, the ice machine will produce 711 pounds by game time. Therefore, the ice machine will indeed produce enough ice (in fact, more than the required 675 pounds) by game time.

5. **Price of Amazon's Kindle**:

(a) Let $t$ be the time in months since February 2009, and let $K$ be the price in dollars. The table below shows the differences over successive time periods. For example, the first difference is $299 - 349 = -50$.

| Change in $t$ | 0 to 5 | 5 to 10 | 10 to 15 |
|---|---|---|---|
| Change in $K$ | $-50$ | $-50$ | $-50$ |

The table of differences shows a common difference of $-50$ dollars every 5 months. This shows that the data can be modeled using a linear function.

(b) The difference in the $t$ values is always 5, and the difference in the $K$ values is always $-50$. This shows that the linear function $K$ has slope $-50/5 = -10$ dollars per month. (Note that the slope is a negative number.) Now $t = 0$ corresponds to February 2009, so (from the data table) the initial value of $K$ is 349, and thus the formula is $K = -10t + 349$.

(c) January 2012 is 35 months after February 2009, so we plug $t = 35$ into the formula from from Part (b): $K = -10 \times 35 + 349 = -1$. Thus, the formula projects a price of $-\$1$ for January 2012.

7. **Tuition at American public universities**:

   (a) Let $d$ be the number of years since 2012 and $T$ the average tuition (in dollars) for public universities. The difference in the $d$ values is always 1, and the difference in the $T$ values is always 277. This shows that the data can be modeled by a linear function with slope 277 dollars per year. Now $d = 0$ corresponds to 2012, so (from the data table) the initial value of $T$ is 8318, and thus the equation is $T = 277d + 8318$.

   (b) The slope for the function from Part (a) is 277 dollars per year.

   (c) The slope for the function from Exercise 6 is 1134 dollars per year.

   (d) Part (b) tells us that the tuition for public institutions increases at a rate of \$277 per year, while Part (c) tells us that the tuition for private institutions increases at a rate of \$1134 per year. The average tuition at private universities is increasing at a greater rate.

   (e) The percentage increase for private universities from 2015 to 2016 is $\dfrac{1134}{31{,}272} = 0.04 = 4\%$, and the percentage increase for public universities from 2015 to 2016 is $\dfrac{277}{9149} = 0.03 = 3\%$. The average tuition for private universities had the larger percentage increase from 2015 to 2016.

9. **Total revenue and profit**:

   (a) The difference in the $N$ values is always 50, and the difference in the $p$ values is always $-0.50$. This shows that the data can be modeled by a linear function with slope $\dfrac{-0.50}{50} = -0.01$ dollar per widget. Thus we have $p = -0.01N + b$, and we use the fact that $p = 43.00$ when $N = 200$ to find $b$. This gives $43.00 = -0.01 \times 200 + b$, and thus $b = 43.00 + 0.01 \times 200 = 45$. Hence the model is $p = -0.01N + 45$.

   (b) Now the total revenue is the price times the number of items, so $R = pN$. Using the formula from Part (a) gives $R = (-0.01N + 45)N$. (This can also be written as $R = -0.01N^2 + 45N$.) This is not a linear function, as can be seen by examining a table of values (or a graph) or by observing that the coefficient of $N$ is not constant.

   (c) We have $P = R - C$. From Part (a) of Exercise 8 we know that $C$ has slope 35 and initial value 900, so its formula is $C = 35N + 900$. Using the formula for $R$ from Part (b) of this exercise, we have $P = (-0.01N + 45)N - (35N + 900)$. (This can also be written as $P = -0.01N^2 + 10N - 900$.) This is not a linear function, as can be seen by examining a table of values (or a graph) or by observing that the coefficient of $N$ is not constant.

11. **The Kelvin temperature scale:**

(a) The change in $F$ is always 36 degrees Fahrenheit for each 20 degree change in $K$ (the temperature in kelvins). This means that $F$ is a linear function of $K$.

(b) The slope is calculated as

$$\frac{\text{Change in } F}{\text{Change in } K} = \frac{36}{20} = 1.8.$$

Thus the slope is 1.8 degrees Fahrenheit per kelvin.

(c) The slope of the function is 1.8, so $F = 1.8K + b$ for some $b$. To find $b$, we use one point on the line; for example, when $K = 200$, then $F = -99.67$. These values give $-99.67 = 1.8 \times 200 + b$, so $b = -459.67$. Thus $F = 1.8K - 459.67$.

(d) Now $F = 98.6$, and we need to find $K$. Thus we have to solve the equation $1.8K - 459.67 = 98.6$:

$$
\begin{aligned}
1.8K - 459.67 &= 98.6 \\
1.8K &= 558.27 \quad \textbf{Add 459.67 to both sides.} \\
K &= 310.15 \quad \textbf{Divide both sides by 1.8.}
\end{aligned}
$$

Thus, body temperature on the Kelvin scale is 310.15 kelvins.

(e) Since the slope of $F$ is 1.8, when temperature increases by one kelvin, the Fahrenheit temperature will increase by 1.8 degrees.

If we solve $F = 1.8K - 459.67$ for $K$ we get $K = \dfrac{1}{1.8}F + \dfrac{459.67}{1.8}$, or $K = 0.56F + 255.37$. So a one degree increase in Fahrenheit will increase the temperature by 0.56 kelvin.

(f) If $K = 0$ then $F = 1.8 \times 0 - 459.67 = -459.67$ degrees Fahrenheit.

13. **Market supply:**

(a) By subtracting entries, we see that for each increase of 0.5 in the quantity $S$, there is an increase of 1.05 in price $P$. Thus $P$ is a linear function of $S$ with slope $\dfrac{1.05}{0.5} = 2.1$. Since the slope is 2.1, we have $P = 2.1S + b$. Now $b$ can be determined using the first data point: $S = 1$ when $P = 1.35$. Substituting, we have $1.35 = 2.1 \times 1.0 + b$, so $b = -0.75$, and therefore $P = 2.1S - 0.75$.

(b) The given table suggests a horizontal span from 0 to 3 billion bushels of wheat and a vertical span from 0 to 5 dollars per bushel. The horizontal axis is billions of bushels produced, and the vertical axis is price per bushel .

(c) If the price increases, then the wheat suppliers will want to produce more wheat. This means that the market supply curve will be going up to the right, so it is increasing.

(d) If the price $P$ is \$3.90 per bushel, then $S$ can be determined using the formula from Part (a): $2.1S - 0.75 = 3.90$. Solving (by adding 0.75 to each side, then dividing by 2.1) yields $S = 2.21$. Thus suppliers would be willing to produce 2.21 billion bushels in a year for a price of \$3.90 per bushel.

15. **Sports car:**

(a) Let $t$ be the time since the car was at rest, in seconds, and $V$ the velocity, in miles per hour. The differences are in the table below:

| Change in $t$ | 0.5 | 0.5 | 0.5 |
|---|---|---|---|
| Change in $V$ | 5.9 | 5.9 | 5.9 |

The ratio of change in $V$ to change in $t$ is always $\dfrac{5.9}{0.5} = 11.8$, so $V$ can be modeled by a linear function of $t$ with slope 11.8.

(b) The slope is 11.8 miles per hour per second. In practical terms, this means that, after every second, the car will be going 11.8 miles per hour faster. This says that the acceleration is constant.

(c) From Part (b), we know that $V = 11.8t + b$ for some $b$. Using the first data point, $t = 2.0$ and $V = 27.9$, we have $27.9 = 11.8 \times 2.0 + b$. Thus $b = 4.3$, and the final formula is $V = 11.8t + 4.3$.

(d) According to the formula in Part (c), the initial velocity is $V(0) = 4.3$ miles per hour. On the other hand, the exercise states that when $t = 0$, the car was at rest, so we would expect the initial velocity to be 0, not 4.3. In practical terms, the formula is valid only after the car has been moving a while since the acceleration is not constant the whole time: acceleration is large initially and decreases afterwards.

(e) We need to find when $V$ is 60. Using the formula, we have $11.8t + 4.3 = 60$, so

$$t = \frac{60 - 4.3}{11.8} = 4.72 \text{ seconds.}$$

This car goes from 0 to 60 mph in 4.72 seconds.

17. **Later high school graduates**: Let $N$ be the number graduating (in millions) and $t$ the number of years since 2001.

(a) The difference in the $t$ values is always 2, and the difference in the $N$ values is always 0.13. This shows that the data can be modeled by a linear function with slope $\dfrac{0.13}{2} = 0.07$ million graduating per year. The slope means that every year the number graduating increases by about 0.07 million (or 70,000).

(b) Now $t = 0$ corresponds to 2001, so (from the data table) the initial value of $N$ is 2.59, and thus the equation is $N = 0.07t + 2.85$.

(c) The number graduating from high school in 2008 is expressed in functional notation as $N(7)$ since 2008 is 7 years since 2001. Its value is $N(7) = 0.07 \times 7 + 2.85 = 3.34$ million graduating.

(d) Since 1994 is 7 years before 2001, we use $t = -7$ in the formula from Part (b). The value is $N(-7) = 0.07 \times (-7) + 2.85 = 2.36$ million graduating. This is much closer to the actual number than the value of 2.02 million graduating given in Part (d) of Exercise 14. The trend in high school graduations was a steady decline from 1985 until the early 1990s followed by a steady increase starting in 1994 or slightly earlier.

19. **Sound speed in oceans**:

(a) Here is a table of differences:

| Change in $S$ | 0.6 | 0.6 | 0.6 | 0.6 |
|---------------|------|------|------|------|
| Change in $c$ | 0.72 | 0.72 | 0.82 | 0.62 |

From this table it is clear that the fourth entry for speed (at the salinity of $S = 36.8$) is in error—it is too high by 0.1. When $S = 36.8$, $c$ should be $1517.62 - 0.1 = 1517.52$. With this choice, all entries are consistent with a linear model.

(b) The linear model has slope

$$\frac{\text{Change in } c}{\text{Change in } S} = \frac{0.72}{0.6} = 1.2 \text{ meters per second per parts per thousand.}$$

The slope means that, at the given depth and temperature, for an increase in salinity of 1 part per thousand the speed of sound increases by 1.2 meters per second.

(c) We use the slope, 1.2, and the last entry in the table. Because

$$\text{Change in } c = \text{Slope} \times \text{Change in } S,$$

we have

$$c(39) - c(37.4) = 1.2 \times (39 - 37.4),$$

and thus

$$c(39) = c(37.4) + 1.2 \times (39 - 37.4) = 1518.24 + 1.2 \times (39 - 37.4) = 1520.16.$$

This means that the speed of sound is 1520.16 meters per second when the salinity is 39 parts per thousand. *Note*: The value of $c(39)$ can also be found by first using the slope and an entry in the table to find the formula for $c$, namely $c = 1.2S + 1473.36$.

21. **Measuring the circumference of the Earth**: Because the rate of travel is 100 stades per day, a trip requiring 50 days is $100 \times 50 = 5000$ stades long. This is estimated to be $\dfrac{1}{50}$ of the Earth's circumference, so Eratosthenes' measure of the circumference of the Earth is $5000 \times 50 = 250{,}000$ stades. This is $250{,}000 \times 0.104 = 26{,}000$ miles.

23. **When data are unevenly spaced**: The average rate of change from 2 to 5 is

$$\frac{\text{Change in } y}{\text{Change in } x} = \frac{9 - 6}{5 - 2} = 1.$$

Similarly, the average rate of change from 5 to 6 is

$$\frac{12 - 9}{6 - 5} = 3.$$

Therefore, the average rate of change from 2 to 5 is not the same as the average rate of change from 5 to 6.

## Skill Building Exercises

S-1. **Testing data for linearity**: Calculating differences, we see that $y$ changes by 5 for each change in $x$ of 2. Thus these data exhibit a constant rate of change and so are linear.

S-3. **Making a linear model**: The data in Exercise S-1 show a constant rate of change of 5 in $y$ per change of 2 in $x$ and therefore the slope is $m = \dfrac{\text{Change in } y}{\text{Change in } x} = \dfrac{5}{2} = 2.5$. We know now that $y = 2.5x + b$ for some $b$. Since $y = 12$ when $x = 2$, $12 = 2.5 \times 2 + b$, so $b = 12 - 2.5 \times 2 = 7$. The equation is $y = 2.5x + 7$.

S-5. **Graphing discrete data**: The figure below shows a plot of the data in Exercise S-2.

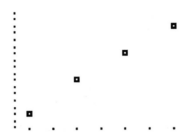

S-7. **Entering and graphing data**: The entered data are shown in the figure on the left below, and the plot of the data is shown in the figure on the right below.

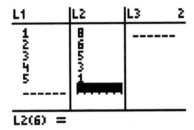
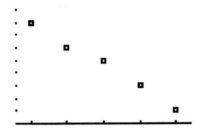

S-9. **Editing data**: In the figure on the left below, we have entered the squares of the data points in the table from Exercise S-7. The plot of the squares is shown in the figure on the right below.

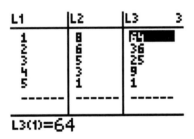
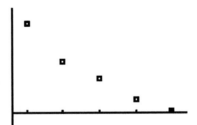

S-11. **Plotting data and functions**: The figure below shows a plot of the data in Exercise S-11.

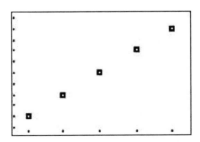

S-13. **Plotting data and functions**: The figure below shows a plot of the data in Exercise S-13.

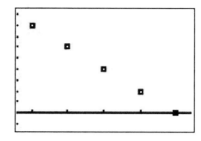

S-15. **Plotting data and functions**: The figure below shows a plot of the data in Exercise S-15.

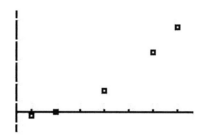

S-17. **Plotting data and functions**: The figure below shows a plot of the data in Exercise S-17.

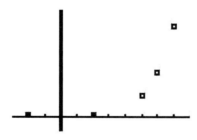

S-19. **Plotting data and functions**: The figure below shows a plot of the data in Exercise S-19.

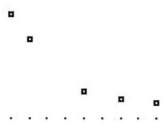

S-21. **Plotting data and functions**: The figure below shows a plot of the data in Exercise S-21.

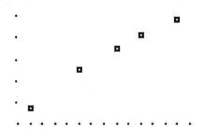

S-23. **Plotting data and functions**: The figure below shows a plot of the data in Exercise S-19 as well as a graph of the function $y = \dfrac{x}{x^2 + 1}$.

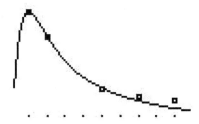

S-25. **Plotting data and functions**: The figure below shows a plot of the data in Exercise S-24 as well as a graph of the function $y = x - \dfrac{5}{x}$.

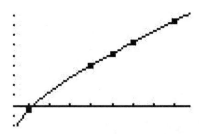

## 3.4  LINEAR REGRESSION

1. **Motor vehicle fatalities**: From the figure below, the equation of the regression line is $R = -0.01t + 1.12$.

```
LinReg
 y=ax+b
 a=-.009
 b=1.12
```

3. **Cost of employee replacement**:

   (a) The plot is shown below.

   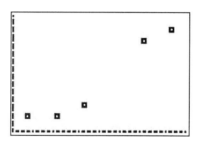

   (b) From the figure on the left below, the equation of the regression line is $P = 0.80S - 6.05$.

   ```
   LinReg
    y=ax+b
    a=.7965116279
    b=-6.046511628
   ```

   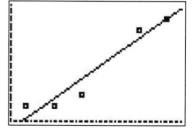

   (c) Adding the regression line to the plot from part a gives the figure on the right above.

   (d) A salary of $50,000 corresponds to $S = 50$, and the regression line gives the percentage $P = 0.80 \times 50 - 6.05 = 33.95$ percent. This is the percentage of the salary spent on the replacement, so the actual cost of replacement is $0.3395 \times 50,000 = 16,975$ dollars.

5. **Personal income:**

(a) From the figure on the left below, the equation of the regression line is $I = 264.93t + 22,884.64$.

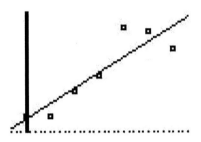

(b) Adding the regression line to the plot of the data gives the figure on the right above.

(c) The slope of the regression line is 264.93 dollars per year. The slope means that, from 1980 on, the average annual income (adjusted for inflation) grew by $264.93 each year.

(d) Examining the plot from part b shows that the last data point is well below the regression line. This point corresponds to $t = 30$, so in 2010 income was much lower than would be expected.

7. **DirecTV subscribers:**

(a) From the figure on the left below, the equation of the regression line is $S = 1.02t + 2.70$.

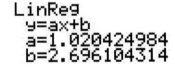
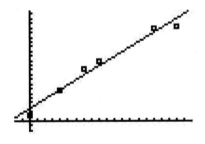

(b) Now 2013 corresponds to $t = 18$, so we plug $t = 18$ into the formula from Part (a): $S = 1.02 \times 18 + 2.70 = 21.06$. Therefore, the equation gives 21.06 million for the number of DirecTV subscribers in 2013.

(c) The slope of the line is 1.02 million subscribers per year, and this means that, according to the model, DirecTV gained 1.02 million subscribers each year.

(d) The graph is shown in the figure on the right above.

9. **Is a linear model appropriate?**

(a) The plotted points are shown below.

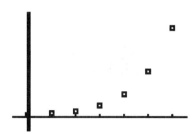

The plot of the data appears to be fit better by a curve rather than by a straight line.

(b) We take the variable to be years since 2004. The plotted points are shown below.

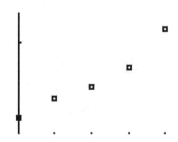

It looks like the points almost fall on a straight line. It appears reasonable to approximate the data with a straight line.

11. **Tourism**: Let $t$ be the number of years since 2010 and $T$ the number of tourists, in millions.

(a) A plot of the data points is shown below.

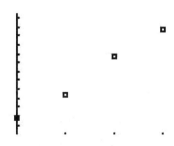

(b) From the left-hand figure below, the equation of the regression line is $T = 3.44t +$
59.46. Adding this line to the plot above gives the right-hand figure below. The
horizontal axis is years since 2010, and the vertical axis is millions of tourists.

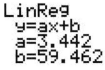

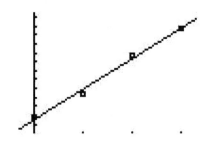

(c) The slope of the regression line is 3.44, and this means that every year the number of international tourists who visited the United States increases by about 3.44
million.

(d) The number of tourists who visited the United States in 2014 is $T(4)$ in functional
notation, since $t = 4$ corresponds to four years after 2010, that is, the year 2014.
Using the regression line, we calculate $T(4) = 3.44 \times 4 + 59.46 = 73.22$ million
tourists.

13. **Long jump**: Let $t$ be the number of years since 1900 and $L$ the length of the winning
long jump, in meters.

(a) Using the calculator, we find the equation of the regression line: $L = 0.034t + 7.197$.

(b) In practical terms the meaning of the slope, 0.034 meter per year, of the regression
line is that each year the length of the winning long jump increased by an average
of 0.034 meter, or about 1.34 inches.

(c) The figure below shows a plot the data points together with a graph of the regression line.

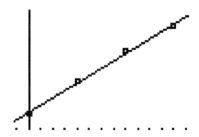

(d) The regression line formula has positive slope and so increases without bound.
Surely there is some length that cannot be attained by a long jump, and therefore

the regression line is not necessarily a good model of the winning length over a long period of time.

(c) The regression line model gives the value $L(20) = 0.034 \times 20 + 7.197 = 7.877$, or about 7.88, meters, which is over two feet longer than the length of the winning long jump in the 1920 Olympic Games. This result is consistent with our answer to Part (d) above.

15. **The effect of sampling error on linear regression**:

(a) The plotted points are shown below.

(b) From the left-hand figure below, the equation of the regression line is $D = 0.88t + 51.98$. In practical terms, the slope of 0.88 foot per hour means that the flood waters are rising by 0.88 foot each hour.

(c) Adding the regression line to the plot gives the right-hand figure below. The horizontal axis is time since flooding began, and the vertical axis is depth.

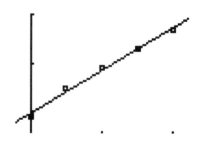

(d) Adding $D = 0.8t + 52$ (thick line) gives the following graph.

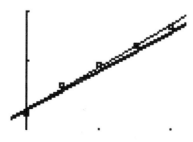

So the data does give a close approximation of the depth function, since the two lines are very close. The regression model shows a water level which is a bit too high.

(e) From the depth function $D = 0.8t + 52$, the actual depth of the water after 3 hours is $0.8 \times 3 + 52 = 54.4$ feet.

(f) The regression line equation predicts the depth after 3 hours to be $0.88 \times 3 + 51.98 = 54.62$, or about 54.6, feet.

17. **Domestic auto sales in the U.S.:** Let $t$ be the number of years since 2005 and $D$ the total U.S. sales of domestic automobiles, in millions.

(a) From the left-hand figure below, the equation of the regression line is $D = -0.32t + 5.68$. The slope is $-0.32$, which means that each year the total U.S. sales of domestic cars decreases by 0.32 million cars, or about 320,000 cars. Note that this *decrease* in sales is reflected by the fact that the slope of the line is negative.

(b) The plot is in the right-hand figure below. The horizontal axis is the number of years since 2005, and the vertical axis is millions of auto sales.

```
LinReg
 y=ax+b
 a=-.32
 b=5.68
```

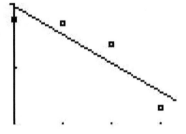

(c) Now 2009 corresponds to $t = 4$, and from the equation of the regression line we expect $D = -0.32 \times 4 + 5.68 = 4.40$ million cars to be sold in 2009. The regression line forecast is larger than the actual figure.

19. **Antimasonic voting**: Using the notation of $M$ as the percentage antimasonic voting and $C$ as the number of church buildings, the statement to be analyzed is "$M$ depends directly on $C$," that is, $M$ is a linear function of $C$. The data plot is shown below.

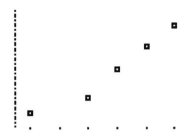

From the left-hand figure below, the equation for the regression line is $M = 5.89C + 45.55$; moreover, plotting that equation with the data points gives the right-hand figure. The horizontal axis is number of churches and the vertical axis is percent antimasonic voting.

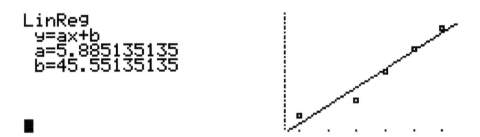

The figure shows that the regression line fits the data well. So the data does support the premise of the stated direct dependence.

21. **Expansion of steam**:

   (a) The linear regression model of volume $V$ as a function of $T$ is $V = 12.57T + 7501.00$.

   (b) If one fire is 100 degrees hotter than another, then $T$ increases by 100, so $V$ increases by $12.57 \times 100 = 1257$; that is, the volume of steam produced by 50 gallons of water in the hotter fire is 1257 cubic feet more.

   (c) The volume $V(420) = 12.57 \times 420 + 7501 = 12{,}780$ cubic feet. In practical terms this answer means that 50 gallons of water applied to a fire of 420 degrees Fahrenheit will produce about 12,780 cubic feet of steam.

(d) If 50 gallons of water expanded to 14,200 cubic feet of steam, then $V = 14{,}200$, and so $14{,}200 = 12.57T + 7501$. Solving for $T$, we find that $12.57T = 14{,}200 - 7501 = 6699$. Thus $T = \dfrac{6699}{12.57} = 532.94$, and therefore the temperature of the fire was 532.94 degrees Fahrenheit.

23. **Whole crop weight versus rice weight**:

(a) Using regression, we find a linear model $B = 0.29W + 0.01$ for $B$ as a function of $W$.

(b) Comparing the predicted rice weight from the linear model of Part (a) with the data, we see that the third sample, with $W = 13.7$ and $B = 3.7$ has a significantly lower rice weight than expected from the whole crop weight. This can also be seen by plotting the data with the regression line.

| $W$ Data | $B$ Data | Predicted $B = 0.29W + 0.01$ |
|----------|----------|------------------------------|
| 6 | 1.8 | 1.75 |
| 11.1 | 3.2 | 3.23 |
| 13.7 | 3.7 | 3.98 |
| 14.9 | 4.3 | 4.33 |
| 17.6 | 5.2 | 5.11 |

(c) An increase in whole crop weight $W$ of one ton per hectare is expected to produce an increase in rice weight $B$ of 0.29 ton per hectare.

25. **Energy cost of running**:

(a) From the figure below, the equation of regression line for $E$ in terms of $v$ is $E = 0.34v + 0.37$.

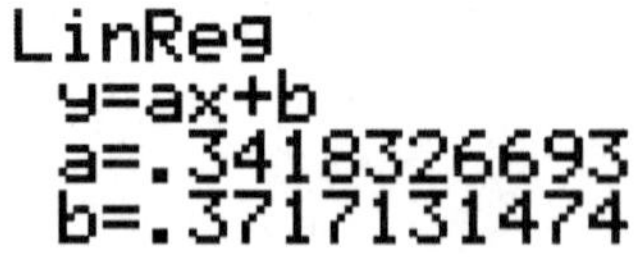

```
LinReg
y=ax+b
a=.3418326693
b=.3717131474
```

(b) The cost of transport is defined to be the slope of the regression line in Part (a), so the cost of transport is 0.34, in units of (unit of $E$) per (unit of $v$), which is milliliter of oxygen per gram per hour per kilometer per hour.

(c) If a rhea weighs 22,000 grams, then $W = 22,000$. The formula for $C$ gives $C = 8.5 \times 22,000^{-0.40} = 0.16$. The figure found for the cost of transport in Part (b) is much higher than what the general formula would predict. Based on this, the rhea is a much less efficient runner that a typical animal of its size.

(d) If the rhea is at rest, then $v = 0$. Using $v = 0$ in the regression equation from Part (a) gives $E = 0.37$ as an estimate for the oxygen consumption for a rhea at rest.

## Skill Building Exercises

S-1. **Sales**:

(a) The slope of 250 dollars per year means that sales for the pertinent period are increasing by $250 each year.

(b) The year 2018 corresponds to $t = 2$, so the regression line gives $S = 250 \times 2 + 2500 = 3000$. Thus, $3000 in sales are indicated for 2018.

S-3. **Employee turnover**: Now 0% employee turnover corresponds to $E = 0$, so we want to solve the equation $10 - \dfrac{S}{500} = 0$. Solving for $S$, we have $10 = \dfrac{S}{500}$ or $S = 500 \times 10 = 5000$. Thus, a monthly salary level of $5000 would result in 0% employee turnover.

S-5. **Getting regression lines**:

(a) The figure below on the left is a plot of the data.

(b) The regression line is $y = -0.26x + 2.54$.

(c) The figure on the right below is the plot of the data with the graph of the regression line added.

S-7. **Getting regression lines**:

    (a) The figure below on the left is a plot of the data.

    (b) The regression line is $y = 1.05x + 2.06$.

    (c) The figure on the right below is the plot of the data with the graph of the regression line added.

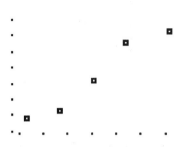 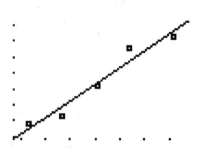

S-9. **Getting regression lines**:

    (a) The figure below on the left is a plot of the data.

    (b) The regression line is $y = -0.32x - 1.59$.

    (c) The figure on the right below is the plot of the data with the graph of the regression line added.

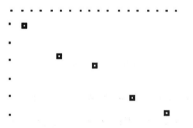 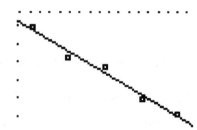

S-11. **Getting regression lines only**: Calculating the regression line we get $y = -x + 4$.

S-13. **Getting regression lines only**: Calculating the regression line we get $y = 0.5x + 2.33$.

S-15. **Getting regression lines only**: Calculating the regression line we get $y = -1.5x + 4$.

S-17. **Correlation coefficient**: The figure on the left below shows the regression line calculation: $y = 2x + 1.4$. The figure on the right shows the plot of the data together with the graph of regression line.

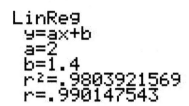
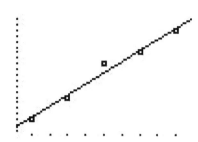

The correlation coefficient is $r = 0.99$.

S-19. **Correlation coefficient**: The figure on the left below shows the regression line calculation: $y = -0.6x + 5.4$. The figure on the right shows the plot of the data together with the graph of regression line.

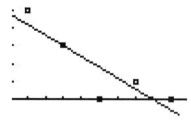

The correlation coefficient is $r = -0.88$.

S-21. **Correlation coefficient**: The figure on the left below shows the regression line calculation: $y = 0.1x - 1$. The figure on the right shows the plot of the data together with the graph of regression line.

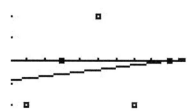

The correlation coefficient is $r = 0.19$.

## 3.5 SYSTEMS OF EQUATIONS

1. **Investing**:

   (a) You have $30,000 to invest, with $M$ dollars in municipal bonds and $T$ dollars in Treasury notes. This gives the equation $M + T = 30,000$. Also, you want to invest twice as much in Treasury notes as in municipal bonds. This gives the equation $T = 2M$. Thus, we have the system of equations

$$
\begin{aligned}
M + T &= 30,000 \\
T &= 2M.
\end{aligned}
$$

   (b) To solve the system of equations in part a, we use the crossing-graphs method. (The answer can also be found by hand calculation.) Because the second equation expresses $T$ in terms of $M$, we solve the first equation for $T$ to get $T = 30,000 - M$. A table of values leads us to choose a horizontal span of 0 to 30,000 and a vertical span of 0 to 60,000. The horizontal axis is the amount, in dollars, invested in municipal bonds, and the vertical axis is the amount, in dollars, invested in Treasury notes. The resulting graphs are in the figure below.

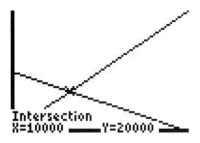

   We see from the figure above that the graphs intersect when $M = 10,000$ and $T = 20,000$. Therefore, we should invest $10,000 in municipal bonds and $20,000 in Treasury notes.

3. **Motherboards**:

   (a) The production line produces 45 items per day, consisting of $M$ motherboards and $C$ chips. This gives the equation $M + C = 45$. Also, each motherboard requires two chips so, for example, if $M = 1$ then we want $C = 2$. In general, we want $C = 2M$. Thus, we have the system of equations

$$M + C = 45$$

$$C = 2M.$$

(b) To solve the system of equations in part a, we use the crossing-graphs method. (The answer can also be found by hand calculation.) Because the second equation expresses $C$ in terms of $M$, we solve the first equation for $C$ to get $C = 45 - M$. A table of values leads us to choose a horizontal span of 0 to 45 and a vertical span of 0 to 90. The horizontal axis is the number of motherboards produced, and the vertical axis is the number of chips produced. The resulting graphs are in the figure below.

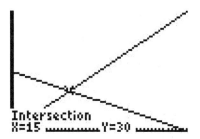

We see from the figure above that the graphs intersect when $M = 15$ and $C = 30$. Therefore, 15 motherboards and 30 chips are produced each day.

5. **A party**: Let $C$ be the number of bags of chips we need to buy and $D$ be the number of drinks we need to buy. We have \$36 to spend, and so

$$\text{Cost of chips} + \text{Cost of drinks} = \text{Available money}$$

$$2C + 0.5D = 36.$$

Additionally, we know that we will buy 5 times as many drinks as bags of chips:

$$\text{Drinks} = 5 \times \text{Chips}$$

$$D = 5C.$$

Thus, we need to solve the system of equations

$$2C + 0.5D = 36$$

$$D = 5C.$$

The first step is to solve each equation for one of the variables. Since the second equation is already solved for $D$, we solve the first for $D$ as well:

$$D = \frac{36 - 2C}{0.5}$$
$$D = 5C.$$

We will certainly buy fewer than 20 bags of chips, and so we use a horizontal span of 0 to 20. The following table of values leads us to choose a vertical span of 0 to 90 drinks. In the right-hand figure below, the thick line is the graph of $D = 5C$. The horizontal axis represents the number of bags of chips, and the vertical axis is the number of drinks.

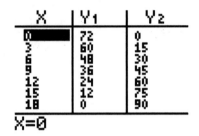
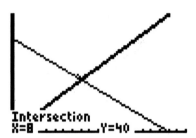

We see that the graphs cross at $C = 8$ and $D = 40$. This means that we should buy 8 bags of chips and 40 drinks.

7. **An order for bulbs**: Let $c$ be the number of crocus bulbs, and let $d$ be the number of daffodil bulbs. We get our first equation from the fact that the total number of bulbs is 55:

$$\text{Crocus} + \text{Daffodil} = \text{Total bulbs}$$
$$c + d = 55.$$

The second equation we need comes from budget constraints:

$$\text{Cost of crocuses} + \text{Cost of daffodils} = \text{Total cost}$$
$$0.35c + 0.75d = 25.65.$$

Thus we need to solve the system of equations

$$c + d = 55$$
$$0.35c + 0.75d = 25.65.$$

Solving each of these equations for $d$, we get

$$d = 55 - c$$
$$d = \frac{25.65 - 0.35c}{0.75}.$$

Since there are certainly no more than 55 crocus bulbs, we use a horizontal span of 0 to 55. From the table below, we choose a vertical span of 0 to 35. In the graph below, the thick line corresponds to $d = \dfrac{25.65 - 0.35c}{0.75}$. The horizontal axis is the number of crocus bulbs, and the vertical axis is the number of daffodil bulbs.

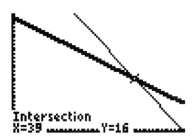

We see that the graphs intersect at the point $c = 39$ and $d = 16$. So we should buy 39 crocus bulbs and 16 daffodil bulbs.

9. **Population growth**: Let $f$ be the number of foxes and $r$ the number of rabbits. Since the rates of change for both foxes and rabbits are constant (33 per year and 53 per year respectively), both are linear functions of time. After $t$ years, there will be $f = 255 + 33t$ foxes and $r = 104 + 53t$ rabbits. We want to know when these values will be the same. That is, we need to find out when $f = r$, or when $255 + 33t = 104 + 53t$. This is a linear equation which we can solve by hand calculation or by graphing. Here is the hand calculation:

$$
\begin{aligned}
255 + 33t &= 104 + 53t & \\
151 + 33t &= 53t & \textbf{Subtract 104 from each side.} \\
151 &= 20t & \textbf{Subtract } 33t \textbf{ from both sides.} \\
7.55 &= t & \textbf{Divide each side by 20.}
\end{aligned}
$$

Hence, the number of foxes will be the same as the number of rabbits in 7.55 years. At that time, there will be $f = 255 + 33 \times 7.55 = 504.15$ foxes (this is better reported as 504 foxes) and the same number of rabbits.

11. **Competition between populations**: The equilibrium point occurs when the per capita growth rate of each population is 0. Thus we have two equations:

$$3(1 - m - n) = 0$$
$$2(1 - 0.7m - 1.1n) = 0.$$

This can solved by hand directly. The first equation is the same as $1 - m - n = 0$, and so $m = 1 - n$. Substituting this into the second equation, we get $2(1 - 0.7(1 - n) - 1.1n) = 0$. Dividing by 2 and expanding, we get $1 - 0.7 + 0.7n - 1.1n = 0$, so $0.3 - 0.4n = 0$. Thus $0.4n = 0.3$, and so $n = \dfrac{0.3}{0.4} = 0.75$ thousand animals. We can substitute this value to find $m$. Now $m = 1 - n = 1 - 0.75 = 0.25$ thousand animals.

13. **Boron uptake**:

   (a) The amount of water-soluble boron $B$ available resulting in the same plant content of boron for Decatur silty clay and Hartsells fine sandy loam is where

   $$33.78 + 37.5B = 31.22 + 71.17B.$$

   You can solve this exercise graphically, using a table, or calculating it by hand. By hand, we move terms to the same side to get

   $$33.78 - 31.22 = 71.17B - 37.5B = (71.17 - 37.5)B.$$

   Thus $B = \dfrac{33.78 - 31.22}{71.17 - 37.5} = 0.08$ part per million.

   (b) The value of boron $B$ from Part (a) gives the same plant content $C$ for either soil type. For the Hartsells fine sandy loam, $C$ increases by 71.17 parts per million with each increase of 1 in $B$, whereas for the Decatur silty clay, $C$ only increases by 37.5 parts per million with each increase of 1 in $B$. Thus if $B$ is larger than 0.08 part per million, the Hartsells fine sandy loam will have the larger plant content of boron.

15. **Fahrenheit and Celsius**: When Fahrenheit is exactly twice as much as Celsius, we have $F = 2C$. That means we need to solve the system of equations

$$F = \frac{9}{5}C + 32$$
$$F = 2C.$$

Since both equations are already solved for $F$, we need only graph both functions and find the intersection. The table of values below leads us to choose a horizontal span of 150 to 200 and a vertical span of 300 to 400. In the graph below, the thick line corresponds to $F = 2C$. The horizontal axis is Celsius temperature, and the vertical axis is Fahrenheit temperature.

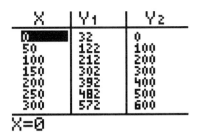 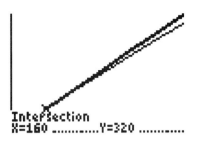

We see that the graphs intersect when $C = 160$ and $F = 320$. Thus the temperature we seek is 160 degrees Celsius and 320 degrees Fahrenheit.

17. **Parabolic mirrors**:

(a) The point $(2, 4)$ is $(a, a^2)$ where $a = 2$, so the light reflected from $(2, 4)$ follows the line $4 \times 2y + (1 - 4 \times 4)x = 2$, that is, $8y - 15x = 2$. The point $(3, 9)$ is $(a, a^2)$ where $a = 3$, so the light reflected from $(3, 9)$ follows the line $4 \times 3y + (1 - 4 \times 9)x = 3$, that is, $12y - 35x = 3$. The light rays meet at the point where these two equations are both satisfied. Using crossing graphs, for example, we first solve each equation for $y$. The first equation is $8y - 15x = 2$, so $8y = 2 + 15x$ and therefore $y = \dfrac{2 + 15x}{8}$, while the second equation is $12y - 35x = 3$, so $12y = 3 + 35x$ and therefore $y = \dfrac{3 + 35x}{12}$. Graphing, we find a common solution at $x = 0$, $y = 0.25$, so the light rays meet at the point $(0, 0.25)$, or $(0, \frac{1}{4})$.

(b) To show that all the reflected light rays pass through the point $(0, 0.25)$, we need only to show that $x = 0$, $y = 0.25$ satisfies the equation $4ay + (1 - 4a^2)x = a$. But plugging in these values, we have $4ay + (1 - 4a^2)x = 4a \times 0.25 + (1 - 4a^2) \times 0 = 4a \times 0.25 + 0 = a \times 1 = a$, so each reflected line does pass through the focal point $(0, 0.25)$.

19. **Figuring taxes**:

(a) The total tax owed is 10% of the amount $T$ on which taxes are calculated, so in millions of dollars the total tax is $0.1T$.

(b) The gross income of 1.47 million dollars equals the total tax plus $B$, and by Part (a) the total tax is $0.1T$. Therefore, $0.1T + B = 1.47$.

(c) The bonus is 20% of the amount $B$ on which the bonus is based, so in millions of dollars the bonus is $0.2B$.

(d) The gross income of 1.47 million dollars equals $T$ plus the bonus, and by Part (c) the bonus is $0.2B$. Therefore, $T + 0.2B = 1.47$.

(e) We need to solve the system of equations

$$0.1T + B = 1.47$$
$$T + 0.2B = 1.47.$$

The first step is to solve each equation for one of the variables. We solve for $T$. The result is

$$T = (1.47 - B)/0.1$$
$$T = 1.47 - 0.2B.$$

In the graph below we use a horizontal span of 0 to 2 million dollars for $B$ and a vertical span of $-5$ to 15 million dollars for $T$.

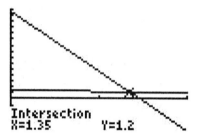

We see that the graphs cross at $B = 1.35$ million dollars and $T = 1.2$ million dollars.

(f) By Part (e) the amount on which the bonus is based is $B = 1.35$ million dollars, so by Part (c) the bonus is $0.2 \times 1.35 = 0.27$ million dollars. By Part (e) the amount on which taxes are calculated is $T = 1.2$ million dollars, so by Part (a) the total tax is $0.1 \times 1.2 = 0.12$ million dollars. The remaining profit after the bonus and taxes are paid is the gross income of 1.47 million dollars minus the bonus and the total tax, or $1.47 - 0.27 - 0.12 = 1.08$ million dollars.

21. **Another interesting system of equations**: Solving both equations for $y$, we see that we need to graph the two functions

$$y = \frac{3 - x}{2}$$
$$y = \frac{-6 + 2x}{-4}.$$

If we do this, we find that both graphs produce the same line, showing that the two functions are in fact identical. If you divide the equation $-2x - 4y = -6$ by $-2$, you get the equation $x + 2y = 3$, which is exactly the first equation. Since the two equations are the same, every point on the line $x + 2y = 3$ is also on the other line, and hence every such point is a solution of the system of equations.

23. **An application of three equations in three unknowns**: Let $N$ be the number of nickels, $D$ the number of dimes, and $Q$ the number of quarters. Since there are 21 coins in the bag, $N + D + Q = 21$. Since there is one more dime than nickel in the bag, $D = N + 1$. Since there is \$3.35 in the bag, $0.05N + 0.10D + 0.25Q = 3.35$. This gives three linear equations in the variables $N$, $D$, and $Q$:

$$
\begin{aligned}
N + D + Q &= 21 \\
D &= N + 1 \\
0.05N + 0.10D + 0.25Q &= 3.35.
\end{aligned}
$$

The second equation is already solved for $D$, so we will substitute that into the other two equations. The first equation becomes $N + (N + 1) + Q = 21$, or $2N + Q + 1 = 21$. Solving for $Q$, we get $Q = 20 - 2N$.

The third equation becomes $0.05N + 0.10(N + 1) + 0.25Q = 3.35$, or $0.05N + 0.10N + 0.10 + 0.25Q = 3.35$. Solving for $Q$ gives

$$
Q = \frac{3.25 - 0.15N}{0.25}.
$$

Graphing this equation together with $Q = 20 - 2N$ using a horizontal span from 0 to 10 nickels and a vertical span from 0 to 30 quarters yields the following picture.

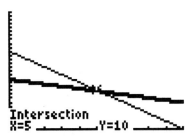

The graphs intersect when $N = 5$ and $Q = 10$.

We know that $D = N + 1$. Since $N = 5$, $D = 5 + 1 = 6$. So the solution to the equation is $N = 5$, $D = 6$, and $Q = 10$: there are 5 nickels, 6 dimes, and 10 quarters.

## Skill Building Exercises

**S-1. An explanation**: Each graph consists of pairs of points that make the corresponding equation true. The solution of the system is the point that makes both true, and that is the common intersection point.

**S-3. Setting up systems of equations**: Because you have $A$ apples and $P$ pears for a total of 40 pieces of fruit, we have the equation $A + P = 40$. Also, there are 3 times as many apples as pears so, for example, if $P = 1$ then we want $A = 3$. In general, we have $A = 3P$. Thus, we have the system of equations

$$
\begin{aligned}
A + P &= 40 \\
A &= 3P.
\end{aligned}
$$

**S-5. Setting up systems of equations**: Because you have \$100 invested, with $A$ dollars in one fund and $B$ dollars in another, we have the equation $A + B = 100$. Further, because the first fund returns 2% and there are $A$ dollars in that fund, the dollar amount of return for that fund is 2% of $A$, or $0.02A$. Similarly, because the second fund returns 5% and there are $B$ dollars in that fund, the dollar amount of return for that fund is 5% of $B$, or $0.05B$. Because the total return is 4% and there are \$100 invested altogether, the dollar amount of the total return is $0.04 \times 100 = 4$ dollars. These three facts give the equation $0.02A + 0.05B = 4$. Thus, we have the system of equations

$$
\begin{aligned}
A + B &= 100 \\
0.02A + 0.05B &= 4.
\end{aligned}
$$

**S-7. Systems that may be solved by hand or with technology**: To use technology we start with

$$
\begin{aligned}
x + y &= 5 \\
x - y &= 1
\end{aligned}
$$

and solve each equation for $y$. For the first equation we have $x + y = 5$, so $y = 5 - x$. For the second equation we have $x - y = 1$, so $-y = 1 - x$ and therefore $y = x - 1$. The figure below shows the graph of both $y = 5 - x$ and $y = x - 1$ using a horizontal span of 2 to 5 and a vertical span of 0 to 3. The intersection point, which is the solution to the system, is at $x = 3, y = 2$.

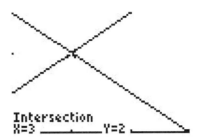

S-9. **Systems that may be solved by hand or with technology**: To use technology we start with

$$2x - y = 0$$
$$3x + 2y = 14$$

and solve each equation for $y$. For the first equation we have $2x - y = 0$, so $y = 2x$. For the second equation we have $3x + 2y = 14$, so $2y = 14 - 3x$ and therefore $y = \dfrac{14 - 3x}{2}$. The figure below shows the graph of both $y = 2x$ and $y = \dfrac{14 - 3x}{2}$ using a horizontal span of 0 to 4 and a vertical span of 0 to 5. The intersection point, which is the solution to the system, is at $x = 2$, $y = 4$.

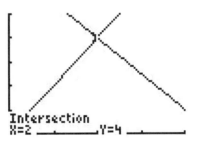

S-11. **Systems that may be solved by hand or with technology**: To use technology we start with

$$3x + 2y = 6$$
$$4x - 3y = 8$$

and solve each equation for $y$. For the first equation we have $3x + 2y = 6$, so $2y = 6 - 3x$ and therefore $y = \dfrac{6 - 3x}{2}$. For the second equation we have $4x - 3y = 8$, so $-3y = 8 - 4x$ and therefore $y = \dfrac{8 - 4x}{-3}$. The figure below shows the graph of both $y = \dfrac{6 - 3x}{2}$ and $y = \dfrac{8 - 4x}{-3}$ using a horizontal span of 0 to 5 and a vertical span of $-5$ to 5. The intersection point, which is the solution to the system, is at $x = 2$, $y = 0$.

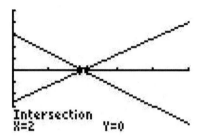

**S-13. Systems that may be solved by hand or with technology**: To use technology we start with

$$3x + y = 0$$
$$x - y = 0$$

and solve each equation for $y$. For the first equation we have $3x + y = 0$, so $y = -3x$. For the second equation we have $x - y = 0$, so $y = x$. The figure below shows the graph of both $y = -3x$ and $y = x$ using a horizontal span of $-5$ to $5$ and a vertical span of $-5$ to $5$. The intersection point, which is the solution to the system, is at $x = 0$, $y = 0$.

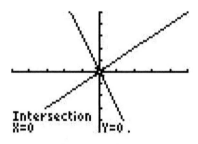

**S-15. Systems that may be solved by hand or with technology**: To use technology we start with

$$x - y = 1$$
$$3x + 2y = 8$$

and solve each equation for $y$. For the first equation we have $x - y = 1$, so $-y = 1 - x$ and therefore $y = -(1 - x) = x - 1$. For the second equation we have $3x + 2y = 8$, so $2y = 8 - 3x$ and therefore $y = \dfrac{8 - 3x}{2}$. The figure below shows the graph of both $y = x - 1$ and $y = \dfrac{8 - 3x}{2}$ using a horizontal span of $0$ to $4$ and a vertical span of $0$ to $2$. The intersection point, which is the solution to the system, is at $x = 2$, $y = 1$.

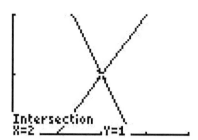

**S-17. Systems that may be solved by hand or with technology**: To use technology we start with

$$x + y = 1$$

$$2x + 3y = 0$$

and solve each equation for $y$. For the first equation we have $x + y = 1$, so $y = 1 - x$. For the second equation we have $2x + 3y = 0$, so $3y = -2x$ and therefore $y = \dfrac{-2x}{3}$. The figure below shows the graph of both $y = 1 - x$ and $y = \dfrac{-2x}{3}$ using a horizontal span of $-5$ to $5$ and a vertical span of $-5$ to $5$. The intersection point, which is the solution to the system, is at $x = 3$, $y = -2$.

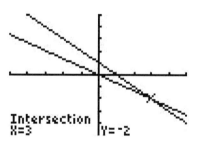

**S-19. Systems that may be solved by hand or with technology**: We start with

$$3x + 4y = 6$$

$$2x - 6y = 5$$

and solve each equation for $y$. For the first equation we have $3x + 4y = 6$, so $4y = 6 - 3x$ and therefore $y = \dfrac{6 - 3x}{4}$. For the second equation we have $2x - 6y = 5$, so $-6y = 5 - 2x$ and therefore $y = \dfrac{5 - 2x}{-6}$. The figure below shows the graph of both $y = \dfrac{6 - 3x}{4}$ and $y = \dfrac{5 - 2x}{-6}$ using a horizontal span of $0$ to $5$ and a vertical span of $-3$ to $3$. The intersection point, which is the solution to the system, is at $x = 2.15$, $y = -0.12$.

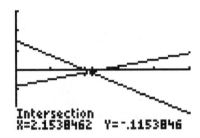

S-21. **Systems that may be solved by hand or with technology**: We start with

$$-7x + 21y = 79$$

$$13x + 17y = 6$$

and solve each equation for $y$. For the first equation we have $-7x + 21y = 79$, so $21y = 79 + 7x$ and therefore $y = \dfrac{79 + 7x}{21}$. For the second equation we have $13x + 17y = 6$, so $17y = 6 - 13x$ and therefore $y = \dfrac{6 - 13x}{17}$. The figure below shows the graph of both $y = \dfrac{79 + 7x}{21}$ and $y = \dfrac{6 - 13x}{17}$ using a horizontal span of $-5$ to $0$ and a vertical span of $0$ to $5$. The intersection point, which is the solution to the system, is at $x = -3.10$, $y = 2.73$.

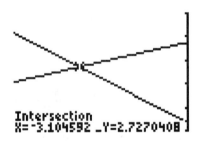

S-23. **Systems not easily solved by hand calculation**: We start with

$$0.7x + 5.3y = 6.6$$

$$5.2x + 2.2y = 1.7$$

and solve each equation for $y$. For the first equation we have $0.7x + 5.3y = 6.6$, so $5.3y = 6.6 - 0.7x$ and therefore $y = \dfrac{6.6 - 0.7x}{5.3}$. For the second equation we have $5.2x + 2.2y = 1.7$, so $2.2y = 1.7 - 5.2x$ and therefore $y = \dfrac{1.7 - 5.2x}{2.2}$. The figure below shows the graph of both $y = \dfrac{6.6 - 0.7x}{5.3}$ and $y = \dfrac{1.7 - 5.2x}{2.2}$ using a horizontal span of $-2$ to $2$ and a vertical span of $0$ to $5$. The intersection point, which is the solution to the system, is at $x = -0.21$, $y = 1.27$.

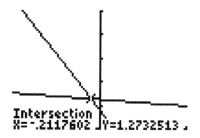

Intersection
X=-.2117602  Y=1.2732513

S-25. **Systems not easily solved by hand calculation**: We start with

$$2.1x + 3.3y = -1.43$$

$$2.2x - 1.1y = 3.78$$

and solve each equation for $y$. For the first equation we have $2.1x + 3.3y = -1.43$, so $3.3y = -1.43 - 2.1x$ and therefore $y = \dfrac{-1.43 - 2.1x}{3.3}$. For the second equation we have $2.2x - 1.1y = 3.78$, so $-1.1y = 3.78 - 2.2x$ and therefore $y = \dfrac{3.78 - 2.2x}{-1.1}$. The figure below shows the graph of both $y = \dfrac{-1.43 - 2.1x}{3.3}$ and $y = \dfrac{3.78 - 2.2x}{-1.1}$ using a horizontal span of $-2$ to $2$ and a vertical span of $-5$ to $5$. The intersection point, which is the solution to the system, is at $x = 1.14$, $y = -1.16$.

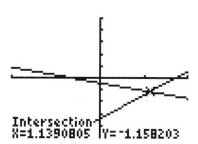

Intersection
X=1.1390805  Y=-1.158203

S-27. **Systems not easily solved by hand calculation**: We start with

$$-4.1x + 5.2y = 29.74$$

$$6.8x - 3.9y = -36.23$$

and solve each equation for $y$. For the first equation we have $-4.1x + 5.2y = 29.74$, so $5.2y = 29.74 + 4.1x$ and therefore $y = \dfrac{29.74 + 4.1x}{5.2}$. For the second equation we have $6.8x - 3.9y = -36.23$, so $-3.9y = -36.23 - 6.8x$ and therefore $y = \dfrac{-36.23 - 6.8x}{-3.9}$. The figure below shows the graph of both $y = \dfrac{29.74 + 4.1x}{5.2}$ and $y = \dfrac{-36.23 - 6.8x}{-3.9}$ using a horizontal span of $-5$ to $0$ and a vertical span of $0$ to $5$. The intersection point, which is the solution to the system, is at $x = -3.74$, $y = 2.77$.

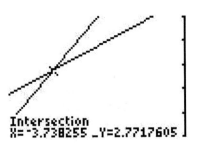

Intersection
X=-3.738255 _Y=2.7717605

## Chapter 3 Review Exercises

1. **Drainage pipe slope:** We have Slope $= \dfrac{\text{Rise}}{\text{Run}}$. In this case, the rise is $-0.5$ foot (since 6 inches make 0.5 foot) for a run of 8 feet. Thus the slope is $\dfrac{-0.5}{8} = -0.0625$ foot per foot. If we measure the rise in inches then we find $\dfrac{-6}{8} = -0.75$ inch per foot for the slope.

2. **Height from slope and horizontal distance:** We have

$$\text{Vertical change} = \text{Slope} \times \text{Horizontal change}.$$

In this case, the slope of the ladder is 3.7, while the horizontal distance is 4.4 feet, so the vertical height is $3.7 \times 4.4 = 16.28$ feet.

3. **A ramp into a van:**

   (a) We have Slope $= \dfrac{\text{Rise}}{\text{Run}}$. In this case, the rise is 18 inches, the distance from the van door to the ground, and the run is 4 feet, since that is the horizontal distance from the ramp to the van. Thus the slope is $\dfrac{18}{4} = 4.5$ inches per foot. If we measure the rise in feet then we find $\dfrac{1.5}{4} = 0.375$ foot per foot for the slope.

   (b) We have

$$\text{Rise} = \text{Slope} \times \text{Run}.$$

In this case we take the slope to be 1.2 inches per foot. The rise is still 18 inches, so we have $18 = 1.2 \times \text{Run}$. Thus the run is $\dfrac{18}{1.2} = 15$ feet. A smaller slope will require a greater distance, so the ramp should rest at least 15 feet from the van.

4. **A reception tent:**

   (a) We have Slope $= \dfrac{\text{Rise}}{\text{Run}}$. The rise from the top of an outside pole to the top of the center pole is $25 - 10 = 15$ feet, and the corresponding run is 30 feet. Thus the slope is $\dfrac{15}{30} = 0.5$ foot per foot.

(b) We have

$$\text{Rise} = \text{Slope} \times \text{Run}.$$

We know from Part (a) that the slope is 0.5 foot per foot. The run is 5 feet, so the rise is $0.5 \times 5 = 2.5$ feet. Thus the height has increased by 2.5 feet from the outside poles. We add the height of an outside pole to get that the height at this point is $2.5 + 10 = 12.5$ feet.

(c) We have

$$\text{Rise} = \text{Slope} \times \text{Run}.$$

We know from Part (a) that the slope is 0.5 foot per foot. The rise from the ground to the top of an outer pole is 10 feet, so we have $10 = 0.5 \times \text{Run}$. Thus the run is $\dfrac{10}{0.5} = 20$ feet. Hence the ropes are attached to the ground 20 feet from the outside poles.

5. **Function value from slope and run**: We have that the slope is $m = -2.2$, so

$$-2.2 = \frac{\text{Change in function}}{\text{Change in variable}} = \frac{g(6.8) - g(3.7)}{6.8 - 3.7} = \frac{g(6.8) - 5.1}{6.8 - 3.7} = \frac{g(6.8) - 5.1}{3.1}.$$

Thus $g(6.8) - 5.1 = 3.1 \times (-2.2)$, and so $g(6.8) = 5.1 + 3.1 \times (-2.2) = -1.72$.

6. **Lanes on a curved track**:

(a) To find the inner radius of the first lane, we put $n = 1$ in the formula and find $R(1) = 100/\pi$. Thus the radius is about 31.83 meters.

(b) The width of a lane is the change in $R$ when we increase $n$ by 1, and that is the slope of the linear function $R$. From the formula we see that the slope is 1.22. Thus the width of a lane is 1.22 meters. This can also be found by subtracting successive values of the radius; for example, $R(2) - R(1) = 1.22$.

(c) We first find for what value of $n$ we have $R(n) = 35$. This means we must solve the linear equation $100/\pi + 1.22(n - 1) = 35$ for $n$. We have

$$\begin{aligned}
100/\pi + 1.22(n - 1) &= 35 \\
1.22(n - 1) &= 35 - 100/\pi \\
n - 1 &= \frac{35 - 100/\pi}{1.22} \\
n &= \frac{35 - 100/\pi}{1.22} + 1 = 3.60.
\end{aligned}$$

Since the number of lanes is a whole number, if you wish to run in a lane with a radius of at least 35 meters you should pick lane 4 or a higher-numbered lane.

7. **Linear equation from two points**: We first compute the slope:

$$m = \frac{\text{Change in function}}{\text{Change in variable}} = \frac{g(1.1) - g(-2)}{1.1 - (-2)} = \frac{3.3 - 7.2}{3.1} = \frac{-3.9}{3.1} = -1.258.$$

(We keep the extra digit for accuracy in the rest of the calculations.) Since $g$ is a linear function with slope $-1.258$, we have $g = -1.258x + b$ for some $b$. Now $g(1.1) = 3.3$, so $3.3 = -1.258 \times 1.1 + b$, and therefore $b = 3.3 + 1.258 \times 1.1 = 4.68$. Thus if we now round the slope to 2 decimal places we get $g = -1.26x + 4.68$.

8. **Working on a commission**:

   (a) Each additional dollar of total sales increases the income by 5% of $1, or 0.05 dollar. This means that the change in $I$ is always the same, 0.05 dollar, for a change of 1 in $S$. Thus $I$ is a linear function of $S$.

   (b) If he sells $1600 in a month then he earns $1000 plus 5% of $1600, or $1000 + 0.05 \times 1600 = 1080$ dollars.

   (c) From Part (a), $I$ is a linear function of $S$ with slope 0.05, since the slope represents the additional income for 1 additional dollar in sales. Since $I$ is a linear function with slope 0.05, we have $I = 0.05S + b$ for some $b$. Now $I(1600) = 1080$ by Part (b), so $1080 = 0.05 \times 1600 + b$, and therefore $b = 1080 - 0.05 \times 1600 = 1000$. Thus $I = 0.05S + 1000$. Alternatively, we note that the initial value (the commission when there are no sales) is $1000, so $b = 1000$.

   (d) We want to find the value of $S$ so that $I = 1350$. That means we need to solve the equation $0.05S + 1000 = 1350$ for $S$. We find:

   $$
   \begin{aligned}
   0.05S + 1000 &= 1350 \\
   0.05S &= 1350 - 1000 \\
   S &= \frac{1350 - 1000}{0.05} = 7000.
   \end{aligned}
   $$

   Thus his monthly sales must be $7000.

9. **Testing data for linearity**:

   (a) There is a constant change of 0.3 in $x$ and a constant change of $-0.6$ in $f$. Thus these data exhibit a constant rate of change and so are linear.

   (b) From Part (a) we know that the data show a constant rate of change of $-0.6$ in $f$ per change of 0.3 in $x$. Hence the slope is $m = \frac{\text{Change in } f}{\text{Change in } x} = \frac{-0.6}{0.3} = -2$. We know now that $f = -2x + b$ for some $b$. Since $f = 8$ when $x = 3$, we have $8 = -2 \times 3 + b$, so $b = 8 + 2 \times 3 = 14$. The formula is $f = -2x + 14$.

10. **An epidemic:**

   (a) There is a constant change of 5 in $d$ and a constant change of 6 in $T$. Thus these data exhibit a constant rate of change and so are linear.

   (b) The slope represents the number of new cases per day.

   (c) The slope is $m = \dfrac{\text{Change in } T}{\text{Change in } d} = \dfrac{6}{5} = 1.2$.

   (d) Now $T$ is a linear function with slope 1.2, and the initial value is 35 (from the first entry in the table). Thus the formula is $T = 1.2d + 35$.

   (e) To predict the total number of diagnosed flu cases after 17 days we put $d = 17$ in the formula from Part (d): $T = 1.2 \times 17 + 35 = 55.4$. Thus we expect 55 flu cases to be diagnosed after 17 days.

11. **Testing and plotting linear data:**

   (a) There is a constant change of 2 in $x$ and a constant change of 0.24 in $f$. Thus these data exhibit a constant rate of change and so are linear.

   (b) The slope is $m = \dfrac{\text{Change in } f}{\text{Change in } x} = \dfrac{0.24}{2} = 0.12$. We know now that $f = 0.12x + b$ for some $b$. Since $f = -1.12$ when $x = -1$, we have $-1.12 = 0.12 \times (-1) + b$, so $b = -1.12 + 0.12 = -1$. The formula is $f = 0.12x - 1$.

   (c) The figure below shows the plot of the data together with the graph of $f = 0.12x - 1$.

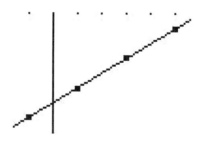

12. **Marginal tax rate:**

   (a) There is a constant change of \$50 in the taxable income and a constant change of \$14 in the tax due. Thus these data exhibit a constant rate of change and so are linear.

   (b) From Part (a) the rate of change in tax due as a function of taxable income is $\dfrac{14}{50} = 0.28$ dollar per dollar. Thus the additional tax due on each dollar is \$0.28.

(c) From the table the tax due on a taxable income of \$97,000 is \$21,913. If the taxable income increases by \$1000 to \$98,000, by Part (b) the tax due will increase by $0.28 \times 1000 = 280$ dollars. Thus the tax due on a taxable income of \$98,000 is $21{,}913 + 280 = 22{,}193$ dollars.

(d) Let $T$ denote the tax due, in dollars, on an income of $A$ dollars over \$97,000. The slope of $T$ is 0.28 by Part (b). The initial value of $T$ is the tax due on 0 dollars over \$97,000, that is, on \$97,000. That tax is \$21,913. Thus the formula is $T = 0.28A + 21{,}913$.

13. **Plotting data and regression lines:**

    (a) The figure below on the left is a plot of the data.

    (b) The regression line is $y = -16.3x + 37.19$.

    (c) The figure on the right below is the plot of the data with the regression line added.

14. **Meaning of slope of regression line:** If the slope of the regression line for number of births per thousand to unmarried 18- to 19-year-old women as a function of year is $-1.13$, then for each year that passes the number of births to unmarried 18- to 19-year old women decreases by about 1.13 births per thousand.

15. **Life expectancy:**

    (a) Let $E$ be life expectancy in years and $t$ the time in years since 2005. Using regression gives the model $E = 0.15t + 77.74$.

    (b) The figure below is the plot of the data with the regression line added.

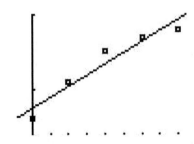

(c) The slope of the regression line is 0.15, and this means that for each increase of 1 in the year of birth the life expectancy increases by 0.15 year, or about 2 months.

(d) Now 2019 corresponds to $t = 14$, so using the regression line we predict the life expectancy to be $0.15 \times 14 + 77.74 = 79.84$ years. If we round to one decimal place, we get 79.8 years.

(e) Now 1580 corresponds to $t = -425$, so from the regression line we expect the life expectancy to be $0.15 \times (-425) + 77.74 = 13.99$ years, or (to one decimal place) 14.0 year. This is nonsensical! The year 2300 corresponds to $t = 295$, so from the regression line we predict the life expectancy to be $0.15 \times 295 + 77.74 = 121.99$ years, or (to one decimal place) 122.0 years. This seems unreasonable. In fact, both of these predictions are far out of the range of applicability of the regression line formula.

16. **XYZ Corporation stock prices**:

(a) Let $P$ be the stock price in dollars and $t$ the time in months since January 2011. Using regression gives the model $P = 0.547t + 43.608$.

(b) The figure below is the plot of the data with the regression line added.

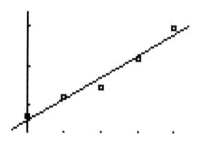

(c) The slope is 0.547 dollar per month, and this means that the stock price increases by about $0.55 each month.

(d) Now January 2012 corresponds to $t = 12$, so using the regression line we predict the stock price to be $0.547 \times 12 + 43.608 = 50.172$, or about \$50.17. Also January 2013 corresponds to $t = 24$, so using the regression line we predict the stock price to be $0.547 \times 24 + 43.608 = 56.736$, or about \$56.74.

17. **Crossing graphs**: We start with

$$4x - 2y = 9$$
$$x + y = 0$$

and solve each equation for $y$. For the first equation we have $4x - 2y = 9$, and therefore $y = \dfrac{9 - 4x}{-2}$. For the second equation we have $x + y = 0$, so $y = -x$. The figure below shows the graph of both $y = \dfrac{9 - 4x}{-2}$ and $y = -x$ using a horizontal span of 0 to 2 and a vertical span of $-5$ to 0. The intersection point, which is the solution to the system, is at $x = 1.5$, $y = -1.5$.

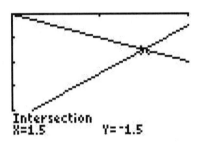

18. **Hand calculation**: To solve Exercise 17 by hand calculation, we take the second equation $x + y = 0$ and solve for $y$. As above for Exercise 17, this yields $y = -x$. Now we take the first equation $4x - 2y = 9$, substitute the expression above for $y$, and then solve for $x$:

$$4x - 2y = 9$$
$$4x - 2(-x) = 9$$
$$6x = 9$$
$$x = \frac{9}{6} = 1.5.$$

Now $y = -x = -1.5$.

19. **Bills**: Let $T$ be the number of $20-dollar bills and $F$ the number of $50-dollar bills. Since the total value is $400, we have $20T + 50F = 400$. Since there are 11 bills, we have $T + F = 11$. The second equation can easily be solved for $F$, giving $F = 11 - T$. Substituting into the first equation, we get

$$
\begin{aligned}
20T + 50F &= 400 \\
20T + 50(11 - T) &= 400 \\
20T + 50 \times 11 - 50T &= 400 \\
-30T &= 400 - 50 \times 11 \\
T &= \frac{400 - 50 \times 11}{-30} = 5.
\end{aligned}
$$

Thus there are 5 $20-dollar bills and 6 $50-dollar bills.

20. **Mixing paint**: Let $B$ be the number of gallons of blue paint and $Y$ the number of gallons of yellow paint. Since the total is 10 gallons, we have $B + Y = 10$. Since there is 3 times as much yellow paint as blue paint, we have $Y = 3B$. We insert the formula for $Y$ from the second equation into the first:

$$
\begin{aligned}
B + Y &= 10 \\
B + 3B &= 10 \\
4B &= 10 \\
B &= \frac{10}{4} = 2.5.
\end{aligned}
$$

Thus you should use 2.5 gallons of blue paint and 7.5 gallons of yellow paint.

## A FURTHER LOOK: PARALLEL AND PERPENDICULAR LINES

1. **Parallel and perpendicular lines**: To determine if the two lines are parallel, perpendicular, or neither, we calculate the slopes of the two lines:

$$
\begin{aligned}
\text{Slope of } \ell_1 &= \frac{\Delta y}{\Delta x} = \frac{10 - 2}{5 - 1} = \frac{8}{4} = 2 \\
\text{Slope of } \ell_2 &= \frac{\Delta y}{\Delta x} = \frac{15 - 3}{7 - 1} = \frac{12}{6} = 2
\end{aligned}
$$

Since the slopes are the same, the lines are parallel.

2. **Parallel and perpendicular lines**: To determine if the two lines are parallel, perpendicular, or neither, we calculate the slopes of the two lines:

$$\text{Slope of } \ell_1 = \frac{\Delta y}{\Delta x} = \frac{10 - 2}{5 - 1} = \frac{8}{4} = 2$$

$$\text{Slope of } \ell_2 = \frac{\Delta y}{\Delta x} = \frac{2 - 3}{3 - 1} = \frac{-1}{2} = -\frac{1}{2}$$

Since the slopes are negative reciprocals, the lines are perpendicular.

3. **Parallel and perpendicular lines**: To determine if the two lines are parallel, perpendicular, or neither, we calculate the slopes of the two lines:

$$\text{Slope of } \ell_1 = \frac{\Delta y}{\Delta x} = \frac{10 - 2}{5 - 1} = \frac{8}{4} = 2$$

$$\text{Slope of } \ell_2 = \frac{\Delta y}{\Delta x} = \frac{4 - 3}{3 - 1} = \frac{1}{2}$$

Since the slopes are neither the same nor negative reciprocals, the lines are neither parallel nor perpendicular.

4. **Parallel and perpendicular lines**: To determine if the two lines are parallel, perpendicular, or neither, we calculate the slopes of the two lines:

$$\text{Slope of } \ell_1 = \frac{\Delta y}{\Delta x} = \frac{10 - 2}{5 - 1} = \frac{8}{4} = 2$$

$$\text{Slope of } \ell_2 = \frac{\Delta y}{\Delta x} = \frac{15 - 3}{(-5) - 1} = \frac{12}{-6} = -2$$

Since the slopes are neither the same nor negative reciprocals, the lines are neither parallel nor perpendicular.

5. **Slope of perpendicular lines**: Since the lines $\ell_1$ and $\ell_2$ are perpendicular, their slopes are negative reciprocals. So if the slope of $\ell_1$ is $-\dfrac{5}{8}$, then the slope of $\ell_2$ is $-\dfrac{1}{-5/8} = -\dfrac{8}{-5} = \dfrac{8}{5}$, or 1.6.

6. **Finding parallel lines**: Parallel lines have the same slope. The line $y = 3x + 4$ has slope 3, so that is the slope of the line whose equation we are seeking. Hence the desired equation has the form $y = 3x + b$. We need to find $b$. The fact that the line passes through the point $(1, 5)$ tells us that when $x = 1$, $y$ is equal to 5. We put these two values into the equation to find $b$:

$$
\begin{aligned}
y &= 3x + b \\
5 &= 3 \times 1 + b \\
5 &= 3 + b \\
2 &= b.
\end{aligned}
$$

Thus $b = 2$, and the equation of the line is $y = 3x + 2$.

7. **Finding parallel lines**: Parallel lines have the same slope. The line $y = 5x + 3$ has slope 5, so that is the slope of the line whose equation we are seeking. Hence the desired equation has the form $y = 5x + b$. We need to find $b$. The fact that the line passes through the point $(-2, 15)$ tells us that when $x = -2$, $y$ is equal to 15. We put these two values into the equation to find $b$:

$$
\begin{aligned}
y &= 5x + b \\
15 &= 5 \times (-2) + b \\
15 &= -10 + b \\
25 &= b.
\end{aligned}
$$

Thus $b = 25$, and the equation of the line is $y = 5x + 25$.

8. **Finding perpendicular lines**: The line $y = 3x + 1$ has slope 3, so any liner perpendicular to the given line has slope $-1/3$. Hence the desired equation has the form $y = -\frac{1}{3}x + b$. We need to find $b$. The fact that the line passes through the point $(9, 1)$ tells us that when $x = 9$, $y$ is equal to 1. We put these two values into the equation to find $b$:

$$
\begin{aligned}
y &= -\frac{1}{3}x + b \\
1 &= -\frac{1}{3} \times 9 + b \\
1 &= -3 + b \\
4 &= b.
\end{aligned}
$$

Thus $b = 4$, and the equation of the line is $y = -\frac{1}{3}x + 4$.

9. **Finding perpendicular lines**: The line $y = -4x + 5$ has slope $-4$, so any line perpendicular to the given line has slope $-1/(-4) = 1/4$. Hence the desired equation has the form $y = \dfrac{1}{4}x + b$. We need to find $b$. The fact that the line passes through the point $(12, 6)$ tells us that when $x = 12$, $y$ is equal to 6. We put these two values into the equation to find $b$:

$$
\begin{aligned}
y &= \frac{1}{4}x + b \\
6 &= \frac{1}{4} \times 12 + b \\
6 &= 3 + b \\
3 &= b.
\end{aligned}
$$

Thus $b = 3$, and the equation of the line is $y = \dfrac{1}{4}x + 3$.

## A FURTHER LOOK: SECANT LINES

1. **Secant lines**: The slope of the secant line is the average rate of change for $x^3$ from $x = 1$ to $x = 3$:

$$
\begin{aligned}
\text{Slope} &= \text{Average rate of change} \\
&= \frac{3^3 - 1^3}{3 - 1} \\
&= 13.
\end{aligned}
$$

Thus the equation of the secant line has the form $y = 13x + b$. When $x = 1$, $y = 1^3 = 1$. We put these two values into the equation to find $b$:

$$
\begin{aligned}
y &= 13x + b \\
1 &= 13 \times 1 + b \\
1 &= 13 + b \\
-12 &= b.
\end{aligned}
$$

The equation of the secant line is $y = 13x - 12$.

2. **Secant lines**: The slope of the secant line is the average rate of change for $\sqrt{x}$ from $x = 4$ to $x = 9$:

$$
\begin{aligned}
\text{Slope} &= \text{Average rate of change} \\
&= \frac{\sqrt{9} - \sqrt{4}}{9 - 4} \\
&= \frac{1}{5}.
\end{aligned}
$$

Thus the equation of the secant line has the form $y = \frac{1}{5}x + b$. When $x = 4$, $y = \sqrt{4} = 2$. We put these two values into the equation to find $b$:

$$
\begin{aligned}
y &= \frac{1}{5}x + b \\
2 &= \frac{1}{5} \times 4 + b \\
2 &= \frac{4}{5} + b \\
\frac{6}{5} &= b.
\end{aligned}
$$

The equation of the secant line is $y = \frac{1}{5}x + \frac{6}{5}$.

3. **Secant lines**: The slope of the secant line is the average rate of change for $x^2 + 1$ from $x = 2$ to $x = 3$:

$$
\begin{aligned}
\text{Slope} &= \text{Average rate of change} \\
&= \frac{3^2 + 1 - (2^2 + 1)}{3 - 2} \\
&= 5.
\end{aligned}
$$

Thus the equation of the secant line has the form $y = 5x + b$. When $x = 2$, $y = 2^2 + 1 = 5$. We put these two values into the equation to find $b$:

$$
\begin{aligned}
y &= 5x + b \\
5 &= 5 \times 2 + b \\
5 &= 10 + b \\
-5 &= b.
\end{aligned}
$$

The equation of the secant line is $y = 5x - 5$.

4. **Secant lines**: The slope of the secant line is the average rate of change for $\frac{15}{x}$ from $x = 3$ to $x = 5$:

$$
\begin{aligned}
\text{Slope} &= \text{Average rate of change} \\
&= \frac{\frac{15}{5} - \frac{15}{3}}{5 - 3} \\
&= -1.
\end{aligned}
$$

Thus the equation of the secant line has the form $y = -x + b$. When $x = 3$, $y = \frac{15}{3} = 5$. We put these two values into the equation to find $b$:

$$
\begin{aligned}
y &= -x + b \\
5 &= -3 + b \\
8 &= b.
\end{aligned}
$$

The equation of the secant line is $y = -x + 8$.

5. **Secant lines**: The slope of the secant line is the average rate of change for $x^2 + x$ from $x = 4$ to $x = 6$:

$$
\begin{aligned}
\text{Slope} \quad &= \quad \text{Average rate of change} \\
&= \quad \frac{6^2 + 6 - (4^2 + 4)}{6 - 4} \\
&= \quad 11.
\end{aligned}
$$

Thus the equation of the secant line has the form $y = 11x + b$. When $x = 4$, $y = 4^2 + 4 = 20$. We put these two values into the equation to find $b$:

$$
\begin{aligned}
y \quad &= \quad 11x + b \\
20 \quad &= \quad 11 \times 4 + b \\
20 \quad &= \quad 44 + b \\
-24 \quad &= \quad b.
\end{aligned}
$$

The equation of the secant line is $y = 11x - 24$.

6. **Approximations**:

(a) The slope of the secant line is the average rate of change for $x^2/2$ from $x = 2$ to $x = 6$:

$$
\begin{aligned}
\text{Slope} \quad &= \quad \text{Average rate of change} \\
&= \quad \frac{6^2/2 - 2^2/2}{6 - 2} \\
&= \quad 4.
\end{aligned}
$$

Thus the equation of the secant line has the form $y = 4x + b$. When $x = 2$, $y = 2^2/2 = 2$. We put these two values into the equation to find $b$:

$$
\begin{aligned}
y \quad &= \quad 4x + b \\
2 \quad &= \quad 4 \times 2 + b \\
2 \quad &= \quad 8 + b \\
-6 \quad &= \quad b.
\end{aligned}
$$

The equation of the secant line is $y = 4x - 6$.

(b) To approximate $f(3)$ using the secant line we put $x = 3$ in the formula $y = 4x - 6$ from Part (a) to get $y = 4 \times 3 - 6 = 6$. Therefore, $f(3)$ is about 6.

7. **Approximations**:

(a) The slope of the secant line is the average rate of change for $1 - x^3$ from $x = 2$ to $x = 4$:

$$
\begin{aligned}
\text{Slope} \quad &= \quad \text{Average rate of change} \\
&= \quad \frac{1 - 4^3 - (1 - 2^3)}{4 - 2} \\
&= \quad -28.
\end{aligned}
$$

Thus the equation of the secant line has the form $y = -28x + b$. When $x = 2$, $y = 1 - 2^3 = -7$. We put these two values into the equation to find $b$:

$$
\begin{aligned}
y \quad &= \quad -28x + b \\
-7 \quad &= \quad -28 \times 2 + b \\
-7 \quad &= \quad -56 + b \\
49 \quad &= \quad b.
\end{aligned}
$$

The equation of the secant line is $y = -28x + 49$.

(b) To approximate $f(3)$ using the secant line we put $x = 3$ in the formula $y = -28x + 49$ from Part (a) to get $y = -28 \times 3 + 49 = -35$. Therefore, $f(3)$ is about $-35$.

8. **Secant lines for graphs that are concave down**: One possible graph is the following.

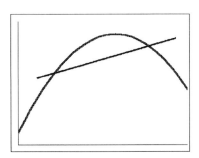

(a) The secant line is below the graph on the interval between the two points that determine the secant line and above the graph outside that interval.

(b) Recall that using average rates of change to approximate function values is the same as approximating the graph by its secant line. Using Part (a), we see that the average rate of change will give an estimate that is too small on the interval between the two points that determine the secant line and too large outside that interval.

9. **Which is best?** If we look above the point $a$ on the horizontal axis, the line that is closest to the point on the graph is secant line number 3. Therefore, line 3 gives the best approximation for $f(a)$.

10. **Difference quotient**: The slope of the secant line is the average rate of change for $x^2$ from $a$ to $x$:

$$\text{Slope} = \text{Average rate of change} = \frac{x^2 - a^2}{x - a}.$$

Therefore, the slope of the secant line is

$$\frac{x^2 - a^2}{x - a}.$$

This expression can be simplified:

$$\frac{x^2 - a^2}{x - a} = \frac{(x - a)(x + a)}{x - a} = x + a.$$

Therefore, the slope of the secant line equals $x + a$.

11. **Derivative**: By the preceding exercise the slope of the secant line from $a$ to $x$ equals $x + a$. To find the derivative of $f(x) = x^2$ at $a$ we let $x$ go to $a$ in this formula for the slope. As $x$ goes to $a$, $x + a$ goes to $a + a = 2a$. Therefore, the derivative of $f$ at $a$ is $2a$.

12. **Difference quotient**: The slope of the secant line is the average rate of change for $x^2 + 1$ from $a$ to $a + h$:

$$
\begin{aligned}
\text{Slope} &= \text{Average rate of change} \\
&= \frac{(a + h)^2 + 1 - (a^2 + 1)}{a + h - a} \\
&= \frac{(a + h)^2 + 1 - (a^2 + 1)}{h}.
\end{aligned}
$$

Therefore the slope of the secant line is

$$\frac{(a + h)^2 + 1 - (a^2 + 1)}{h}.$$

This expression can be simplified:

$$
\begin{aligned}
\frac{(a + h)^2 + 1 - (a^2 + 1)}{h} &= \frac{a^2 + 2ah + h^2 + 1 - a^2 - 1}{h} \\
&= \frac{2ah + h^2}{h} \\
&= 2a + h.
\end{aligned}
$$

Therefore, the slope of the secant line equals $2a + h$.

13. **Derivative**: By the preceding exercise the slope of the secant line from $a$ to $a + h$ equals $2a + h$. To find the derivative of $g(x) = x^2 + 1$ at $a$ we let $h$ go to 0 in this formula for the slope. As $h$ goes to 0, $2a + h$ goes to $2a$. Therefore, the derivative of $g$ at $a$ is $2a$.

# Solution Guide for Chapter 4: Exponential Functions

## 4.1 EXPONENTIAL GROWTH AND DECAY

1. **A violin string:**

   (a) In order to make the next higher note, the current length is multiplied by $0.944$, so the decay factor for $L$ is $0.944$. The length of an unstopped string is 32 centimeters, and that is the initial value for $L$. Therefore, $L = 32 \times 0.944^n$.

   (b) Making a note 4 notes higher than an unstopped string corresponds to taking $n = 4$. By part a, the resulting length is $L = 32 \times 0.944^4 = 25.41$. Thus, the string must be stopped to a length of 25.41 centimeters in order to make a note that is 4 notes higher than an unstopped string.

3. **Solar panels:**

   (a) The yearly growth factor for $P$ is 1.45, and the initial value for $P$ is 3.8 gigawatts (the installations in 2007). Therefore, $P = 3.8 \times 1.45^t$.

   (b) Because 2006 is 1 year before 2007, the year 2006 corresponds to $t = -1$. By part a, we find $P(-1) = 3.8 \times 1.45^{-1} = 2.62$. Thus, the installations in 2006 were about 2.62 gigawatts.

5. **A weight-gain program:**

   (a) Each week the woman's weight is multiplied by 1.03, so $w$ is an exponential function with weekly growth factor 1.03. The initial value is her current weight of 104 pounds. Therefore, $w = 104 \times 1.03^t$.

   (b) We want to solve the equation

   $$104 \times 1.03^t = 135.$$

   We solve this using the crossing-graphs method. After consulting a table of values, we use a horizontal span of 0 to 10 weeks and a vertical span of 0 to 160 pounds. In the figure below, we have graphed the function $w$ along with the constant function

135 (thick line). We see that they intersect when $t = 8.83$. Therefore, she will reach 135 pounds after 8.83 weeks.

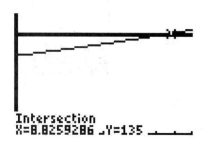

Intersection
X=8.8259286  Y=135

7. **Unit conversion with exponential decay**:

   (a) Because $m$ is measured in months, inspection of the formula for $P$ shows that the monthly decay factor is $a = 0.86$. A week is $1/4$ of a month, so the weekly decay factor is $0.86^{1/4} = 0.96$. The initial value is still 300, so the formula in terms of weeks is $P = 300 \times 0.96^w$.

   (b) The monthly decay factor is 0.86, and each year is 12 months, so the yearly decay factor is $0.86^{12} = 0.16$. The formula in terms of years is $P = 300 \times 0.16^y$.

9. **Half-life of tritium**: If we start with 100 grams of tritium then, by the same reasoning as in Example 4.1, the decay is modeled by

$$A = 100 \times 0.945^t.$$

As in Example 4.1, the half-life of tritium is obtained by solving $100 \times 0.945^t = 50$. The result is $t = 12.25$ years. So, after 12.25 years we will have 50 grams of tritium left. This is the same length of time as found in Example 4.1, Part 4, for 50 grams of tritium to decay to 25 grams.

Because the amount we start with decays by half every 12.25 years, we see that after starting with 100 grams of tritium we will have 50 grams after 12.25 years, 25 grams after $2 \times 12.25 = 24.5$ years, 12.5 grams after $3 \times 12.25 = 36.75$ years, and 6.25 grams after $4 \times 12.25 = 49$ years. This is four half-lives.

11. **The MacArthur-Wilson Theory of biogeography**: According to the theory, stabilization occurs when the rate of immigration equals the rate of extinction, that is, $I = E$. This is equivalent to solving the equation $4.2 \times 0.93^t = 1.5 \times 1.1^t$. This can be solved using a table or a graph. The graph below, which uses a horizontal span of 0 to 10 and a vertical span of 0 to 5, shows that the solution occurs when $t = 6.13$ years, at which time the immigration and extinction rates are each 2.69 species per year.

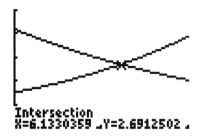

Intersection
X=6.1330359 _Y=2.6912502 .

13. **Gold prices**:

(a) Every 3 years gold prices doubled, so they grew by a factor of 2 every 3 years. Therefore, in 1 year gold prices grew by a factor of $2^{1/3} = 1.26$. Therefore, the yearly growth factor was 1.26.

(b) The yearly growth factor was 1.26, so to find next year's gold price we multiply this year's price by 1.26.

15. **Grains of wheat on a chess board**: The growth factor here is 2, and the initial amount is 1. Thus the function we need is $1 \times 2^t$, or simply $2^t$, where $t$ is measured in days since the first day. Since we count the first square as being on the initial day ($t = 0$), the second day corresponds to $t = 1$, and so on; thus the 64th day corresponds to $t = 63$. Hence the king will have to put $2^{63} = 9.22 \times 10^{18}$ grains of wheat on the 64th square. If we divide $9.22 \times 10^{18}$ by the number of grains in a bushel, namely $1,100,000$, we get $8.38 \times 10^{12}$ bushels. If each bushel is worth \$4.25, then the value of the wheat on the 64th square was

$$4.25 \times 8.38 \times 10^{12} = 3.56 \times 10^{13} = 35,600,000,000,000 \text{ dollars.}$$

This is over 35.6 trillion dollars, a sum that may trouble even a king.

17. **Mobile phones**:

(a) The amount of data passing through mobile phone networks grows by a factor of 2 each year. This amount is an exponential function of time because the amount is growing by constant multiples.

(b) The yearly growth factor is 2, so $D = D_0 \times 2^t$.

(c) The amount of data will be 100 times its initial value when $D = 100D_0$. This occurs at the time $t$ when $D_0 \times 2^t = 100D_0$ or

$$2^t = 100.$$

We want to solve this equation. We solve it using the crossing-graphs method. After consulting a table of values, we use a horizontal span of 0 to 8 years and a vertical span of 0 to 300. In the figure below, we have graphed the function $2^t$ along with the constant function 100. We see that they intersect when $t = 6.64$ years. Therefore, it will be 6.64 years before the amount of data is 100 times its initial value.

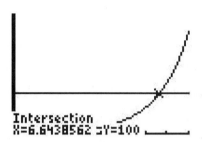

19. **Headway on four-lane highways**:

(a) A flow of 500 vehicles per hour is $\dfrac{500}{3600} = 0.14$ vehicle per second, so $q = 0.14$. Thus the probability that the headway exceeds 15 seconds is given by $P = e^{-0.14 \times 15} = 0.12$, or about 12%.

(b) On a four-lane highway carrying an average of 500 vehicles per hour, $q = 0.14$, and so $P = e^{-0.14t}$. Using the law of exponents, this can be written as $P = e^{-0.14t} = (e^{-0.14})^t$. In this form it is clear that the decay factor is $a = e^{-0.14} = 0.87$.

## Skill Building Exercises

S-1. **Exponential growth with given initial value and growth factor**: The initial value is $P = 23$, and the growth factor is $a = 1.4$, so the formula for the function is

$$N(t) = 23 \times 1.4^t.$$

The table of values below leads us to choose a horizontal span of 0 to 5 and a vertical span of 0 to 130. The graph is shown below.

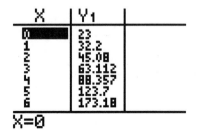
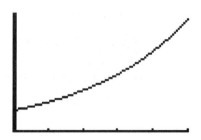

S-3. **Exponential growth**: To find next year's amount we multiply this year's amount by 3, so $A$ is an exponential function with yearly growth factor 3. Because $A$ is an exponential function, we can find a formula for $A$ in the form $Pa^t$, where $a$ is the yearly growth factor 3 and $P$ is the initial value 10. Therefore, $A = 10 \times 3^t$.

S-5. **Exponential decay**: To find the next day's amount we multiply this day's amount by 0.6, so $A$ is an exponential function with daily decay factor 0.6. Because $A$ is an exponential function, we can find a formula for $A$ in the form $Pa^t$, where $a$ is the daily decay factor 0.6 and $P$ is the initial value 7. Therefore, $A = 7 \times 0.6^t$.

S-7. **Exponential change**: To find the next week's amount we divide this week's amount by 7, which means that we multiply by $1/7$. Therefore, $A$ is an exponential function with weekly decay factor $1/7$. Because $A$ is an exponential function, we can find a formula for $A$ in the form $Pa^t$, where $a$ is the weekly decay factor $1/7$ and $P$ is the initial value 8. Therefore, $A = 8 \times (1/7)^t$.

S-9. **Function value from initial value and growth factor**: Since the growth factor is 2.4 and the initial value is $f(0) = 3$, $f(2) = 3 \times 2.4 \times 2.4 = 17.28$. Since $f$ is an exponential function, we can get a formula for $f(x)$ in the form $Pa^x$, where $a$ is the growth factor 2.4 and $P$ is the initial value 3. Thus $f(x) = 3 \times 2.4^x$.

S-11. **Finding the growth factor**: The growth factor is the factor by which $f$ changes for a change of 1 in the variable. If $a$ is the growth factor, then

$$a = \frac{f(5)}{f(4)} = \frac{10}{8} = 1.25$$

S-13. **Rate of change**: A graph showing exponential decay is concave up—see, for example, Figure 4.2 of the text.

S-15. **Rate of change**: The rate of change of an exponential function is proportional to the function value.

S-17. **Changing units**: If the yearly growth factor is 1.17, then, since a month is $1/12$th of a year, the monthly growth factor is $1.17^{1/12} = 1.013$, or about 1.01.

S-19. **Changing units**: If the decade growth factor is 3.4, then, since a year is $1/10$th of a decade, the yearly growth factor is $3.4^{1/10} = 1.13$.

S-21. **Changing units**: If the century decay factor is 0.2, then, since a year is $1/100$th of a century, the yearly decay factor is $0.2^{1/100} = 0.984$, or about 0.98.

## 4.2  CONSTANT PERCENTAGE CHANGE

1. **Mittenwalde is rich**:

    (a) Let $D$ denote the debt, in guilders, at the time $t$ years after 1562. Because the debt increased by 6% each year, the yearly growth factor for the function $D$ is $a = 1 + 0.06 = 1.06$. The initial value is 400 guilders, so an exponential model is $D = 400 \times 1.06^t$. Because 2012 is $2012 - 1562 = 450$ years after 1562, we put $t = 450$ into this formula. The result is $D(450) = 400 \times 1.06^{450} = 9.766 \times 10^{13}$. Now one trillion equals $10^{12}$, so the debt in 2012 was 97.66 trillion guilders.

    (b) If a 1562 guilder is about 0.45 dollar, the value of the debt in trillions of dollars is $0.45 \times 97.66 = 43.947$. Thus, the debt was about 43.95 trillion dollars.

    (c) Answers will vary.

3. **Ponzi schemes**:

    (a) Because the return must increase by 5% each month, the monthly growth factor for the function $R$ is $a = 1 + 0.05 = 1.05$. The initial value is 20,000 dollars, so an exponential model is $R = 20{,}000 \times 1.05^t$.

    (b) Because 3 years is 36 months, we put $t = 36$ into the formula from part a to get $R(36) = 20{,}000 \times 1.05^{36} = 115{,}836.32$. Thus, at the end of 3 years the operator must pay a return of \$115,836.32.

    (c) If each new investor pays \$2000, to cover the amount calculated in part b the number of new investors must be $115{,}836.32/2000 = 58$, where we have rounded to the nearest whole number.

5. **Theater productions**:

    (a) Because the number of performances increased by 10% each year, the yearly growth factor for the function $P$ is $a = 1 + 0.10 = 1.10$. The initial number of performances is 56.61 thousand, so an exponential model is $P = 56.61 \times 1.10^t$.

    (b) Now 2007 is $2007 - 1995 = 12$ years after 1995, so we put $t = 12$ into the formula for $P$:
    $$P(12) = 56.61 \times 1.10^{12} = 177.67.$$

    Therefore, the model gives the value of 177.67 thousand for the number of performances in 2007.

7. **A population with given per capita growth rate**:

   (a) Since the population grows at a rate of 2.3% per year, the yearly growth factor for the function is $a = 1 + .023 = 1.023$. Since initially the population is $P = 3$ million, the exponential function is

   $$N(t) = 3 \times 1.023^t.$$

   Here $N$ is the population in millions and $t$ is time in years.

   (b) The population after 4 years is expressed by $N(4)$ using functional notation. Its value, using the formula from Part (a), is $N(4) = 3 \times 1.023^4 = 3.29$ million.

9. **Unit conversion with exponential decay**:

   (a) Now $N(2) = 500 \times 0.68^2 = 231.2$, which means that after 2 years there are 231.2 grams of the radioactive substance present.

   (b) The yearly decay factor is $0.68 = 1 - 0.32$, so $r = 0.32$. Thus the yearly percentage decay rate is 32%.

   (c) Now $t = \dfrac{1}{12}$ corresponds to one month, so the monthly decay factor is

   $$0.68^{1/12} = 0.968.$$

   The monthly percentage decay rate is found using the same method as in Part (b): $0.968 = 1 - 0.032$, so $r = 0.032$. Thus the monthly percentage decay rate is 3.2%.

   (d) There are 31,536,000 seconds in a year, so one second corresponds to $\dfrac{1}{31{,}536{,}000}$ year. Thus the decay factor per second is

   $$0.68^{1/31,536,000} = 0.9999999878 = 1 - 0.0000000122.$$

   Thus $r = 0.0000000122$, and so the percentage decay rate per second is only 0.00000122%.

11. **Gray wolves in Idaho**:

   (a) Because the population decreases by 19% each year, the yearly decay factor for the function $W$ is $a = 1 - 0.19 = 0.81$. The initial value of $W$ is the population in 2009, which is 870. Therefore, an exponential model is $W = 870 \times 0.81^t$.

   (b) We want to solve the equation

   $$870 \times 0.81^t = 20.$$

   We solve this using the crossing-graphs method. After consulting a table of values, we use a horizontal span of 0 to 20 years and a vertical span of 0 to 880 wolves.

In the figure below, we have graphed the function $W$ along with the constant function 20 (thick line). We see that they intersect when $t = 17.90$ years. Therefore, the rate of decline must continue for 17.90 years for the population to reach 20.

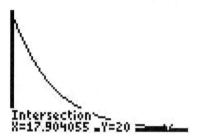

13. **Inflation**: The yearly growth factor for prices is given by $1 + 0.03 = 1.03$. We want to find the value of $t$ for which $1.03^t = 2$, since that is when the prices will have doubled. We solve this using the crossing graphs method. The table of values below leads us to choose a horizontal span of 0 to 25 years and a vertical span of 0 to 3. In the graph below, the horizontal axis is years, and the vertical axis is the factor by which prices increase. We see that the two graphs cross at $t = 23.45$ years. Thus, prices will double in 23.45 years. Note that we did not use the amount of the fixed income. The answer, 23.45 years, is the same regardless of the income.

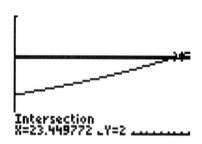

15. **The population of Canada**:

    (a) The yearly growth factor for the population is $1 + 0.011 = 1.011$, and the initial population is 34.34 million. The population $N$ (in millions) at the time $t$ years after 2011 is

    $$N = 34.34 \times 1.011^t.$$

    (b) The population of Canada in 2015 is expressed as $N(4)$ in functional notation, since 2015 is $t = 4$ years after 2011. The value of $N(4)$ is $34.34 \times 1.011^4 = 35.88$ million.

(c) The population of Canada is 37.5 million when $34.34 \times 1.011^t = 37.5$. We solve this equation using the crossing-graphs method. A table leads us to choose a horizontal span of 0 to 10 and a vertical span of 30 to 40. In the figure below, the curve is the graph of $34.34 \times 1.011^t$, and the thick line is the graph of 37.5. We see that the two cross at 8.05 years after 2011. That is sometime in 2019.

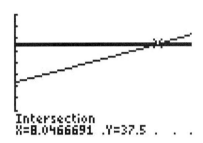

```
Intersection
X=8.0466691 .Y=37.5 . . .
```

17. **Tsunami waves in Crescent City**:

(a) The probability $P$ is an exponential function of $Y$ because for each change in $Y$ of 1 year, the probability $P$ decreases by 2%, so $P$ exhibits constant percentage change.

(b) The decay factor for $P$ is 1 minus the decay percentage, so $a = 1 - r = 1 - 0.02 = 0.98$ per year.

(c) The initial value of $P$, $P(0)$, represents the probability of no tsunami wave of height 15 feet or more striking over a period of 0 years. Since this is a certainty (as a probability), the initial value of $P$ is 1.

(d) Since $P$ is an exponential function of $Y$ (by Part (a)) with a decay factor of 0.98 (by Part (b)) and an initial value of 1 (by Part (c)) the formula is $P = 1 \times 0.98^Y$, or simply $P = 0.98^Y$.

(e) The probability of no tsunami waves 15 feet or higher striking Crescent City over a 10-year period is $P(10) = 0.98^{10} = 0.82$. Thus the probability of no such tsunami wave striking over a ten-year period is 0.82 or 82%. Over a 100-year period, the probability is $P(100) = 0.98^{100} = 0.13$. Thus the probability of no such tsunami wave striking over a 100-year period is 0.13 or only 13%.

(f) To find the probability $Q$ that at least one wave 15 feet or higher will strike Crescent City over a period of $Y$ years, we first see that $P + Q = 1$ since $P + Q$ represents all possibilities. Thus $Q = 1 - P = 1 - 0.98^Y$.

19. **The photic zone:** Near Cape Cod, Massachusetts, the depth of the photic zone is about 16 meters. We know that the photic zone extends to a depth where the light intensity is about 1% of surface light, that is, $I = 0.01I_0$. Since the **Beer-Lambert-Bouguer law** implies that $I$ is an exponential function of depth, $I = I_0 \times a^d$. Since $I(16) = 0.01I_0$, we have $I_0 \times a^{16} = 0.01I_0$, so $a^{16} = 0.01$. Thus the decay factor is $a = 0.01^{1/16} = 0.750$, representing a 25.0% decrease in light intensity for each additional meter of depth.

21. **A diet:**

   (a) Because the difference decreases by 10% each month, the monthly decay factor for the function $D$ is $a = 1 - 0.10 = 0.90$. The initial value of $D$ is the difference at the beginning, which is $260 - 200 = 60$ pounds. Therefore, an exponential model is $D = 60 \times 0.9^t$.

   (b) When he weighs 210 pounds the difference is $210 - 200 = 10$ pounds. Therefore, we want to solve the equation

   $$60 \times 0.9^t = 10.$$

   We solve this using the crossing-graphs method. After consulting a table of values, we use a horizontal span of 0 to 20 years and a vertical span of 0 to 60 pounds. In the figure below, we have graphed the function $D$ along with the constant function 10 (thick line). We see that they intersect when $t = 17.01$ months. Therefore, it will take 17.01 months for his weight to reach 210 pounds.

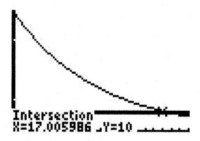

23. **Continuous compounding:**

   (a) The limiting value of the function given by the formula

   $$\left(1 + \frac{0.1}{n}\right)^n$$

   can be found using a table or a graph. Scanning down a table, for example, shows that the limiting value is 1.1052 to four decimal places.

   (b) Since the APR is 0.1, $e^{APR} = e^{0.1} = 1.1052$ to four decimal places..

(c) Since the answers to Parts (a) and (b) are the same, this shows that $e^{\text{APR}}$ is the yearly growth factor for continuous compounding in the case when the APR is 10%.

## Skill Building Exercises

**S-1. Making models:** The growth rate is 5%, so the growth factor for the function $y$ is $a = 1 + 0.05 = 1.05$. Because the initial value is 300, an exponential model is $y = 300 \times 1.05^t$.

**S-3. Making models:** The decay rate is 20%, so the decay factor for the function $y$ is $a = 1 - 0.2 = 0.80$. Because the initial value is 700, an exponential model is $y = 700 \times 0.80^t$.

**S-5. Percentage growth rate from growth factor:** The growth factor of $a = 1.06$ corresponds to a percentage growth rate of $r = a - 1 = 1.06 - 1 = 0.06$ as a decimal or 6%.

**S-7. Percentage decay rate from decay factor:** A decay factor of $a = 0.70$ corresponds to a percentage decay rate of $r = 1 - a = 1 - 0.70 = 0.30$ as a decimal or 30%.

**S-9. Function value from initial value and percentage growth rate:** The percentage growth rate is $r = 0.05$ as a decimal, so the growth factor is $a = 1 + r = 1 + 0.05 = 1.05$. We can find a formula for $f$ in the form $Pa^x$, where $a$ is the growth factor 1.05 and $P$ is the initial value $f(0) = 8$. Therefore, $f(x) = 8 \times 1.05^x$.

**S-11. Function value from initial value and percentage decay rate:** The percentage decay rate is $r = 0.05$ as a decimal, so the decay factor is $a = 1 - r = 1 - 0.05 = 0.95$. We can find a formula for $f$ in the form $Pa^x$, where $a$ is the decay factor 0.95 and $P$ is the initial value $f(0) = 6$. Therefore, $f(x) = 6 \times 0.95^x$.

**S-13. Percentage growth:** If a function has an initial value of 30 and grows at a rate of 4% per year, then it is an exponential function with $P = 30$ and $a = 1 + r = 1 + 0.04 = 1.04$. Thus an exponential function which describes this is $30 \times 1.04^t$, with $t$ in years.

**S-15. Percentage decay:**If a function has an initial value of 10 and decays by 4% per year, then it is an exponential function with $P = 10$ and $a = 1 - r = 1 - 0.04 = 0.96$. Thus an exponential function which describes this is $10 \times 0.96^t$, with $t$ in years.

**S-17. Percentage decay:**If a function has an initial value of 5 and decays by 3% per day, then it is an exponential function with $P = 5$ and $a = 1 - r = 1 - 0.03 = 0.97$. Thus an exponential function which describes this is $5 \times 0.97^t$, with $t$ in days.

S-19. **Percentage change**: Because the quantity increases by 5% each year, the yearly percentage growth rate is $r = 0.05$ as a decimal. Therefore, the yearly growth factor is $a = 1 + r = 1 + 0.05 = 1.05$. The 10-year growth factor is then $1.05^{10} = 1.629$. This represents a percentage increase of $1.629 - 1 = 0.629$ as a decimal or 62.9% over the 10-year period.

S-21. **Percentage change**: Because the quantity decreases by 5% each year, the yearly percentage decay rate is $r = 0.05$ as a decimal. Therefore, the yearly decay factor is $a = 1 - r = 1 - 0.05 = 0.95$. The 10-year decay factor is then $0.95^{10} = 0.599$. This represents a percentage decrease of $1 - 0.599 = 0.401$ as a decimal or 40.1% over the 10-year period.

S-23. **Percentage change**: Because the quantity decreases by 3% each year, the yearly percentage decay rate is $r = 0.03$ as decimal. Therefore, the yearly decay factor is $a = 1 - r = 1 - 0.03 = 0.97$. The 5-year decay factor is then $0.97^5 = 0.859$. This represents a percentage decrease of $1 - 0.859 = 0.141$ as a decimal or 14.1% over the 5-year period.

S-25. **Percentage change**: If a population grows by 15% each decade, then the decade growth factor is $a = 1 + r = 1 + 0.15 = 1.15$. The yearly growth factor, since a year is 1/10th of a decade, is $1.15^{1/10} = 1.014$, which is a growth of 1.4% per year.

S-27. **Percentage change**: If a population declines by 2% each month, then the monthly decay factor is $a = 1 - r = 1 - 0.02 = 0.98$. Since a year is 12 months, the yearly decay factor is $0.98^{12} = 0.785$, which represents a decay factor of 21.5% per year.

S-29. **Percentage change**: If the investment grows by 1% each year, then the yearly growth factor is $a = 1 + r = 1 + 0.01 = 1.01$. The 50-year growth factor is $1.01^{50} = 1.645$. Now $r = a - 1 = 0.645$, which is a growth rate of 64.5% per 50 years.

## 4.3 MODELING EXPONENTIAL DATA

1. **Making an exponential model**: To show that this is an exponential function we must show that the successive ratios are the same.

| $t$ increment | 0 to 1 | 1 to 2 | 2 to 3 | 3 to 4 | 4 to 5 |
|---|---|---|---|---|---|
| $\dfrac{\text{new}}{\text{old}}$ ratio of $f(t)$ | $\dfrac{3.95}{3.80} = 1.04$ | $\dfrac{4.11}{3.95} = 1.04$ | $\dfrac{4.27}{4.11} = 1.04$ | $\dfrac{4.45}{4.27} = 1.04$ | $\dfrac{4.62}{4.45} = 1.04$ |

Note that the ratios are not precisely equal, but they are equal when rounded to 2 decimal places. The table in the exercise tells us that the initial value is $P = 3.80$ (this corresponds to $t = 0$). Since $t$ increases in units of 1, we don't need to adjust the units, and so the growth factor is $a = 1.04$. The formula for the function is

$$f(t) = 3.80 \times 1.04^t.$$

3. **Data that are not exponential**: To show that these data are not exponential, we show that some of the successive ratios are not equal. It suffices to compute the first two ratios:

$$\frac{26.6}{4.9} = 5.43 \text{ and } \frac{91.7}{26.6} = 3.45.$$

Since these ratios are not equal and yet the changes in $t$ are the same, $h$ is not an exponential function.

5. **Making guitars**:

   (a) Because $n$ increases by 1 each time, to show that the data are exponential, we must show that the successive ratios (rounded to two decimal places) are the same.

| $n$ increment | New/Old ratio |
|---|---|
| 0 to 1 | $23.60/25.00 = 0.94$ |
| 1 to 2 | $22.27/23.60 = 0.94$ |
| 2 to 3 | $21.02/22.27 = 0.94$ |
| 3 to 4 | $19.84/21.02 = 0.94$ |

   Because all of the ratios are equal, the data are exponential.

   (b) Because $n$ increases by 1 each time, the common ratio, 0.94, is the decay factor for $D$ as a function of $n$. The initial value is 25.00 inches according to the table. Therefore, an exponential model is

   $$D = 25.00 \times 0.94^n.$$

   (c) The distance from the fifth fret to the sixth fret is $D(5) - D(6)$. By the model in part b, the value is

   $$D(5) - D(6) = 25.00 \times 0.94^5 - 25.00 \times 0.94^6 = 1.10.$$

   Therefore, the distance from the fifth fret to the sixth fret is 1.10 inches according to the model.

7. **A coin collection**:

(a) To show that these data are exponential, we must show that the successive ratios (rounded to two decimal places) are the same.

| $t$ increment | $\dfrac{\text{New}}{\text{Old}}$ ratio |
|---|---|
| 0 to 1 | $156.00/130.00 = 1.20$ |
| 1 to 2 | $187.20/156.00 = 1.20$ |
| 2 to 3 | $224.64/187.20 = 1.20$ |
| 3 to 4 | $269.57/224.64 = 1.20$ |

Because all of the ratios are equal, the data are exponential.

(b) Because $t$ increases by 1 each time, the common ratio, 1.20, is the yearly growth factor. The initial value is 130.00 dollars according to the table. Therefore, an exponential model is

$$V = 130 \times 1.20^t.$$

(c) We want to solve the equation

$$130 \times 1.20^t = 500.$$

We solve this using the crossing-graphs method. After consulting a table of values, we use a horizontal span of 0 to 10 years and a vertical span of 0 to 800 dollars. In the figure below, we have graphed the function $V$ along with the constant function 500 (thick line). We see that they intersect when $t = 7.39$ years. Therefore, 7.39 years after the start the collection have a value of \$500.

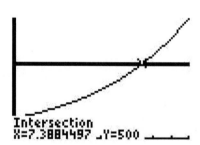

Intersection
X=7.3884497   Y=500

9. **Magazine sales**: The calculations of differences and ratios for the years in the data set are in the table below.

| Interval | Differences | Ratios |
|---|---|---|
| 2009–2010 | $8.82 - 7.76 = 1.06$ | $8.82/7.76 = 1.137$ |
| 2010–2011 | $9.88 - 8.82 = 1.06$ | $9.88/8.82 = 1.120$ |
| 2011–2012 | $10.94 - 9.88 = 1.06$ | $10.94/9.88 = 1.107$ |
| 2012–2013 | $12.00 - 10.94 = 1.06$ | $12.00/10.94 = 1.097$ |
| 2013–2014 | $13.08 - 12.00 = 1.08$ | $13.08/12.00 = 1.090$ |
| 2014–2015 | $14.26 - 13.08 = 1.18$ | $14.26/13.08 = 1.090$ |
| 2015–2016 | $15.54 - 14.26 = 1.28$ | $15.54/14.26 = 1.090$ |

From 2009 to 2013, the magazine sales exhibit a constant growth rate of 1.06 thousand dollars per year. From 2013 to 2016, sales grew at a constant proportional rate, that is, each year's sales are 1.09 times those of the previous year—this is a growth rate of 9% per year.

11. **An investment**:

(a) The amount of money originally invested is the balance when $t = 0$. Hence the original investment was \$1750.00.

(b) Each successive ratio of new/old is 1.012, which shows the data is exponential. Furthermore, since $t$ increases by single units, the common ratio 1.012 is the monthly growth factor. The formula for an exponential model is

$$B = 1750.00 \times 1.012^t.$$

Here $t$ is time in months, and $B$ is the savings balance in dollars.

(c) The growth factor is $1.012 = 1 + 0.012$, so $r = 0.012$. The monthly interest rate is 1.2%.

(d) To find the yearly interest rate we first have to find the yearly growth factor. There are 12 months in a year, so the yearly growth factor is $1.012^{12} = 1.154$. Now $1.154 = 1 + 0.154$, so $r = 0.154$, and thus the yearly interest rate is 15.4%.

(e) The account will grow for 18 years or 216 months, so when she is 18 years old, her college fund account balance will be

$$B(216) = 1750 \times 1.012^{216} = \$23{,}015.94.$$

(f) The account will double the first time when it reaches \$3500 and double the second time when it reaches \$7000. That is, we need to solve the equations

$$1750 \times 1.012^t = 3500 \text{ and } 1750 \times 1.012^t = 7000.$$

We do this using the crossing graphs method. The vertical span we want is 0 to 8000, and the table of values below for $B$ leads us to choose 0 to 120 months for a horizontal span. In the right-hand figure below we have graphed $B$ along with 3500 and 7000 (thick lines). The horizontal axis is months since the initial investment, and the vertical axis is account balance.

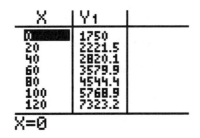

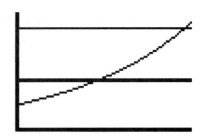

In the left-hand figure below, we have calculated the intersection of $B$ with 3500, and we find that the balance doubles in 58.11, or about 58.1, months. In the right-hand figure below, we have calculated the intersection with 7000, and we see the account reaches \$7000 after 116.22 months. Thus the second doubling occurs 58.11 months after the first one. (In fact, the account doubles every 58.11 months.)

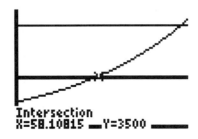

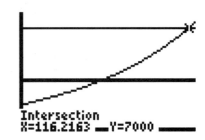

13. **A skydiver**:

(a) The successive ratios in $D$ values are always 0.42, so the data can be modeled with an exponential function. Since the $t$ values are measured every 5 seconds, 0.42 is the decay factor every 5 seconds. To find the decay factor every second we compute

$$0.42^{1/5} = 0.84.$$

The initial value is 176.00, and so the exponential model is

$$D = 176.00 \times 0.84^t.$$

(b) The decay factor per second is $0.84 = 1 - 0.16$, so $r = 0.16$, and the percentage decay rate is 16% per second. This means that the difference between the terminal velocity and the skydiver's velocity decreased by 16% each second.

(c) Now $D$ represents the difference between terminal velocity of 176 feet per second and the skydiver's velocity, $V$. Thus $D = 176 - V$, and so $V = 176 - D$. Using the formula from Part (a) for $D$, we find that

$$V = 176.00 - 176.00 \times 0.84^t.$$

(d) Now 99% of terminal velocity is $0.99 \times 176 = 174.24$, and we want to know when the velocity reaches this value. That is, we want to solve the equation

$$176 - 176 \times 0.84^t = 174.24.$$

We do this using the crossing graphs method. We know that the velocity starts at 0 and increases toward 176, and so we use a vertical span of 160 to 190. The table below leads us to choose a horizontal span of 0 to 35. In the right-hand figure below, we have graphed velocity and the target velocity of 174.24 (thick line). We see that the intersection occurs at $t = 26.41$. The skydiver reaches 99% of terminal velocity after 26.41 seconds.

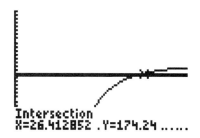

15. **An inappropriate linear model for radioactive decay**:

(a) As seen in the figure below, the data points do seem to fall on a straight line.

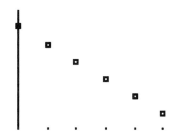

(b) In the left-hand figure below, we have calculated the regression line parameters, and the equation of the regression line is $U = -0.027t + 0.999$. In the right-hand figure, we have added the graph of the regression line to the data plot.

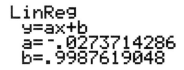

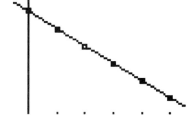

(c) Using the regression line equation for the decay, we see that the time required for decay to 0.5 gram is given by solving $-0.027t + 0.999 = 0.5$. This shows that the time required would be 18.48 minutes.

(d) The regression line predicts that after 60 minutes there will be $-0.621$ gram of the uranium-239 left, which is impossible.

Two items are worth noting here. The first is that the data has a linear appearance because we have sampled over too short a time period. Sampling over a longer time period would have clearly shown that the data is not linear. It is crucial in any experiment to gather enough data to allow for sound conclusions. Second, it is always dangerous to propose mathematical models based solely on the appearance of data. In the case of radioactive decay, physicists have good scientific reasons for believing that radioactive decay is an exponential phenomenon. Collected data serves to verify this, and to allow for the calculation of a constant such as the half-life for specific substances.

17. **Rates vary**: To determine the interest rates, we calculate the ratios of new balances to old balances, as shown in the table below.

| Time interval | Ratios of New/Old Balances |
|---|---|
| $t = 0$ to $t = 1$ | $262.50/250.00 = 1.050$ |
| $t = 1$ to $t = 2$ | $275.63/262.50 = 1.050$ |
| $t = 2$ to $t = 3$ | $289.41/275.63 = 1.050$ |
| $t = 3$ to $t = 4$ | $302.43/289.41 = 1.045$ |
| $t = 4$ to $t = 5$ | $316.04/302.43 = 1.045$ |
| $t = 5$ to $t = 6$ | $330.26/316.04 = 1.045$ |

Thus the yearly interest rate was 5.0% for the first 3 years, then 4.5% after that.

19. **Stochastic population growth**:

(a) The table below shows the Monte Carlo method applied by rolling a die ten times. The faces showing for these ten rolls are: 5, 2, 6, 3, 6, 1, 2, 1, 5, and 4. Your answer will presumably be different since it is highly unlikely that your die will roll the same values as these(!). Since the numbers are the size of a population, we will only use whole numbers.

| $t$ | Die face | Population % Change | Population change | Population |
|---|---|---|---|---|
| 0 | | | | 500 |
| 1 | 5 | Up 4% | +20 | 520 |
| 2 | 2 | Down 1% | −5 | 515 |
| 3 | 6 | Up 9% | +46 | 561 |
| 4 | 3 | No change | 0 | 561 |
| 5 | 6 | Up 9% | +50 | 611 |
| 6 | 1 | Down 2% | −12 | 599 |
| 7 | 2 | Down 1% | −6 | 593 |
| 8 | 1 | Down 2% | −12 | 581 |
| 9 | 5 | Up 4% | +23 | 604 |
| 10 | 4 | Up 2% | +12 | 616 |

(b) The exponential model is $500 \times 1.02^t$ since the initial value is 500 and the growth rate is 2%. A graph of the data points from Part (a) together with this exponential model is shown below. Although the points are scattered about and do not lie on the exponential curve, they do follow, very roughly, the general upward trend of the curve.

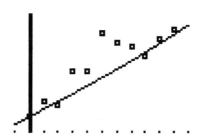

## Skill Building Exercises

S-1. **Finding an exponential formula**: Since $N$ is multiplied by 8 if $t$ is increased by 1, the growth factor $a$ is 8. Since the initial value $P$ is 7, a formula for $N$ is $N = P \times a^t = 7 \times 8^t$.

S-3. **Finding an exponential formula**: Since $N$ is multiplied by 62 if $t$ is increased by 7, the growth factor $a$ satisfies the equation $a^7 = 62$. Thus $a = 62^{1/7} = 1.803$. Since the initial value $P$ is 12, a formula for $N$ is $N = P \times a^t = 12 \times 1.803^t$.

S-5. **Finding an exponential formula**: Since $N = Pa^t$ and we have $N(1) = 4$ and $N(3) = 16$, taking the ratio gives $a^2 = a^{3-1} = 16/4 = 4$, so $a = 4^{1/2} = 2$ and so $N = P \times 2^t$. We find the initial value using the value $N(1) = 4$, so $P \times 2^1 = 4$ and therefore $P = 4/2 = 2$. The formula is $N = 2 \times 2^t$.

S-7. **Finding an exponential formula**: Since $N$ is divided by 8 if $t$ is increased by 3, the decay factor $a$ is $\left(\dfrac{1}{8}\right)^{1/3} = \dfrac{1}{2}$. Since the initial value $P$ is 4, a formula for $N$ is $N = P \times a^t = 4 \times \left(\dfrac{1}{2}\right)^t$, which can also be written as $N = 4 \times 0.5^t$.

S-9. **Testing exponential data**: Calculating the ratios of successive terms of $y$, we get $\dfrac{10}{5} = 2$, $\dfrac{20}{10} = 2$, and $\dfrac{40}{20} = 2$. Since the $x$ values are evenly spaced and these ratios show a constant value of 2, the table does show exponential data; moreover since $x$ increases by 2's the growth factor $a$ satisfies $a^2 = 2$ and so $a = 2^{1/2} = 1.41$. The initial value of $P = 5$ comes from the first entry in the table. Thus an exponential model for the data is $y = 5 \times 1.41^x$.

S-11. **Testing exponential data**: Calculating the ratios of successive terms of $y$, we get $\dfrac{18}{6} = 3$, $\dfrac{54}{18} = 3$, and $\dfrac{162}{54} = 3$. Since the $x$ values are evenly spaced and these ratios show a constant value of 3, the table does show exponential data; moreover since $x$ increases by 2's the growth factor $a$ satisfies $a^2 = 3$ and so $a = 3^{1/2} = 1.73$. The initial value of $P = 6$ comes from the first entry in the table. Thus an exponential model for the data is $y = 6 \times 1.73^x$.

S-13. **Testing exponential data**: Calculating the ratios of successive terms of $y$, we get $\dfrac{90}{100} = 0.90$, $\dfrac{80}{90} = 0.89$, and $\dfrac{70}{80} = 0.88$. Since the $x$ values are evenly spaced and these ratios do not show a constant value, the table does not show exponential data.

S-15. **Testing exponential data**: Calculating the ratios of successive terms of $y$, we get $\dfrac{10}{5} = 2$, $\dfrac{20}{10} = 2$, and $\dfrac{40}{20} = 2$. Since the $x$ values are evenly spaced and these ratios show a constant value of 2, the table does show exponential data; moreover since $x$ increases by 3's the growth factor $a$ satisfies $a^3 = 2$ and so $a = 2^{1/3} = 1.26$. Since $y = Pa^x$, we now know that $y = P \times 1.26^x$. To find $P$, we use the first entry in the table, when $x = 3$, $y = 5$: $P \times 1.26^3 = 5$ and so $P = 5/(1.26^3) = 2.50$. Thus an exponential model for the data is $y = 2.50 \times 1.26^x$.

S-17. **Growth rate**: Since $N$ is multiplied by $a$ if $t$ is increased by 2, the growth factor is $a^{1/2}$, which can also be written as $\sqrt{a}$.

S-19. **Growth rate**: Since $N$ is multiplied by $a^4$ if $t$ is increased by 1, then when $t$ is increased by 5, $N$ will be multiplied by $a^4$ five times, so by $(a^4)^5 = a^{20}$.

S-21. **Growth rate**: Since $N = Pa^t$ and $N = 6$ when $t = 4$ and $N = 48$ when $t = 7$, then the ratio is $a^3 = a^{7-4} = 48/6 = 8$ and so the growth factor is $a = 8^{1/3} = 2$. To find the value for $N$ when $t = 8$, we start from the value $N = 48$ when $t = 7$ and multiply by the growth factor, so the value for $N$ is $48 \times 2 = 96$.

## 4.4 MODELING NEARLY EXPONENTIAL DATA

1. **Choose the model**: Because the speed of the car is constant, the rate of change of the distance traveled is constant. Therefore, a linear model for the distance traveled is better.

3. **First down!**

   (a) The regression parameters for the exponential model are calculated in the figure on the left below. We find the exponential model $P = 74.955 \times 0.893^D$.

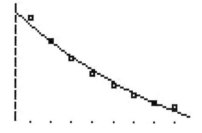

   (b) The plot on the right above shows the data together with the exponential model from part a.

   (c) Fourth down and ten yards to go corresponds to $D = 10$, and by the model from part a that value is $P(10) = 74.955 \times 0.893^{10} = 24.17$. Thus, the probability is 24.17%, or about 24%.

5. **Rare coins**:

   (a) The regression parameters for the exponential model are calculated in the figure below. We find the exponential model $C = 29.490 \times 1.076^t$.

   ```
   ExpReg
    y=a*b^x
    a=29.4896357
    b=1.07572582
   ```

   ∎

   (b) By inspection of the model from Part (a), the yearly growth factor is $1.076 = 1 + 0.076$, so $r = 0.076$. Thus the yearly percentage growth rate is 7.6%, and this is the percentage by which the value of the 1877 cent increases from year to year.

7. **Population growth**: Let $t$ be years since 2012 and $N$ the population, in thousands. The regression parameters for the exponential model are calculated in the figure below. We find the exponential model $N = 2.300 \times 1.090^t$.

   ```
   ExpReg
    y=a*b^x
    a=2.300181189
    b=1.090147741
   ```

   ∎

9. **Cell phones**: Let $t$ be years since 2010 and $C$ the number of cell phone subscribers, in millions.

   (a) The plot of the data points is in the figure below. The horizontal axis is years since 2010, and the vertical axis is the number of cell phone subscribers, in millions.

   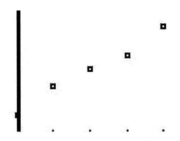

(b) The regression parameters for the exponential model are calculated in the left-hand figure below. The exponential model is given by

$$C = 298.914 \times 1.043^t.$$

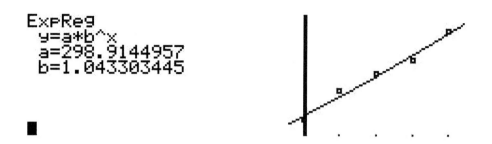

(c) We have added the graph of the exponential model in the right-hand figure above.

(d) Since the yearly growth factor is $1.043 = 1 + 0.043$, we have that $r = 0.043$. Thus the yearly percentage growth rate is 4.3%.

(e) Since 2016 is 6 years after 2010, we would predict $298.914 \times 1.043^6 = 384.82$ million, or about 384.8 million, cell phone subscribers in 2016. So it appears that the executive's plan was reasonable.

11. **National health care spending**: Let $t$ be years since 1970 and $H$ the costs in billions of dollars.

   (a) The plot of the data points is in the figure on the left below. The horizontal axis is years since 1970, and the vertical axis is the cost, in billions of dollars.

   (b) The regression parameters for the exponential model are calculated in the right-hand figure below. The exponential model is given by

$$H = 94.414 \times 1.091^t.$$

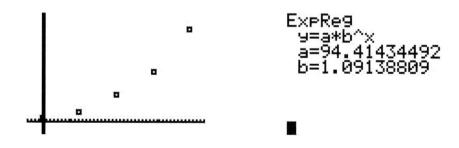

   (c) The yearly growth factor is $1.091 = 1 + 0.091$, so $r = 0.091$. Thus the yearly percentage increase is 9.1%.

(d) The year 2011 is 41 years after 1970, so $t = 41$. Thus the amount spent in 2011 on health care is expressed in functional notation as $H(41)$. We would estimate that

$$H(41) = 94.414 \times 1.091^{41} = 3356.23 \text{ billion dollars.}$$

Thus, the estimate is that about 3356 billion dollars (or 3.356 trillion dollars) was spent on health care in the year 2011.

13. **Grazing rabbits**:

(a) The plot of the data points is shown below. The horizontal axis is the vegetation level $V$, and the vertical axis is the difference $D$.

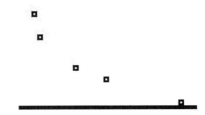

(b) We use regression to calculate the exponential model and find

$$D = 0.211 \times 0.988^V.$$

(c) The satiation level for the rabbit is 0.18 pound per day, and thus we have

$$D = \text{satiation level} - A = 0.18 - A.$$

Solving for $A$ yields $A = 0.18 - D$, and, using the formula from Part (b), we find that

$$A = 0.18 - 0.211 \times 0.988^V.$$

(d) Now 90% of the satiation level is $0.90 \times 0.18 = 0.162$, so we need to find when $A = 0.162$. By the formula in Part (c), this means

$$0.18 - 0.211 \times 0.988^V = 0.162.$$

We solve this using the crossing graphs method. The table of values for $A$ below shows that we reach 0.162 somewhere between 150 and 250. We use this for a horizontal span, and we use 0.1 to 0.2 for a vertical span. In the graph below, the horizontal axis is the vegetation level, and the vertical axis is the amount eaten. We have added the graph of 0.162 (thick line) and calculated the intersection point. Thus, the amount of food eaten by the rabbit will be 90% of satiation level when the vegetation level is $V = 203.89$ pounds per acre.

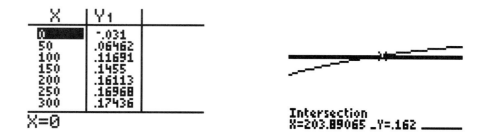

15. **Nearly linear or exponential data**: The left-hand figure below is the plot of the data from Table A. This appears to be linear. Computing the regression line gives $f = 19.842t - 16.264$, and adding it to the plot (right-hand figure) shows how very close to linear $f$ is. We conclude that Table A is approximately linear, with a linear model of $f = 19.842t - 16.264$.

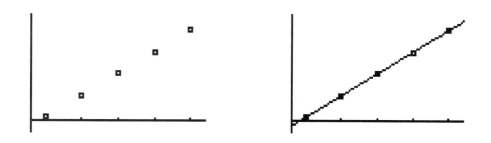

The plot of data from Table B in the figure below looks more like exponential data than linear data. We use regression to calculate the exponential model and find

$$g = 2.330 \times 1.556^t.$$

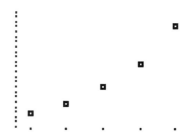

17. **Sound pressure**:

(a) A plot of the data is shown below.

(b) We use regression to calculate the exponential model and find $P = 0.0003 \times 1.116^D$.

(c) When loudness $D$ is increased by one decibel, the pressure $P$ is multiplied by a factor of 1.116, that is, it increases by 11.6%.

19. **Growth in length of haddock**:

(a) We use regression to calculate the exponential model and find $D = 40.909 \times 0.822^t$.

(b) The difference $D$ is the difference between 53 centimeters and the length at age $t$, $L = L(t)$. Thus $D = 53 - L$. Solving for $L$ and substituting the expression from Part (a) for $D$, we have

$$
\begin{aligned}
D &= 53 - L \\
D + L &= 53 \\
L &= 53 - D \\
L &= 53 - 40.909 \times 0.822^t.
\end{aligned}
$$

(c) The figure below shows a plot of the experimentally gathered data for the length $L$ at ages 2, 5, 7, 13, and 19 years along with the graph of the model for $L$ from Part (b). This graph shows that the 5-year-old haddock is a bit shorter than would be expected from the model for $L$.

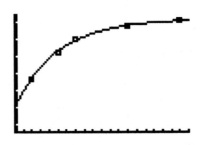

(d) If a fisherman has caught a haddock which measures 41 centimeters, then $L = 41$ and so its age $t$ satisfies the equation $41 = 53 - 40.909 \times 0.822^t$. Solving for $t$, we find that $t = 6.26$. Thus the haddock is about 6.3 years old.

21. **Injury versus speed**:

(a) We use regression to calculate the exponential model and find $N = 16.726 \times 1.021^s$.

(b) Using the formula from Part (a), we find that $N(70) = 16.726 \times 1.021^{70} = 71.65$. In practical terms this means that if vehicles are traveling at 70 miles per hour, then we expect 71.65 persons injured (on average) per 100 vehicles involved in accidents.

(c) An increase in 1 mile per hour of speed $s$ causes the number of people injured per 100 accident-involved vehicles $N$ to be multiplied by 1.021, and so to increase by 2.1%.

23. **Walking in Seattle**:

(a) We use regression to calculate the exponential model and find $P = 84.726 \times 0.999^D$.

(b) The percentage of pedestrians who walk at least $D = 200$ feet from parking facilities is $P = 84.726 \times 0.999^{200} = 69.36\%$.

(c) For $D = 0$, the exponential model indicates that $P = 84.726$ and so only 84.726% of pedestrians walk at least 0 feet. The correct percentage is 100%. Thus this model is not appropriate to use for very small distances $D$.

25. **Frequency of earthquakes**:

(a) We use regression to calculate the exponential model and find $N = 24{,}670{,}000 \times 0.125^M$.

(b) To find the number of earthquakes per year of magnitude at least 5.5, we put $M = 5.5$ in the formula from Part (a) and find $N = 24{,}670{,}000 \times 0.125^{5.5} = 266.18$, or about 266 earthquakes per year.

(c) The number of earthquakes per year of magnitude 8.5 or greater predicted by the model is $N = 24{,}670{,}000 \times 0.125^{8.5} = 0.52$, or about 0.5 earthquake per year.

(d) The limiting value for $N$ is seen to be 0 using a table or graph. In practical terms this means that earthquakes of large magnitude are very rare.

(e) In the situation described, the number of earthquakes predicted is multiplied by 0.125, which is a reduction of 87.5%. Thus there are 87.5% fewer earthquakes of magnitude $M + 1$ or greater than of magnitude $M$ or greater.

27. **Economic growth of the United States**: Let $t$ be years since 2005 and $G$ the GDP in trillions of dollars. The regression parameters for the exponential model are calculated in the figure below. We find the exponential model $G = 12.492 \times 1.029^t$.

```
ExpReg
 y=a*b^x
 a=12.49165953
 b=1.028765313
```

29. **Research project**: Answers will vary.

## Skill Building Exercises

S-1. **Population growth**: Because the population has abundant resources, an exponential model should be appropriate.

S-3. **Running speed versus length**: Since a 1-unit increase in length always results in the same increase in running speed, the rate of change is constant, and therefore we should use a linear model.

S-5. **Exponential regression**: The exponential model is $y = 51.01 \times 1.04^x$.

S-7. **Exponential regression**: The exponential model is $y = 2.22 \times 1.96^x$. A plot of the exponential model, together with the data, is shown below.

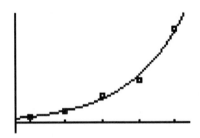

S-9. **Exponential regression**: The exponential model is $y = 6.35 \times 1.03^x$. A plot of the exponential model, together with the data, is shown below.

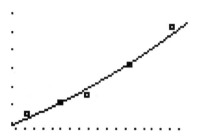

S-11. **Exponential regression**: The exponential model is $y = 2.39 \times 1.40^x$. A plot of the exponential model, together with the data, is shown below.

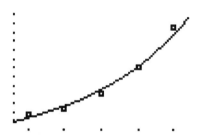

S-13. **Exponential regression**: The exponential model is $y = 3.17 \times 1.15^x$. A plot of the exponential model, together with the data, is shown below.

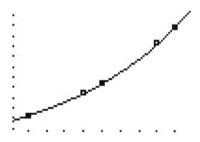

S-15. **Using exponential models**: The model from Exercise S-5 is $y = 51.01 \times 1.04^x$. To find $y(3.3)$, we plug in 3.3 for $x$: $y(3.3) = 51.01 \times 1.04^{3.3} = 58.06$. To solve $y = 65$ for $x$, we need to find $x$ such that $51.01 \times 1.04^x = 65$. The solution is $x = 6.18$; this can be calculated using a table or crossing-graphs, for example.

S-17. **Using exponential models**: The model from Exercise S-7 is $y = 2.22 \times 1.96^x$. To find $y(7)$, we plug in 7 for $x$: $y(7) = 2.22 \times 1.96^7 = 246.69$. To solve $y = 21$ for $x$, we need to find $x$ such that $2.22 \times 1.96^x = 21$. The solution is $x = 3.34$; this can be calculated using a table or crossing-graphs, for example.

S-19. **Using exponential models**: The model from Exercise S-9 is $y = 6.35 \times 1.03^x$. To find $y(10)$, we plug in 10 for $x$: $y(10) = 6.35 \times 1.03^{10} = 8.53$. To solve $y = 9$ for $x$, we need to find $x$ such that $6.35 \times 1.03^x = 9$. The solution is $x = 11.80$; this can be calculated using a table or crossing-graphs, for example.

S-21. **Using exponential models**: The model from Exercise S-11 is $y = 2.39 \times 1.40^x$. To find $y(2.2)$, we plug in 2.2 for $x$: $y(2.2) = 2.39 \times 1.40^{2.2} = 5.01$. To solve $y = 8$ for $x$, we need to find $x$ such that $2.39 \times 1.40^x = 8$. The solution is $x = 3.59$; this can be calculated using a table or crossing-graphs, for example.

S-23. **Using exponential models**: The model from Exercise S-13 is $y = 3.17 \times 1.15^x$. To find $y(8.8)$, we plug in 8.8 for $x$: $y(8.8) = 3.17 \times 1.15^{8.8} = 10.84$. To solve $y = 10$ for $x$, we need to find $x$ such that $3.17 \times 1.15^x = 10$. The solution is $x = 8.22$; this can be calculated using a table or crossing-graphs, for example.

S-25. **Linear or exponential**:

(a) The linear model is $y = 1.92x + 1.06$. A plot of the linear model, together with the data, is shown on the left below.

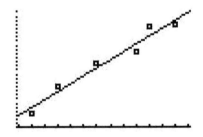 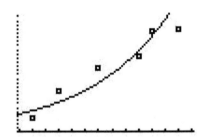

(b) The exponential model is $y = 3.77 \times 1.18^x$. A plot of the exponential model, together with the data, is shown on the right above.

(c) The linear model is a better fit.

S-27. **Linear or exponential**:

(a) The linear model is $y = -27.97x + 467.66$. A plot of the linear model, together with the data, is shown on the left below.

(b) The exponential model is $y = 521.29 \times 0.91^x$. A plot of the exponential model, together with the data, is shown on the right above.

(c) There is not much difference between the linear fit and the exponential fit.

## 4.5 LOGARITHMIC FUNCTIONS

1. **Earthquakes in Alaska and Chile**:

   (a) The New Madrid earthquake of Example 4.12 had a Richter magnitude of 8.8, whereas the 1964 Alaska earthquake had a Richter magnitude of 9.2. The Alaska earthquake was therefore $10^{9.2-8.8} = 10^{0.4} = 2.51$ times as powerful as the New Madrid earthquake.

   (b) The Chilean earthquake was 2 times as powerful as the Alaska quake. We want to write 2 as a power of 10. Solving the equation $2 = 10^t$, we find that $t = \log 2 = 0.30$, so the Chilean earthquake was $10^{0.3}$ times as powerful as the Alaska quake. Thus the Richter scale reading for the Chilean earthquake was 0.3 more than that of the Alaska earthquake, so it was $9.2 + 0.3 = 9.5$.

3. **Recent earthquakes**:

   (a) The first quake had a magnitude of 7.8, and the aftershock had a magnitude of 7.3. The first quake was therefore $10^{7.8-7.3} = 10^{0.5} = 3.16$ times as powerful as the aftershock.

   (b) We want to find the magnitude of a quake that is 2 times as powerful as the first quake. First we write 2 as a power of 10. Solving the equation $2 = 10^t$, we find that $t = \log 2 = 0.30$. Therefore, the larger quake would be $10^{0.3}$ times as powerful as the first quake. Thus the Richter scale reading for the larger quake would be 0.3 more than that of the first quake, so it would be $7.8 + 0.3 = 8.1$.

5. **Doubling time**:

   (a) Now an APR of 5% corresponds to $r = 0.05$, and we calculate

   $$D(0.05) = \frac{\log 2}{\log(1 + 0.05)} = 14.21.$$

   Thus, the doubling time for this investment is 14.21 years.

   (b) We use a horizontal span of 0 to 0.1. After consulting a table of values, we use a vertical span of 0 to 30 years. The graph is below.

   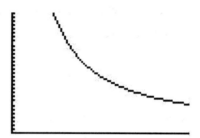

   (c) Because the graph of $D$ is concave up, a small change in the interest rate has a greater effect on the doubling time if interest rates are low.

7. **Moore's law**:

   (a) If a chip were introduced in the year 2022, then that chip would have $10^{0.15 \times 7} = 11$ times as many transistors as a chip released in 2015, such as the 6th generation Core.

   (b) The number of transistors will be 200 times as many as the number of transistors of the 6th generation Core when $200 = 10^{0.15t}$, where $t$ is years since 2015, according to Moore's law. Solving gives $\log 200 = 0.15t$, and so $t = \dfrac{\log 200}{0.15} = 15.34$. Thus that limit will be reached in the year 2030. (The equation $200 = 10^{0.15t}$ can also be solved using the crossing-graphs method.)

   (c) If the fastest speed corresponds to about $10^{40}$ times the number of transistors as that of the 6th generation Core, then, according to Moore's law, this limit will be reached when $10^{40} = 10^{0.15t}$, where $t$ is years since 2015. Thus $40 = 0.15t$, so $t = \dfrac{40}{0.15} = 267$. Therefore, that limit will be reached in the year 2282, or about the year 2300.

9. **The pH scale**:

   (a) Rain in the eastern United States has a pH level of 3.8, which is 1.8 less than the 5.6 pH level of normal rain. Thus the eastern United States rain is $10^{1.8} = 63.10$ times as acidic as normal rain.

(b) A pH of 5 is 0.6 less than a pH of 5.6, so such water is $10^{0.6} = 3.98$ times as acidic as normal water.

11. **Whispers**: Because there are 10 whisperers, the relative intensity is multiplied by 10. Multiplying by $10 = 10^1$ increases the logarithm by 1 unit. Because the decibel is 10 times the logarithm of the relative intensity, we are adding $10 \times 1 = 10$ units on the decibel scale. (In terms of the formula for decibels, multiplying the relative intensity by 10 adds $10 \log 10 = 10$ to the decibel level.) Therefore, the effect of the 10 whisperers together on the decibel level is to raise it by 10 decibels from the level for one whisperer, 30 decibels, for a total of 40 decibels.

13. **Weight gain**:

(a) The figure below shows a graph of $G$ against $M$ using a horizontal span of 0 to 0.4 and a vertical span of 0 to 0.05.

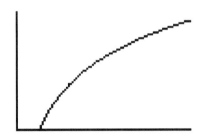

(b) If $M = 0.3$, then $G = 0.067 + 0.052 \log 0.3 = 0.04$ unit. This could also have been solved using a table or graph.

(c) A zookeeper wants $G = 0.03$, so $0.03 = 0.067 + 0.052 \log M$. This can be solved directly: $0.052 \log M = 0.03 - 0.067 = -0.037$, so $\log M = \dfrac{-0.037}{0.052}$ and therefore $M = 10^{-0.037/0.052} = 0.19$ unit should be the daily milk-energy intake.

(d) The graph is increasing and concave down; thus higher values of $M$ produce smaller and smaller effects on the value of $G$, supporting the quotation from the study.

15. **Age of haddock**:

(a) The figure below shows a graph of age $T$ versus length $L$ using a horizontal span of 25 to 50 and a vertical span of 0 to 15.

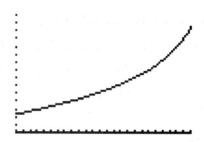

(b) In functional notation the age of a haddock that is 35 centimeters long is expressed by $T(35)$. The value of $T(35)$ is given by the formula, for example, as $T(35) = 19 - 5 \ln{(53 - 35)} = 4.55$ years old.

(c) If a haddock is 10 years old, then $T = 10$ and so $10 = 19 - 5 \ln{(53 - L)}$. Solving for $L$ directly, we see that $-5 \ln{(53 - L)} = 10 - 19 = -9$, so $\ln{(53 - L)} = \dfrac{-9}{-5} = 1.8$. Thus $53 - L = e^{1.8}$, so $53 = e^{1.8} + L$ and finally $L = 53 - e^{1.8} = 46.95$ centimeters.

17. **Stand density:**

(a) If the stand has $N = 500$ trees per acre, and the diameter of a tree of average size is $D = 7$ inches, then, according to the formula, $\log SDI = \log 500 + 1.605 \log 7 - 1.605 = 2.45$ and therefore the stand-density index is $SDI = 10^{2.45} = 281.84$, or about 282.

(b) If $D$ remains the same and $N$ is increased by a factor of 10, then $\log N$ will increase by 1 and so the logarithm of the stand-density index will increase by 1. Since the logarithm increases by 1, the $SDI$ will increase by a factor of $10^1 = 10$.

(c) If the diameter of a tree of average size is 10 inches, then

$$\log SDI = \log N + 1.605 \log 10 - 1.605 = \log N + 1.605 \times 1 - 1.605 = \log N,$$

and therefore $SDI = N$.

19. **Spectroscopic parallax:**

(a) If the distance to the star Kaus Astralis is $D = 38.04$ parsecs, then its spectroscopic parallax is $S = 5 \log 38.04 - 5 = 2.90$.

(b) If the spectroscopic parallax for the star Rasalhague is $S = 1.27$, then $1.27 = 5 \log D - 5$, so $5 \log D = 1.27 + 5 = 6.27$, and therefore $\log D = \dfrac{6.27}{5}$. Thus the distance to Rasalhague is $D = 10^{6.27/5} = 17.95$ parsecs.

(c) If distance $D$ is multiplied by 10, then $\log D$ is increased by 1, and so the spectroscopic parallax $S$ is increased by $5 \times 1 = 5$ units.

(d) Because the star Shaula is 3.78 times as far away as the star Atria, the distance $D$ for Atria in multiplied by 3.78. As in Part (c), the spectroscopic parallax $S$ is increased by $5 \log 3.78 = 2.89$, so that the spectroscopic parallax for Shaula is 2.89 units more than that of Atria.

21. **Rocket staging**:

(a) Since $R_1 = R_2 = 3.4$ and $c = 3.7$, the total velocity attained by this two-stage craft is $v = 3.7 \times \ln 3.4 + 3.7 \times \ln 3.4 = 9.06$ kilometers per second.

(b) Since the velocity $v$ from Part (a) is greater than 7.8 kilometers per second, this craft can achieve a stable orbit.

23. **Relative abundance of species**:

(a) Since $S = 197$ and $N = 6814$, $\dfrac{S}{N} = \dfrac{197}{6814} = 0.0289$ for this collection.

(b) The figure below shows a graph of the function $\dfrac{x-1}{x} \ln(1-x)$ using a horizontal span of 0 to 1 and a vertical span of 0 to 1.

(c) Using crossing graphs, we can determine the value of $x$ for which $\dfrac{x-1}{x} \ln(1-x) = 0.0289$. We find that $x = 0.9945$.

(d) For this collection $\alpha = \dfrac{N(1-x)}{x} = \dfrac{6814(1-0.9945)}{0.9945} = 37.68$.

(e) The number of species of moth in the collection represented by 5 individuals is $\alpha \left( \dfrac{x^5}{5} \right) = 37.68 \left( \dfrac{0.9945^5}{5} \right) = 7.33$, or about 7 species of moths.

(f) The figure below shows a graph of the number of species represented by $n$ individuals, that is $\alpha \left( \dfrac{x^n}{n} \right) = 37.68 \left( \dfrac{0.9945^n}{n} \right)$ against $n$, using a horizontal span of 1 to 20 and a vertical span of 0 to 40.

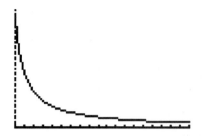

25. **Magazine circulation**: Below is a table that shows the natural logarithm of the circulation data. (Here $t$ is the time in years since 2012, and $C$ is the circulation in thousands.) For example, $\ln 2.64 = 0.97$ (rounded to 2 decimal places).

| $t$ | 0 | 1 | 2 | 3 | 4 |
|---|---|---|---|---|---|
| $C$ | 2.64 | 2.78 | 2.93 | 3.08 | 3.25 |
| $\ln C$ | 0.97 | 1.02 | 1.08 | 1.12 | 1.18 |

The figure below shows a plot of $\ln C$ (the third row of the table) versus $t$ (the first row). The points nearly fall on a straight line, so an exponential model for the circulation $C$ may be appropriate.

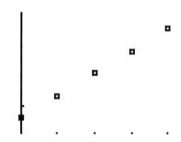

## Skill Building Exercises

S-1. **Richter scale**: If one earthquake reads 4.2 on the Richter scale and another reads 7.2, then the 7.2 earthquake is $10^{7.2-4.2} = 10^3 = 1000$ times as powerful as the 4.2 earthquake.

S-3. **Richter scale**: The second earthquake is 100 times as strong as the earthquake with a Richter scale reading of 6.5. Since $100 = 10^2$, the Richter scale reading of the more powerful earthquake is $6.5 + 2 = 8.5$.

S-5. **Richter scale**: If the relative intensity of a quake is multiplied by $10^t$, then the change is Richter scale is $\log 10^t = t$, so $t$ units are added to the Richter reading.

S-7. **The decibel scale**: If one sound has a relative intensity of 1000 times that of another, then the common logarithm of the more intense sound is 3 more than that of the other, since $1000 = 10^3$. Since the decibel level is 10 times the logarithm, the decibel level is increased by $10 \times 3 = 30$ decibels. Thus the more intense sound has a decibel level 30 decibels more than the other.

S-9. **The decibel scale**: If one sound has a decibel reading of 5, and another has a decibel reading of 7, then the difference in the decibel levels is 2. Since the decibel level is 10 times the logarithm of the relative intensity, then the difference in the logarithms is $\frac{2}{10} = 0.2$. Thus the relative intensity of the higher decibel sound is $10^{0.2} = 1.58$ times that of the lower decibel sound.

S-11. **Calculating common logarithms**: Since $1000 = 10^3$, $\log 1000 = 3$.

S-13. **Calculating common logarithms**: Since $\frac{1}{10} = 10^{-1}$, $\log \frac{1}{10} = -1$.

S-15. **Calculating natural logarithms by hand**: Since $e = e^1$, $\ln e = \ln e^1 = 1$.

S-17. **Calculating natural logarithms by hand**: Since $\frac{1}{e} = e^{-1}$, $\ln \frac{1}{e} = \ln e^{-1} = -1$.

S-19. **Calculating natural logarithms by hand**: $\ln e^e = e$.

S-21. **How the natural logarithm changes**: If $x$ is multiplied by $e^t$, then $\ln x$ becomes $\ln(x \times e^t) = \ln x + \ln e^t = \ln x + t$, so $t$ is added to the logarithm $\ln x$.

S-23. **Solving logarithmic equations**: If $\log x = 1$, then $x = 10^1 = 10$.

S-25. **Solving logarithmic equations**: If $\ln x = 3$, then $x = e^3$, or 20.09.

S-27. **Solving logarithmic equations**: If $\ln x = 0$, then $x = e^0 = 1$.

S-29. **How the logarithm increases**: If $\log x = 8.3$ and $\log y = 10.3$, then $\log y$ is 2 more than $\log x$, so $y$ is 100 times $x$, that is, $y = 100x$.

S-31. **Solving exponential equations**: If $e^t = 5$, then $\ln e^t = \ln 5$. Now $\ln e^t = t$, so $t = \ln 5$.

S-33. **Solving exponential equations**: If $e^t = a$, then $\ln e^t = \ln a$. Now $\ln e^t = t$, so $t = \ln a$.

S-35. **Logarithmic regression**: The regression parameters for the model are calculated in the figure on the left below. We find the model $y = 17.45 - 3.19 \ln x$. The plot on the right below shows the data together with the graph of the model.

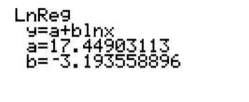
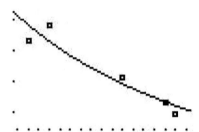

S-37. **Logarithmic regression**: The regression parameters for the model are calculated in the figure on the left below. We find the model $y = 12.78 - 1.17 \ln x$. The plot on the right below shows the data together with the graph of the model.

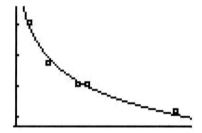

## Chapter 4 Review Exercises

1. **Percentage growth**: The population grows by 4.5% each year, so the yearly growth factor is $a = 1 + r = 1 + 0.045 = 1.045$. Since a month is 1/12th of a year, the monthly growth factor is $1.045^{1/12} = 1.0037$. That is a growth of 0.37% per month.

2. **Percentage decline**: The population declines by 0.5% each year, so the yearly decay factor is $a = 1 - r = 1 - 0.005 = 0.995$. Since a decade is 10 years, the decade decay factor is $0.995^{10} = 0.9511$. Since $1 - 0.9511 = 0.0489$, that is a decline of 4.89% per decade.

3. **Credit cards**:

(a) The decay factor is $a = (1 + 0.02)(1 - 0.05) = 0.969$.

(b) Because the decay factor is $a = 0.969$, we have $r = 1 - 0.969 = 0.031$. Hence the balance decreases by 3.1% each month.

(c) Because 3 years is 36 months, the balance after 3 years of payments will be $1000 \times 0.969^{36} = 321.85$ dollars.

4. **More on credit cards**:

   (a) We have $r = 0.015$ and $m = 0.04$, so the base is $a = (1 + 0.015)(1 - 0.04) = 0.9744$.

   (b) Because the decay factor is $a = 0.9744$, we have $r = 1 - 0.9744 = 0.0256$. Hence the balance decreases by 2.56% each month.

   (c) We have $r = 0.015$ and $m = 0.01$, so the base is $a = (1 + 0.015)(1 - 0.01) = 1.00485$. This number is greater than 1, so the balance will increase each month. This is not a realistic situation.

5. **Testing exponential data**: Calculating the ratios of successive terms of $y$, we get $\dfrac{30.0}{25.0} = 1.2$, $\dfrac{36.0}{30.0} = 1.2$, and $\dfrac{43.2}{36.0} = 1.2$. Since the $x$ values are evenly spaced and these ratios show a constant value of 1.2, the table does show exponential data.

6. **Modeling exponential data**: The data from Exercise 5 show a constant ratio of 1.2 for $x$ values increasing by 1's, and so the growth factor is $a = 1.2$. The initial value of $P = 25.0$ comes from the first entry in the table. Thus an exponential model for the data is $y = 25 \times 1.2^x$.

7. **Credit card balance**:

   (a) Each successive ratio of new/old is 0.97, so this is exponential data. Because $n$ increases by single units, the decay factor is $a = 0.97$. The initial value of 500.00 comes from the first entry in the table. Thus an exponential model for the data is $B = 500 \times 0.97^n$.

   (b) From the first entry in the table the initial charge is $500.00.

   (c) Because 2 years is 24 months, the balance after 2 years of payments will be $500 \times 0.97^{24} = 240.71$ dollars.

8. **Inflation**:

   (a) Each successive ratio of new/old is 1.03, so this is exponential data.

   (b) Let $t$ be the time in years since the start of 2013 and $P$ the price in dollars. Because $t$ increases by single units, by Part (a) the growth factor is $a = 1.03$. The initial value of 265.50 comes from the first entry in the table. Thus an exponential model for the data is $P = 265.5 \times 1.03^t$.

(c) We want to find the first whole number $t$ for which $P$ is greater than 325. From a table of values or a graph we find the number to be $t = 7$. Thus at the beginning of 2020 the price will surpass \$325.

9. **Exponential regression**: The exponential model is $y = 20.97 \times 1.34^x$.

10. **Exponential regression**: The exponential model is $y = 10.96 \times 0.80^x$. A plot of the exponential model, together with the original data, is shown below.

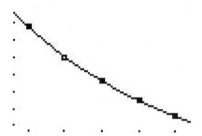

11. **Sales**:

(a) The exponential model for $A$ is $A = 8.90 \times 1.02^t$, and the exponential model for $B$ is $B = 2.30 \times 1.27^t$.

(b) By Part (a) the growth factor for $A$ is 1.02, so the sales for *Alpha* are growing by 2% per year. By Part (a) the growth factor for $B$ is 1.27, so the sales for *Beta* are growing by 27% per year.

(c) We want to find the first whole number $t$ for which $B$ is greater than $A$. From a table of values or a graph we find the number to be $t = 7$. Thus at the beginning of 2020 sales for *Beta* will overtake sales for *Alpha*.

12. **Credit card payments**:

(a) The exponential model for $B$ is $B = 520 \times 0.9595^n$.

(b) From Part (a) the initial charge was \$520.00.

(c) From Part (a) the decay factor is 0.9595. Because $m = 0.05$ we know that

$$(1 + r)(1 - 0.05) = 0.9595,$$

and thus

$$1 + r = \frac{0.9595}{1 - 0.05}.$$

Hence

$$r = \frac{0.9595}{1 - 0.05} - 1 = 0.01.$$

Thus the monthly finance charge is 1%.

13. **Solving logarithmic equations**: If $\log x = -1$ then $x = 10^{-1} = 0.1$.

14. **Comparing logarithms**: If $\log x = 3.5$ and $\log y = 2.5$ then $\log y$ is 1 less than $\log x$. Hence $y$ is $10^{-1}$ times $x$, so $y = 0.1x$.

15. **Comparing earthquakes**: The Double Spring Flat quake was $6.0 - 5.5 = 0.5$ higher on the scale than the Little Skull Mountain quake. Hence the Double Spring Flat quake was $10^{0.5} = 3.16$ times as powerful as the Little Skull Mountain quake.

16. **Population growth**:

    (a) We put $N = 200$ in the formula and get $T = 25 \log 200 - 50 = 7.53$. Thus the population reaches a size of 200 after 7.53 years.

    (b) We want to find the value of $N$ for which $T = 0$. Thus we need to solve the equation $25 \log N - 50 = 0$ for $N$. We find $25 \log N = 50$ or $\log N = \dfrac{50}{25} = 2$. Hence $N = 10^2 = 100$. Thus the initial population size was 100.


# A FURTHER LOOK: SOLVING EXPONENTIAL EQUATIONS

1. **Using the laws of logarithms**: Using the quotient law and then the product law, we have
$$\ln \frac{AB}{C} = \ln AB - \ln C = \ln A + \ln B - \ln C = 3 + 4 - 5 = 2.$$

2. **Using the laws of logarithms**: Using the quotient law and then the product law, we have
$$\ln \frac{A}{BC} = \ln A - \ln BC = \ln A - (\ln B + \ln C) = 3 - (4 + 5) = -6.$$

3. **Using the laws of logarithms**: Using the power law, we have
$$\ln \sqrt{C} = \ln C^{1/2} = \frac{1}{2} \ln C = \frac{1}{2} \times 5 = \frac{5}{2}.$$

4. **Using the laws of logarithms**: Using the product law and then the power law, we have
$$\ln A^2 B^3 = \ln A^2 + \ln B^3 = 2 \ln A + 3 \ln B = 2 \times 3 + 3 \times 4 = 18.$$

5. **Using the laws of logarithms**: Using the quotient law, we have

$$\ln \frac{1}{A} = \ln 1 - \ln A = 0 - 3 = -3.$$

6. **Using the laws of logarithms**: Using the quotient law and then the power and product laws, we have

$$\ln \frac{A^2}{BC} = \ln A^2 - \ln BC = 2\ln A - (\ln B + \ln C) = 2 \times 3 - (4 + 5) = -3.$$

7. **Solving exponential equations**: We can solve for $t$ by taking logarithms and using their properties.

$$
\begin{aligned}
3 \times 4^t &= 21 &&\text{original equation}\\
\ln(3 \times 4^t) &= \ln 21 &&\text{taking ln of each side}\\
\ln 3 + \ln 4^t &= \ln 21 &&\text{product law}\\
t \ln 4 &= \ln 21 - \ln 3 &&\text{power law}\\
t &= \frac{\ln 21 - \ln 3}{\ln 4} = 1.40.
\end{aligned}
$$

8. **Solving exponential equations**: We can solve for $t$ by taking logarithms and using their properties.

$$
\begin{aligned}
\frac{5^t}{4} &= 9 &&\text{original equation}\\
\ln\left(\frac{5^t}{4}\right) &= \ln 9 &&\text{taking ln of each side}\\
\ln 5^t - \ln 4 &= \ln 9 &&\text{quotient law}\\
t \ln 5 &= \ln 9 + \ln 4 &&\text{power law}\\
t &= \frac{\ln 9 + \ln 4}{\ln 5} = 2.23.
\end{aligned}
$$

9. **Solving exponential equations**: We can solve for $t$ by taking logarithms and using their properties.

$$
\begin{aligned}
7^{t-1} &= 4^{t+1} \quad &&\text{original equation} \\
\ln 7^{t-1} &= \ln 4^{t+1} \quad &&\text{taking ln of each side} \\
(t-1)\ln 7 &= (t+1)\ln 4 \quad &&\text{power law} \\
t\ln 7 - \ln 7 &= t\ln 4 + \ln 4 \\
t(\ln 7 - \ln 4) &= \ln 4 + \ln 7 \\
t &= \frac{\ln 4 + \ln 7}{\ln 7 - \ln 4} = 5.95.
\end{aligned}
$$

10. **Solving exponential equations**: We can solve for $t$ by taking logarithms and using their properties.

$$
\begin{aligned}
2 \times 5^t &= 3^{2t} \quad &&\text{original equation} \\
\ln(2 \times 5^t) &= \ln 3^{2t} \quad &&\text{taking ln of each side} \\
\ln 2 + t\ln 5 &= 2t\ln 3 \quad &&\text{product and power laws} \\
t(\ln 5 - 2\ln 3) &= -\ln 2 \\
t &= \frac{-\ln 2}{\ln 5 - 2\ln 3} = 1.18.
\end{aligned}
$$

11. **Solving exponential equations**: We can solve for $t$ by taking logarithms and using their properties.

$$
\begin{aligned}
5 \times 4^t &= 7 \quad &&\text{original equation} \\
\ln(5 \times 4^t) &= \ln 7 \quad &&\text{taking ln of each side} \\
\ln 5 + t\ln 4 &= \ln 7 \quad &&\text{product and power laws} \\
t\ln 4 &= \ln 7 - \ln 5 \\
t &= \frac{\ln 7 - \ln 5}{\ln 4} = 0.24.
\end{aligned}
$$

12. **Solving exponential equations**: We can solve for $t$ by taking logarithms and using their properties.

$$
\begin{aligned}
2^{t+3} &= 7^{t-1} \quad \text{original equation} \\
\ln 2^{t+3} &= \ln 7^{t-1} \quad \text{taking ln of each side} \\
(t+3)\ln 2 &= (t-1)\ln 7 \quad \text{power law} \\
t\ln 2 + 3\ln 2 &= t\ln 7 - \ln 7 \\
t(\ln 2 - \ln 7) &= -\ln 7 - 3\ln 2 \\
t &= \frac{-\ln 7 - 3\ln 2}{\ln 2 - \ln 7} = 3.21.
\end{aligned}
$$

13. **Solving exponential equations**: We can solve for $t$ by taking logarithms and using their properties.

$$
\begin{aligned}
\frac{12^t}{9^t} &= 7 \times 6^t \quad \text{original equation} \\
\ln \frac{12^t}{9^t} &= \ln(7 \times 6^t) \quad \text{taking ln of each side} \\
t\ln 12 - t\ln 9 &= \ln 7 + t\ln 6 \quad \text{quotient, power and product laws} \\
t\ln 12 - t\ln 9 - t\ln 6 &= \ln 7 \\
t &= \frac{\ln 7}{\ln 12 - \ln 9 - \ln 6} = -1.29.
\end{aligned}
$$

14. **Solving exponential equations**: We can solve for $t$ by taking logarithms and using their properties.

$$
\begin{aligned}
a^{2t} &= 3 \times a^{4t} \quad \text{original equation} \\
\ln a^{2t} &= \ln(3 \times a^{4t}) \quad \text{taking ln of each side} \\
2t\ln a &= \ln 3 + 4t\ln a \quad \text{power and product laws} \\
t2\ln a - t4\ln a &= \ln 3 \\
t &= \frac{\ln 3}{-2\ln a} = -\frac{\ln 3}{2\ln a}.
\end{aligned}
$$

15. **Solving exponential equations**: We can solve for $t$ by taking logarithms and using their properties.

$$
\begin{aligned}
ab^t &= c && \text{original equation} \\
\ln ab^t &= \ln c && \text{taking ln of each side} \\
\ln a + t \ln b &= \ln c && \text{power and product laws} \\
t \ln b &= \ln c - \ln a \\
t &= \frac{\ln c - \ln a}{\ln b}.
\end{aligned}
$$

16. **Solving exponential equations**: We can solve for $t$ by taking logarithms and using their properties.

$$
\begin{aligned}
a^{t+1} &= b^{t-1} && \text{original equation} \\
\ln a^{t+1} &= \ln b^{t-1} && \text{taking ln of each side} \\
(t+1)\ln a &= (t-1)\ln b && \text{power and product laws} \\
t \ln a - t \ln b &= -\ln b - \ln a \\
t &= \frac{-\ln b - \ln a}{\ln a - \ln b}.
\end{aligned}
$$

Since $a \neq b$, $\ln a - \ln b \neq 0$, so this last division is valid. The answer can also be written as $t = \dfrac{-\ln a - \ln b}{\ln a - \ln b}$ or $t = \dfrac{\ln a + \ln b}{\ln b - \ln a}$.

17. **Solving logarithmic equations**: We can solve for $t$ by taking exponentials and using their properties.

$$
\begin{aligned}
\ln(3t-1) &= 4 && \text{original equation} \\
e^{\ln(3t-1)} &= e^4 && \text{taking } e \text{ of each side} \\
3t - 1 &= e^4 && \text{since } e^{\ln A} = A \\
3t &= e^4 + 1 \\
t &= \frac{e^4 + 1}{3} = 18.53.
\end{aligned}
$$

18. **Solving logarithmic equations**: We can solve for $t$ by taking exponentials and using their properties.

$$
\begin{aligned}
\log(6t+4) &= -2 && \text{original equation} \\
10^{\log(6t+4)} &= 10^{-2} && \text{taking 10 of each side} \\
6t+4 &= 10^{-2} && \text{since } 10^{\log A} = A \\
6t &= 10^{-2} - 4 \\
t &= \frac{10^{-2}-4}{6} = -0.665.
\end{aligned}
$$

19. **Solving logarithmic equations**: We can solve for $t$ by taking exponentials and using their properties.

$$
\begin{aligned}
\ln(at+b) &= c && \text{original equation} \\
e^{\ln(at+b)} &= e^c && \text{taking } e \text{ of each side} \\
at+b &= e^c && \text{since } e^{\ln A} = A \\
at &= e^c - b \\
t &= \frac{e^c - b}{a}.
\end{aligned}
$$

20. **Solving logarithmic equations**: We can solve for $t$ by taking exponentials and using their properties.

$$
\begin{aligned}
\log(at-b) &= c && \text{original equation} \\
10^{\log(at-b)} &= 10^c && \text{taking 10 of each side} \\
at-b &= 10^c && \text{since } 10^{\log A} = A \\
at &= 10^c + b \\
t &= \frac{10^c + b}{a}.
\end{aligned}
$$

21. **Solving logarithmic equations**: We can solve for $t$ by taking exponentials and using their properties.

$$
\begin{aligned}
\ln(2^t + 1) &= 3 && \text{original equation} \\
e^{\ln(2^t+1)} &= e^3 && \text{taking } e \text{ of each side} \\
2^t + 1 &= e^3 && \text{since } e^{\ln A} = A \\
2^t &= e^3 - 1 && \text{which is exponential, so we take } \ln \\
\ln 2^t &= \ln(e^3 - 1) && \text{taking } \ln \text{ of each side} \\
t \ln 2 &= \ln(e^3 - 1) && \text{power and product laws} \\
t &= \frac{\ln(e^3 - 1)}{\ln 2} = 4.25.
\end{aligned}
$$

22. **Solving logarithmic equations**: We can solve for $t$ by taking exponentials and using their properties.

$$
\begin{aligned}
\log(3^t - 2) &= 2 && \text{original equation} \\
10^{\log(3^t-2)} &= 10^2 && \text{taking } 10 \text{ of each side} \\
3^t - 2 &= 10^2 && \text{since } 10^{\log A} = A \\
3^t &= 10^2 + 2 && \text{which is exponential, so we take } \ln \\
\ln 3^t &= \ln(10^2 + 2) && \text{taking } \ln \text{ of each side} \\
t \ln 3 &= \ln(10^2 + 2) && \text{power and product laws} \\
t &= \frac{\ln(10^2 + 2)}{\ln 3} = 4.21.
\end{aligned}
$$

23. **Decibels**:

(a) If we double the relative intensity, so multiply it by 2, the effect on the decibels is

$$
\begin{aligned}
\text{New decibels} &= 10\log(2 \times \text{old Relative intensity}) \\
&= 10(\log 2 + \log(\text{old Relative intensity})) \\
&= 10\log 2 + 10\log(\text{old Relative intensity}) \\
&= 10\log 2 + \text{Old decibels} \\
&= 3.01 + \text{Old decibels}
\end{aligned}
$$

(b) Since $D = 10\log R$, $\log R = D/10$ and so $10^{\log R} = 10^{D/10}$ and thus $R = 10^{D/10}$.

24. **Spectroscopic parallax preview**:

(a) If $D$ is doubled, we have

$$
\begin{aligned}
\text{New S} &= 5\log 2D - 5 \\
&= 5(\log 2 + \log D) - 5 \\
&= 5\log 2 + 5\log D - 5 \\
&= 5\log 2 + \text{Old S}
\end{aligned}
$$

Therefore, $S$ increases by $5\log 2$, or about 1.51.

(b) Since $5\log D - 5 = S$, $5\log D = S + 5$ and so $\log D = \dfrac{S+5}{5}$ and thus $D = 10^{\log D} = 10^{(S+5)/5}$.

25. **Doubling time**:

(a) $B$ will double when it reaches $5000 \times 2 = 10{,}000$ dollars.

$$
\begin{aligned}
5000 \times 1.005^t &= 10{,}000 \\
1.005^t &= 10{,}000/5000 = 2 \\
\ln 1.005^t &= \ln 2 \\
t \ln 1.005 &= \ln 2 \\
t &= \frac{\ln 2}{\ln 1.005} = 138.98
\end{aligned}
$$

So $B$ doubles in about 139 months, or 11.58 years.

(b) The doubling time is the value of $t$ for which $y(t) = 2a$:

$$
\begin{aligned}
ab^t &= 2a \\
b^t &= 2a/a = 2 \\
\ln b^t &= \ln 2 \\
t \ln b &= \ln 2 \\
t &= \frac{\ln 2}{\ln b}
\end{aligned}
$$

26. **Moore's law**:

(a) Applying the common logarithm, we solve for $k$:

$$
\begin{aligned}
10^k &= 2^{1/2} \\
\log\left(10^k\right) &= \log\left(2^{1/2}\right) \\
k &= \log\left(2^{1/2}\right) = 0.15.
\end{aligned}
$$

(b) If the yearly growth factor is $a = 2^{1/2}$, then the 2-year growth factor is $a^2 = \left(2^{1/2}\right)^2 = 2^{2/2} = 2$, so the function doubles every 2 years.

(c) According to Part (b), every year the speed of computer chips increases by a factor of $a = 2^{1/2}$, which equals $10^k$ if $k$ is the solution of the equation in Part (a).

# Solution Guide for Chapter 5: A Survey of Other Common Functions

## 5.1 LOGISTIC FUNCTIONS

1. **Ebola**:

   (a) Because the data points appear to be leveling off, they show a health crisis that is under control.

   (b) By the formula for $E$, the limiting value (corresponding to the carrying capacity) is 27,841.42. The disease was spreading at the fastest rate when the number of diseases was half of this value, so when the total number of cases was 13,920 (to the nearest whole number).

   (c) By part b, we want to solve the equation $E(t) = 13{,}920$. Using the crossing-graphs method, we find the solution $t = 7.6$, where we have rounded to one decimal place. Thus, the disease was growing at the fastest rate 7.6 months after April 1, 2014.

3. **Multiple myeloma**:

   (a) We use a horizontal span of 0 to 1200 days. After consulting a table of values, we use a vertical span of 0 to 1100 billion cells. The graph is below.

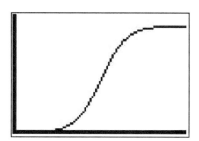

(b) The size at the time of diagnosis is $N(0) = 1$ (measured in billions of cells). We want to find when the size is 100 times that, so we want to solve the equation $N(t) = 100$. Using the crossing-graphs method, we find the solution $t = 413.12$. Thus, it takes 413.12 days for the tumor to reach 100 times its size at the time of diagnosis.

5. **Magazine sales:** The formula for a logistic model is $N = \dfrac{K}{1 + be^{-rt}}$ with $t$ measured in months. We are given that $N(0) = 300$ and $K = 1200$, so $b = \dfrac{K}{N(0)} - 1 = \dfrac{1200}{300} - 1 = 3$. Since the monthly growth would be 20% if the market were unlimited, $a = 1 + 0.20 = 1.20$ and so $r = \ln a = \ln 1.20 = 0.182$. Thus our logistic model is $N = \dfrac{1200}{1 + 3e^{-0.182t}}$.

7. **African bees:** The formula for a logistic model is $N = \dfrac{K}{1 + be^{-rt}}$. We are given that $N(0) = 50$ and $K = 3600$, so $b = \dfrac{K}{N(0)} - 1 = \dfrac{3600}{50} - 1 = 71$. Since the annual growth would be 30% if the growth were unlimited, $a = 1 + 0.30 = 1.30$ and so $r = \ln a = \ln 1.30 = 0.262$. Thus our logistic model is $N = \dfrac{3600}{1 + 71e^{-0.262t}}$. To determine when $N = 1800$ hives are affected, we solve $\dfrac{3600}{1 + 71e^{-0.262t}} = 1800$, finding that $t = 16.27$ years.

9. **PTA participation:**

   (a) Because the school board believes their plan can eventually lead to an attendance level of 50 parents, we take the carrying capacity to be $K = 50$.

   (b) We are given that $P(0) = 25$, and in Part (a) we found $K = 50$, so $b = \dfrac{K}{P(0)} - 1 = \dfrac{50}{25} - 1 = 1$.

   (c) In the absence of limiting factors the school board believes its plan can increase participation by 10% each month, so we take $a = 1 + 0.10 = 1.10$. Therefore, $r = \ln a = \ln 1.10 = 0.095$.

   (d) Using the answers from the preceding parts, we find the logistic model $P = \dfrac{50}{1 + e^{-0.095m}}$.

11. **Reliability curves:**

(a) The regression parameters for the logistic model are calculated in the figure on the left below. We find the logistic model $P = \dfrac{102.55}{1 + 5.13e^{-0.349m}}$.

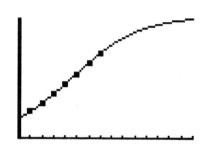

(b) The graph of the logistic model and the data is in the figure on the right above. We used a horizontal span of 0 to 15 months and a vertical span of 0 to 100 percent.

(c) We want to find the value of $m$ for which $P(m) = 95$, so we want to solve the equation $\dfrac{102.55}{1 + 5.13e^{-0.349m}} = 95$. The crossing-graphs method gives the solution $m = 12$, where we have rounded to the nearest whole number. Therefore, the model predicts that the chip will be ready for production after 12 months.

13. **Paramecium cells:**

(a) Entering the data, we see that logistic regression yields (see figure below on the left) $N = \dfrac{708.42}{1 + 80.12e^{-0.532t}}$.

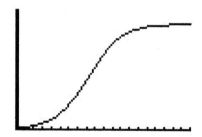

(b) The graph of the logistic model is the figure above on the right using a horizontal span of 0 to 20 and a vertical span of 0 to 800.

(c) Using crossing-graphs or tables, for example, we see that the population reaches 450 when $t = 9.29$ days.

15. **Northern Yellowstone elk**:

    (a) Entering the data, we see that logistic regression yields (see figure below) $N = \dfrac{12{,}937.03}{1 + 3.14e^{-0.447t}}$ if $t$ is time in years since 1968.

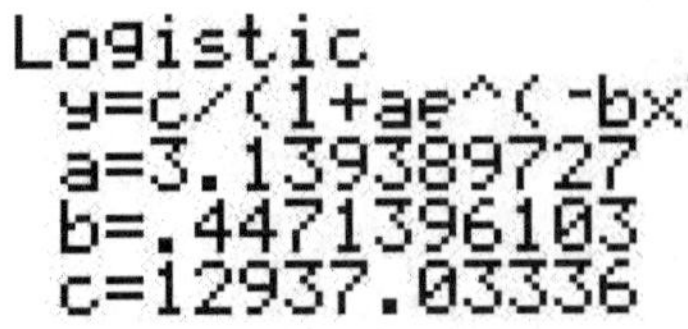

$$
\begin{aligned}
&\texttt{Logistic}\\
&\texttt{y=c/(1+ae\^{}(-bx))}\\
&\texttt{a=3.139389727}\\
&\texttt{b=.4471396103}\\
&\texttt{c=12937.03336}
\end{aligned}
$$

    (b) Using crossing-graphs or tables, for example, we see that the population reaches half the carrying capacity, that is $12{,}937.03/2 = 6468.52$, when $t = 2.56$ years.

17. **World population**:

    (a) Entering the data, we see that logistic regression yields (see the figure below) $N = \dfrac{12.14}{1 + 3.84e^{-0.027t}}$, with $t$ measured in years since 1950.

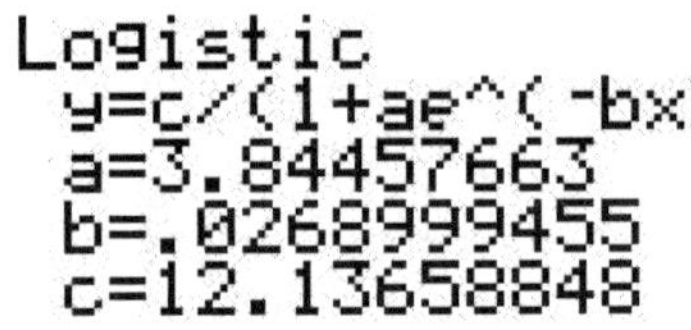

$$
\begin{aligned}
&\texttt{Logistic}\\
&\texttt{y=c/(1+ae\^{}(-bx))}\\
&\texttt{a=3.84457663}\\
&\texttt{b=.0268999455}\\
&\texttt{c=12.13658848}
\end{aligned}
$$

    (b) The $r$ value is 0.027 per year.

    (c) The carrying capacity $K$ is 12.14 billion people.

    (d) Using crossing-graphs or tables, for example, we see that the population reaches 90% of the carrying capacity, that is, 10.926 billion, when $t = 131$ years after 1950, so in the year 2081.

19. **More on the Pacific sardine**:

    (a) Note first that 50,000 tons equals 0.05 million tons. Following the suggestion, we make a new logistic formula using $K = 2.4$, $r = 0.338$, and $N(0) = 0.05$ and then ask when this formula equals 1.2 million tons. We want $b = \dfrac{K}{N(0)} - 1 = \dfrac{2.4}{0.05} - 1 = 47$, so the new logistic formula is

$$
N = \frac{2.4}{1 + 47e^{-0.338t}}.
$$

We then solve the equation $N = 1.2$. We do this using the crossing graphs method.

Based on the table below, we used a horizontal span of 0 to 15 and a vertical span of 0 to 2 in the graph below. (The horizontal axis is time in years, and the vertical axis is population in millions of tons.)

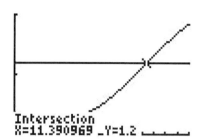

We see that the desired time $t$ is 11.39 years. This is the time required to recover from a level of 50,000 tons to 1.2 million tons.

(b) Here we need to adjust the $r$ value in the formula found in Example 5.1 to get $N = \dfrac{2.4}{1 + 239e^{-0.215t}}$. We then proceed as in the solution to Part (a) above by entering this new formula for $N$ and finding $t_1$ and $t_2$ where $N(t_1) = 0.05$ and $N(t_2) = 1.2$, so then the time to recover from 0.05 to 1.2 million tons is $t_2 - t_1$.

To find $t_1$ we make the graph using a horizontal span of 0 to 15 and a vertical span of 0 to 0.2. The left-hand figure below shows that, with the reduced growth rate, the level of 50,000 tons is reached at $t_1 = 7.564$ years.

To find $t_2$ we put 1.2 in place of 0.05 on the function entry screen and make the graph using a horizontal span of 0 to 40 and a vertical span of 0 to 2.5 as in Example 5.1. The right-hand figure below shows that, with the reduced growth rate, the level of 1.2 million tons is reached at $t_2 = 25.472$ years. Then the difference is $t_2 - t_1 = 25.472 - 7.564$, or about 17.91 years. This is the time required to recover from a level of 50,000 tons to 1.2 million tons with the reduced growth rate. Note that this is considerably longer than the time of 11.39 years found in Part (a).

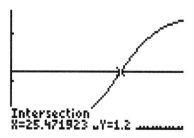

21. **Eastern Pacific yellowfin tuna:**

(a) Since the equation given in the exercise matches the integral form of the logistic equation, all we have to do is read the $r$ value from the formula: The $r$ value for Eastern Pacific yellowfin tuna is 2.61 per year.

(b) The constant in the numerator is 148, so the carrying capacity for the tuna is 148 thousand tons.

(c) The optimum yield level is half of the carrying capacity, so it is 74 thousand tons.

(d) Based on a table of values for $N$, we used a horizontal span of 0 to 5 and a vertical span of 0 to 160 in the left-hand graph below. (The horizontal axis is time in years, and the vertical axis is population in thousands of tons.)

(e) The population was growing the fastest when the optimum yield level of 74 thousand tons was reached. To determine this time, we use the crossing graphs method. From the right-hand graph below we see that the population was growing the fastest at the time $t = 0.49$ year.

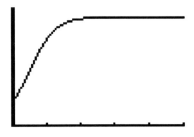  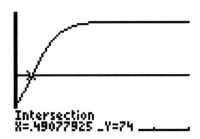

23. **An inverted logistic curve:**

(a) Because 2000 is $2000 - 1875 = 125$ years since 1875, we use a horizontal span of 0 to 125 years. After consulting a table of values, we use a vertical span of 0 to 0.13. The graph is in the figure below.

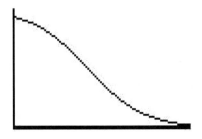

(b) Scanning down a table of values for $T$ using a calculator gives the limiting value of 0. (This can also be found from the graph.) This means that in the long run there will be very few deaths due to tuberculosis.

(c) We want to locate the inflection point on the graph because at that point the rate of decrease is the largest. (Before then, the function is decreasing at an increasing rate.) Tracing the graph indicates that the inflection point occurs at about $t = 55$, which corresponds to the year 1930. Therefore, the fraction of deaths due to tuberculosis was decreasing most rapidly in about 1930.

## Skill Building Exercises

S-1. **Estimating optimum yield**: To answer this we locate the inflection point. The optimum yield level is about 200, and this occurs when the time is about 8.

S-3. **Logistic growth**: If the notion of carrying capacity is added to the basic assumptions leading to the exponential model, then we get the logistic model.

S-5. **Harvesting**: The maximum sustainable yield model says that a renewable resource that grows logistically should be harvested at half the carrying capacity.

S-7. **Harvesting continued**: The point on the graph of $N$ as a function of $t$ corresponding to $N = \dfrac{K}{2}$ is the inflection point on the logistic curve; in particular, it is the point of greatest growth in $N$.

S-9. **Finding logistic parameters**: The parameter $b$ equals $\dfrac{K}{N(0)} - 1$. The optimal yield level is $400 = K/2$, so $K = 800$. Since the initial population is 200, $b = \dfrac{800}{200} - 1 = 3$.

S-11. **Finding logistic parameters**: Since $b = \dfrac{K}{N(0)} - 1$, if $b = 5$ and $N(0) = 150$, then $\dfrac{K}{150} - 1 = 5$. Thus $\dfrac{K}{150} = 6$ and so $K = 6 \times 150 = 900$, and therefore the optimal yield level is $K/2 = 900/2 = 450$.

S-13. **Finding logistic parameters**: Since $b = \dfrac{K}{N(0)} - 1$, if $K = 1000$ and $b = 3$, then $\dfrac{1000}{N(0)} - 1 = 3$. Thus $\dfrac{1000}{N(0)} = 4$ and so the initial population is $N(0) = 1000/4 = 250$.

S-15. **Finding logistic parameters**: In the absence of constraints, the growth factor is $a = e^r = e^{0.3} = 1.35$, and so the percentage growth would be 35%.

S-17. **Finding logistic parameters**: In the absence of constraints, the growth factor is $a = e^r = e^{0.7} = 2.01$, and so the percentage growth would be 101%.

S-19. **Finding logistic parameters**: In the absence of constraints, the $r$ value is $r = \ln a$. Since the population has 11% growth, the growth factor is $a = 1 + 0.11 = 1.11$, and so $r = \ln 1.11 = 0.104$.

S-21. **Finding logistic parameters**: In the absence of constraints, the $r$ value is $r = \ln a$. Since the population has 64% growth, the growth factor is $a = 1 + 0.64 = 1.64$, and so $r = \ln 1.64 = 0.495$.

S-23. **Finding the logistic formula**: The formula for a logistic model is $N = \dfrac{K}{1 + be^{-rt}}$, so since $K = 2300$, $r = 0.77$ and $b = 7$, we have $N = \dfrac{2300}{1 + 7e^{-0.77t}}$.

S-25. **Finding the logistic formula**: The formula for a logistic model is $N = \dfrac{K}{1 + be^{-rt}}$. In this case $K = 400$, $r = 0.44$ and $b = \dfrac{K}{N(0)} - 1 = \dfrac{400}{10} - 1 = 39$, so we have $N = \dfrac{400}{1 + 39e^{-0.44t}}$.

S-27. **Finding the logistic formula**: The formula for a logistic model is $N = \dfrac{K}{1 + be^{-rt}}$. In this case the optimum yield is $K/2 = 400$, so $K = 800$. The initial value is 100, so $b = \dfrac{K}{N(0)} - 1 = \dfrac{800}{100} - 1 = 7$. The population, in the absence of constraints, would grow at 22% per year, so $a = 1 + 0.22 = 1.22$ and so $r = \ln a = \ln 1.22 = 0.20$. Thus we have $N = \dfrac{800}{1 + 7e^{-0.20t}}$.

S-29. **Using logistic regression**: Entering the data, we find that logistic regression yields (see figure below) $N = \dfrac{220.42}{1 + 3.95e^{-0.764t}}$.

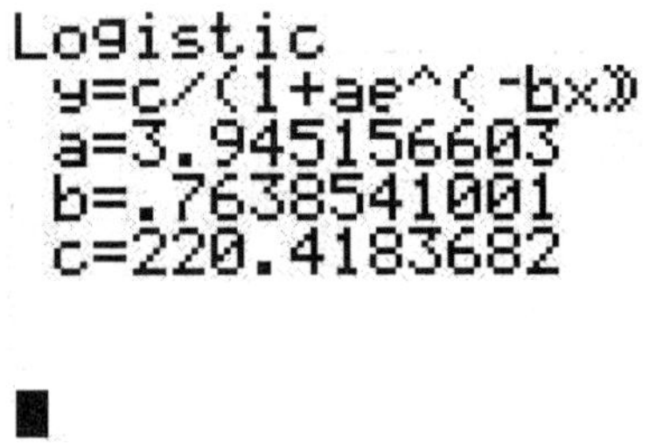

```
Logistic
y=c/(1+ae^(-bx)
a=3.945156603
b=.7638541001
c=220.4183682
```

S-31. **Using logistic regression**:

(a) The carrying capacity is $K = 329.65$.

(b) The population will reach 125 when $125 = \dfrac{329.65}{1 + 9.28e^{-0.537t}}$. Solving using crossing-graphs, tables, or logarithms yields $t = 3.23$.

S-33. **Using logistic regression**:

(a) The carrying capacity is $K = 2384.95$.

(b) The population will reach 1625 when $1625 = \dfrac{2384.95}{1 + 20.18e^{-0.211t}}$. Solving using crossing-graphs, tables, or logarithms yields $t = 17.84$.

S-35. **Using regression and plotting**: Entering the data, we find that logistic regression yields (see the figure on the left below) $N = \dfrac{344.17}{1 + 240.87e^{-0.513t}}$.

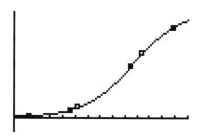

In the figure on the right above, we have plotted the data along with the regression model.

S-37. **Using regression and plotting**: Entering the data, we find that logistic regression yields (see the figure on the left below) $N = \dfrac{926.55}{1 + 294.67e^{-0.267t}}$.

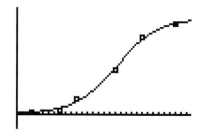

In the figure on the right above, we have plotted the data along with the regression model.

## 5.2  POWER FUNCTIONS

1. **The role of the constant term in power functions with positive power:** We use a horizontal span of 0 to 5 as suggested, and the table of values below leads us to choose a vertical span of 0 to 100. In the right-hand figure, $x^2$ is the bottom graph, $2x^2$ is the next one up, and $4x^2$ is the top graph.

| X | Y3 | Y4 |
|---|-----|-----|
| 0 | 0 | 0 |
| 1 | 3 | 4 |
| 2 | 12 | 16 |
| 3 | 27 | 36 |
| 4 | 48 | 64 |
| 5 | 75 | 100 |
| 6 | 108 | 144 |

Y4=0

Larger values of $c$ make the function increase faster.

3. **Math and the city:**

   (a) We use the homogeneity property of power functions. If the population $N$ is multiplied by 2, the number $G$ of gas stations is multiplied by

   $$2^{\text{Function power}} = 2^{0.77} = 1.71.$$

   Thus, if one city is twice as large as another, it has about 1.71 times as many gas stations.

   (b) The formula for $G$ is $G = cN^{0.77}$, and we know that if $N = 2.2$ then $G = 1239$. Thus, $1239 = c \times 2.2^{0.77}$. Solving for $c$ gives $c = \dfrac{1239}{2.2^{0.77}} = 675.16$.

   (c) By part a we have $c = 675.16$, so the formula for $G$ is $G = 675.16N^{0.77}$. Because the population of Los Angeles is about 3.9 million, to estimate the number of gas stations in Los Angeles we put $N = 3.9$ into the formula for $G$:

   $$G = 675.16 \times 3.9^{0.77} = 1925,$$

   where we have rounded to the nearest whole number. Thus, the number of gas stations in Los Angeles is about 1925.

5. **Kleiber's law:**

(a) We use the homogeneity property of power functions. The 200-pound man weighs 200 times as much as the 1-pound squirrel. If the weight $W$ is multiplied by 200, the metabolic rate $M$ is multiplied by

$$200^{\text{Function power}} = 200^{3/4} = 53.18.$$

Thus, the man's metabolic rate is 53.18 times that of the squirrel.

(b) The 200-pound man weighs $\dfrac{200}{130}$ times as much as the 130-pound woman. If the weight $W$ is multiplied by $\dfrac{200}{130}$, the metabolic rate $M$ is multiplied by

$$\left(\frac{200}{130}\right)^{\text{Function power}} = \left(\frac{200}{130}\right)^{3/4} = 1.38.$$

Thus, the man's metabolic rate is 1.38 times that of the woman.

(c) The man has a higher metabolic rate. A higher metabolic rate means more energy is expended in a fixed period of time. Thus, a larger part of caloric intake from food is used in energy production. Thus, overeating would be more likely to lead to weight gain for the woman.

7. **Speed and stride length**: We use the homogeneity property of power functions to solve this. If one animal has a stride length 3 times that of another, then it will run faster by a factor of

$$3^{\text{Function power}} = 3^{1.7} = 6.47$$

.

9. **Stevens's power law**: We use the homogeneity property of power functions. If the force $f$ is multiplied by 2, the perceived pressure $p$ is multiplied by

$$2^{\text{Function power}} = 2^{1.1} = 2.14.$$

Therefore, if the force is doubled the perceived pressure increases by a factor of 2.14.

11. **Mosteller formula for body surface area:**

(a) We put $h = 188$ and $w = 86$ into the formula for $B$:

$$B = \frac{1}{60} \times 188^{1/2} \times 86^{1/2} = 2.12.$$

Therefore, the body surface area for a man who is 188 centimeters tall and weighs 86 kilograms is about 2.12 square meters.

(b) Because the height $h$ stays the same, we may take $B$ to be a power function of $w$ with power $1/2$. We use the homogeneity property of power functions. If the weight increases by 10% or 0.1 as a decimal, $w$ is multiplied by $1 + 0.1 = 1.1$. Therefore, $B$ is multiplied by

$$1.1^{\text{Function power}} = 1.1^{1/2} = 1.0488,$$

where we have rounded to four decimal places. Multiplying by 1.0488 corresponds to an increase by a factor of $1.0488 - 1 = 0.0488$ or 4.88%. Therefore, the body surface area will increase by 4.88%.

13. **Length of skid marks versus speed:**

(a) We are on dry concrete, so the value of the friction coefficient is $h = 0.85$. Thus the function we are dealing with is

$$L = \frac{1}{30 \times 0.85}S^2 = \frac{S^2}{25.5}.$$

If the speed is $S = 55$ miles per hour, then the skid marks will have length

$$L = \frac{55^2}{25.5} = 118.63 \text{ feet.}$$

(b) Again we are on dry concrete pavement, so we use the same function as in Part (a). We need to know what speed will result in skid marks which are 230 feet long. That is, we need to solve the equation

$$\frac{S^2}{25.5} = 230.$$

We will solve this using the crossing graphs method. We use a horizontal span of 0 to 80 miles per hour, and the table of values for $L$ below leads us to choose a vertical span of 0 to 300. We have graphed $L$ and the target length of 230 (thick line). We see that the intersection occurs at a speed of $S = 76.58$ miles per hour. The driver would appear to be in danger of receiving a citation.

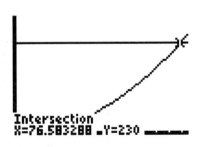

| X | Y1 | Y2 |
|---|---|---|
| 0 | 0 | 230 |
| 20 | 15.686 | 230 |
| 40 | 62.745 | 230 |
| 60 | 141.18 | 230 |
| 80 | 250.98 | 230 |
| 100 | 392.16 | 230 |
| 120 | 564.71 | 230 |

X=0

Intersection
X=76.583288  Y=230

(c) Since we now do not know the value of $h$, we must use the homogeneity property to solve this problem. We want to change the distance by a factor of 0.5. Since the power for the function is 2, changing the speed by a factor of $f$ changes the distance $L$ by a factor of $f^2$. We want to cut the speed in half, so we want to choose $f$ so that

$$f^2 = 0.5.$$

This may be solved by the crossing graphs method or simply by noting that the solution is

$$f = \sqrt{0.5} = 0.7071.$$

Hence we need to change our speed from 60 miles per hour to

$$60 \times 0.7071 = 42.43 \text{ miles per hour.}$$

The new speed should be between 42 and 43 miles per hour.

15. **Life expectancy of stars**:

(a) Since $E$ is a power function with negative exponent, as $M$ increases, $E$ decreases, so a more massive star has a shorter life expectancy than a less massive star.

(b) Since Spica has a mass of 7.3 solar masses, $M = 7.3$, and therefore $E = M^{-2.5} = 7.3^{-2.5}$, which equals 0.0069, or about 0.007, of a solar lifetime. Since Earth's life expectancy is about 10 billion years, Spica's life expectancy is about $10 \times 0.0069 = 0.069$ billion years or about 70 million years.

(c) In functional notation, the life expectancy of a main-sequence star with mass $M = 0.5$ is $E(0.5)$. Its value is $E = 0.5^{-2.5}$ or about 5.66 solar lifetimes. This is about 56.6 billion years.

(d) Because Vega has a life expectancy of 6.36 billion years and 1 solar lifetime is 10 billion years, Vega has a life expectancy of 0.636 of a solar lifetime. Since $E = 0.636$, we have the equation $M^{-2.5} = 0.636$, which we can solve for $M$. This yields $M = 1.20$, so the mass of Vega is 1.20 solar masses.

(e) If one main-sequence star is twice as massive as another, then $M$ is changed by a factor of 2, and therefore $E$ is changed by a factor of $2^{-2.5} = 0.18$. Thus the life expectancy of the larger star is 0.18 times that of the smaller.

17. **Tsunami waves and breakwaters**: The wave height $h$ beyond the breakwater is a power function of the width ratio $R$ with power 0.5.

   (a) We measure $H$, $W$, $w$, and $h$ in feet. Since the wave has height 8 feet in a channel of width 5000 before the breakwater, $H = 8$ and $W = 5000$. Since the breakwater has width 3000 feet, $w = 3000$, and so $R = \dfrac{w}{W} = \dfrac{3000}{5000} = 0.6$. The height of the wave beyond the breakwater is $h = HR^{0.5} = 8 \times 0.6^{0.5} = 6.20$ feet.

   (b) If the channel width is cut in half, then $R$ is multiplied by $\dfrac{1}{2}$. Thus the new height is $\left(\dfrac{1}{2}\right)^{0.5} = 0.71$ times the original height.

19. **Terminal velocity**: Terminal velocity $T$ is a power function with power 0.5, so if the length $L$ changes by a factor of $t$, then $T$ changes by a factor of $t^{0.5}$.

   (a) The man is 36 times as long as the mouse, so $t = 36$. The terminal velocity of the man is therefore $36^{0.5} = 6$ times that of the mouse.

   (b) If the terminal velocity of the man is 120 miles per hour, then, from Part (a), that of the mouse is $\dfrac{120}{6} = 20$ miles per hour.

   (c) Since a squirrel is 7 inches long, it is only $\dfrac{7}{72}$ times the length of the 6-foot man, so the terminal velocities differ by a factor of $\left(\dfrac{7}{72}\right)^{0.5}$. Thus the terminal velocity of a squirrel is $120 \times \left(\dfrac{7}{72}\right)^{0.5} = 37.42$ miles per hour.

21. **Newton's law of gravitation**:

   (a) If the distance $d$ is changed by a factor of 0.5, then the gravitational force will be changed by a factor of

$$0.5^{\text{Function power}} = 0.5^{-2} = 4.$$

   Hence if the distance between the asteroids is halved, the force of gravity will be 4 times as strong.

   If the distance is reduced to one quarter of its original value, it is changed by a factor of 0.25. That gives a change in force by a factor of

$$0.25^{\text{Function power}} = 0.25^{-2} = 16.$$

   Thus if the distance is reduced to one quarter of its original value, the force of gravity will be 16 times stronger.

   (b) We want to know the value of $c$ that will make

$$2{,}000{,}000 = c \times 300^{-2}.$$

This is a linear equation in $c$ which we can solve by dividing each side by $300^{-2}$:

$$c = \frac{2{,}000{,}000}{300^{-2}} = 1.8 \times 10^{11}.$$

Hence we know that, for these asteroids, the gravitational force is given by

$$F = 1.8 \times 10^{11} \times d^{-2}.$$

When $d = 800$, we get that the force is

$$F = 1.8 \times 10^{11} \times 800^{-2} = 281{,}250 \text{ newtons.}$$

This could also be solved used the homogeneity property of power functions. The first step is to note that increasing the distance from 300 to 800 kilometers is an increase by a factor of $800/300 = 2.67$.

(c) We use a horizontal span of 0 to 1000 as suggested. The table of values below for $F = 1.8 \times 10^{11} \times d^{-2}$ leads us to choose a vertical span of 0 to $20{,}000{,}000 = 2 \times 10^{7}$. In the graph below, the horizontal axis is distance between the centers, and the vertical axis is force of gravity.

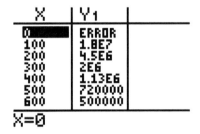 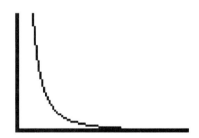

The graph shows that when the asteroids are relatively close, gravitational force is very strong. (It is in fact so strong that when bodies of planetary size get too close, the force of gravity will tear one or both of them to pieces.) As they move farther apart, the gravitational force decreases toward zero.

23. **Giant ants and spiders**:

(a) Weight is a power function of length, with power 3. Thus, if the length of the ant is increased by a factor of 500, then its weight is increased by a factor of $500^3 = 125{,}000{,}000$.

(b) Cross-sectional area of limbs is a power function of length, with power 2. Thus, if the length of the ant is increased by a factor of 500, then the cross-sectional area of its legs is increased by a factor of $500^2 = 250{,}000$.

(c) Using the hint and the results of Parts (a) and (b), we see that the pressure on a leg increases by a factor of $\dfrac{125{,}000{,}000}{250{,}000} = 500$. The poor ant will be flattened.

25. **Time to failure**: If the building collapses in 30 seconds at one intensity, put $t_1 = 30$ and denote that intensity as $I_1$. Tripling that intensity means putting $I_2 = 3I_1$, so

$$\frac{t_1}{t_2} = \left(\frac{I_2}{I_1}\right)^2 = \left(\frac{3I_1}{I_1}\right)^2 = 9.$$

Since $t_1 = 30$, we have $\dfrac{30}{t_2} = 9$. Solving for $t_2$ yields 3.33 seconds for the building to collapse.

### Skill Building Exercises

S-1. **Graphs of power functions:**

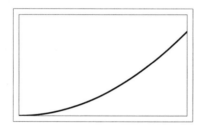

S-3. **Graphs of power functions:**

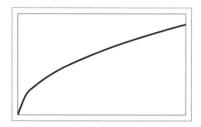

S-5. **Graphs of power functions:**

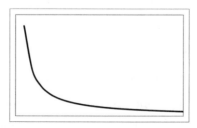

S-7. **Homogeneity**: In this case $f$ is power function with power $k = 4$. By the homogeneity property, if $x$ is doubled, then $f$ is increased by a factor of $2^4 = 16$.

S-9. **Homogeneity**: In this case $f$ is a power function with power $k = 2.53$. By the homogeneity property, if $x$ is increased by a factor of $t$, then $f$ is increased by a factor of $t^{2.53}$. To find the $t$ tripling the value of $f$, we need to solve the equation $t^{2.53} = 3$. Solving by the crossing-graphs method or algebraically, we see that $t = 1.54$, so $x$ must be increased by a factor of 1.54.

S-11. **Homogeneity**: In this case $f$ is a power function with power $k = 3.11$. By the homogeneity property, if $z$ is increased by a factor of $t$, then $f$ is increased by a factor of $t^{3.11}$. To find the $t$ increasing the value of $f$ by a factor of 9, we need to solve the equation $t^{3.11} = 9$. Solving by the crossing-graphs method or algebraically, we see that $t = 2.03$, so $z$ must be increased by a factor of 2.03. Thus $y$ is 2.03 times $z$.

S-13. **Homogeneity**: In this case $f$ is a power function with power $k$. By the homogeneity property, if $x$ is increased by a factor of $t$, then $f$ is increased by a factor of $t^k$. Since $x$ increases from 1.76 to 6.6, 1.76 is increased by a factor of $t = \dfrac{6.6}{1.76}$. Thus $f$ is increased by a factor of $t^k = \left( \dfrac{6.6}{1.76} \right)^k$. Since $f(6.6)$ is 6.2 times as large as $f(1.76)$,

$$6.2 = t^k = \left( \frac{6.6}{1.76} \right)^k.$$

Solving for $k$ yields $k = 1.38$.

S-15. **Constant term**: Since $f(x) = cx^{-1.32}$ and $f(5) = 11$, $11 = c \times 5^{-1.32}$. Thus $c = \dfrac{11}{5^{-1.32}}$, which equals 92.05, so $c = 92.05$.

S-17. **Making power formulas**: We want to find $c$ and $k$ in the formula $f(x) = cx^k$. Now $f(1) = 7$ says $c \times 1^k = 7$, so $c = 7$. Then $f(25) = 35$ says $7 \times 25^k = 35$, so $25^k = 5$. Thus, $k = 1/2$, and we have the formula $f(x) = 7x^{1/2}$.

S-19. **Making power formulas**: We want to find $c$ and $k$ in the formula $f(x) = cx^k$. First we use the fact that, if $x$ is multiplied by 2, then $f(x)$ is multiplied by 8. By the homogeneity property of power functions with power $k$, if $x$ is multiplied by 2, then $f(x)$ is multiplied by $2^k$. Thus, $2^k = 8$. We see that $k = 3$. Now $f(2) = 16$ says $c \times 2^3 = 16$, so $c = 2$. Thus, we have the formula $f(x) = 2x^3$.

## 5.3  MODELING DATA WITH POWER FUNCTIONS

1. **Zipf's law**:

    (a) The regression parameters for the power model are calculated in the figure on the left below. We find the power model $N = 7552.85 \times r^{-0.85}$.

    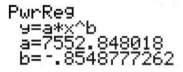

    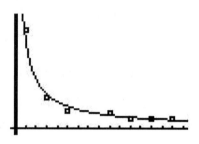

    (b) The plot on the right above shows the data together with the power model.

    (c) We use $r = 6$ in the formula from part a. We find that $N(6) = 1647$, so the model estimates the population of Phoenix to be 1647 thousand. (We have rounded the answer, in thousands, to the nearest whole number.)

3. **Stopping distance**:

    (a) The regression parameters for the power model are calculated in the figure on the left below. We find the power model $D = 0.89 \times S^{1.42}$.

    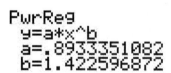

    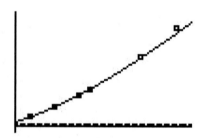

    (b) We use the homogeneity property of power functions. If the speed $S$ is multiplied by 2, the stopping distance $D$ is multiplied by

    $$2^{\text{Function power}} = 2^{1.42} = 2.68.$$

    (c) The plot on the right above shows the data together with the power model.

5. **Hydroplaning**:

(a) The regression parameters for the power model are calculated in the figure below. We find the power model $V = 10.41p^{0.5}$.

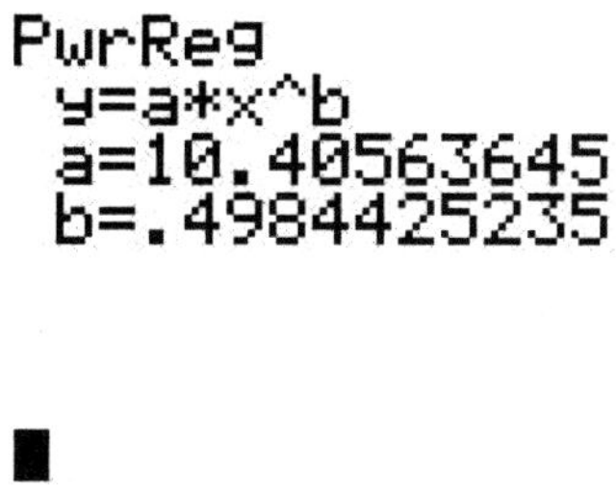

```
PwrReg
 y=a*x^b
 a=10.40563645
 b=.4984425235
```

■

(b) We want to compare $V(35)$ with $V(60)$. From the formula in Part (a) we get $V(35) = 10.41 \times 35^{0.5} = 61.59$, while $V(60) = 10.41 \times 60^{0.5} = 80.64$. Thus the critical speed for hydroplaning is about 62 mph for the car and about 81 mph for the bus. Since both are traveling at 65 mph, the car is in danger of hydroplaning, while the bus is not. Thus, if both apply their brakes, the car might hydroplane and hit the bus from the rear.

7. **Mass-luminosity relation**:

(a) The regression parameters for the power model are calculated in the figure below. We find the power model $L = M^{3.5}$, where we have rounded to one decimal place.

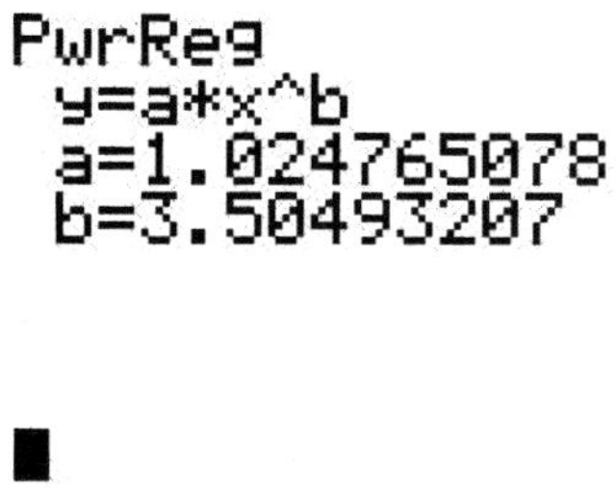

```
PwrReg
 y=a*x^b
 a=1.024765078
 b=3.50493207
```

■

(b) Since Kruger 60 has a mass of 0.11 solar mass, its relative luminosity is $L(0.11)$ in functional notation. The value of $L(0.11)$ is $0.11^{3.5} = 4.41 \times 10^{-4}$ or 0.000441, or about 0.0004.

(c) Since Wolf 359 has a relative luminosity of about 0.0001, for that star $L = 0.0001$ and so $M^{3.5} = 0.0001$. Solving for $M$, for example by $M = 0.0001^{1/3.5}$, shows that $M = 0.072$, or about 0.07, of a solar mass.

(d) If one star is 3 times as massive an another, then $M$ changes by a factor of 3 and so, by the homogeneity property with power 3.5, $L$ changes by a factor of $3^{3.5} = 46.77$. Thus the more massive star has a relative luminosity 46.77, or about 47, times that of the less massive star.

9. **Speed in flight versus length:**

(a) Table 5.3 indicates that larger animals fly faster, since as $L$ increases so does $F$.

(b) The regression parameters for the power model are calculated in the figure on the left below. We find the power model $F = 20.61L^{0.34}$.

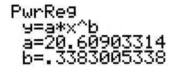

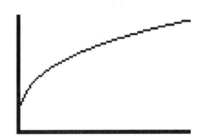

(c) The graph of $F = 20.61L^{0.34}$ is shown above on the right. The horizontal span is 0 to 300, and the vertical span is 0 to 150.

(d) The graph is concave down and increasing. This means that the flying speed $F$ increases as the length $L$ increases, but at a decreasing rate.

(e) If $L$ is increased by a factor of 10, then $F$ is increased by a factor of $10^{0.34} = 2.19$. Thus the longer bird should fly 2.19 times as fast.

11. **Metabolism:**

(a) The regression parameters for the power model are calculated in the figure below. We find the power model $B = 39.94W^{0.75}$.

PwrReg
y=a*x^b
a=39.93735076
b=.7450302162

■

(b)  i. By Part (a), the basic energy needed for survival ($B$ in our notation) is proportional not to the weight $W$ but to $W^{0.75}$. Thus to see how an animal is eating relative to its energy requirements, we should divide its intake by $W^{0.75}$, not by $W$.

ii. The metabolic weight of a 2.76-pound animal is $2.76^{0.75} = 2.14$, and the metabolic weight of a 126.8-pound animal is $126.8^{0.75} = 37.79$.

iii. We use the metabolic weights found in Part ii above. For the rabbit the ratio is $\dfrac{0.18}{2.14} = 0.08$, and for the sheep it is $\dfrac{2.8}{37.79} = 0.07$. So the daily consumption levels of the two animals are about the same on this basis.

13. **Proportions of trees**:

(a) The regression parameters for the power model are calculated in the figure below. We find the power model $h = 35.91 d^{0.66}$.

```
PwrReg
y=a*x^b
a=35.91136989
b=.6608072672
```

(b) First we find what the model from Part (a) gives for these two trees. For the plains cottonwood, the model gives $h = 35.91 \times 2.9^{0.66} = 72.51$. For the willow, the model gives $h = 35.91 \times 6.2^{0.66} = 119.73$. Therefore, the actual height of 80 feet for the plains cottonwood is greater than the height of about 73 feet predicted by the model, but the actual height of 95 feet for the willow is less than the height of about 120 feet predicted by the model. We conclude that the plains cottonwood is taller for its diameter.

This answer can also be found by plotting the data points together with a graph of the model. The data point for the plains cottonwood is above the graph of the model, but the data point for the willow is below the graph of the model.

(c)  i. The powers of $d$ for the trees and the critical height function are about the same, since $\dfrac{2}{3} = 0.67$ to two decimal places. However, the coefficient is 35.91 for the trees, while the critical height function has a coefficient of 140, which is about four times as large.

ii. No tree from the table is taller than its critical buckling height. This can be determined by looking at a table of values for $140 d^{2/3}$.

15. **Self-thinning:**

   (a) As time went by, the density decreased, which means that there were fewer plants in the plot.

   (b) The regression parameters for the power model are calculated in the figure below. We find the power model $w = 4766.06p^{-1.48}$.

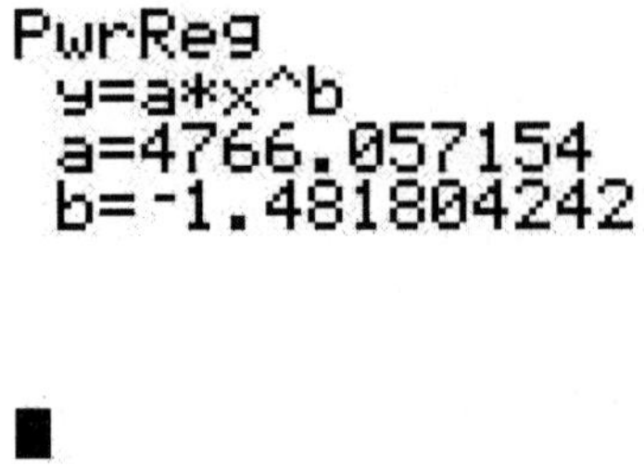

   ```
   PwrReg
   y=a*x^b
   a=4766.057154
   b=-1.481804242
   ```

   ■

   (c) We use the homogeneity property of power functions. The power here is $-1.48$, so if the density $p$ changes by a factor of $\dfrac{1}{2}$ then the weight $w$ changes by a factor of $\left(\dfrac{1}{2}\right)^{-1.48} = 2.79$.

   (d) The total yield is $y = w \times p = 4766.06p^{-1.48} \times p = 4766.06p^{-0.48}$. The power is negative, so $y$ decreases as $p$ increases. On the other hand, as time increases, $p$ decreases, as noted in Part (a). As $p$ decreases, $y$ increases; so, as time increases, the yield also increases.

   Multiplying the corresponding entries in the two columns of the table confirms this trend (though the line for January 30 represents an exception).

17. **Cost of transport:**

   (a) In general, as the weight increases, the cost of transport decreases. The only exception is the white rat.

   (b) The regression parameters for the power model are calculated in the figure below. We find the power model $C = 8.58W^{-0.40}$.

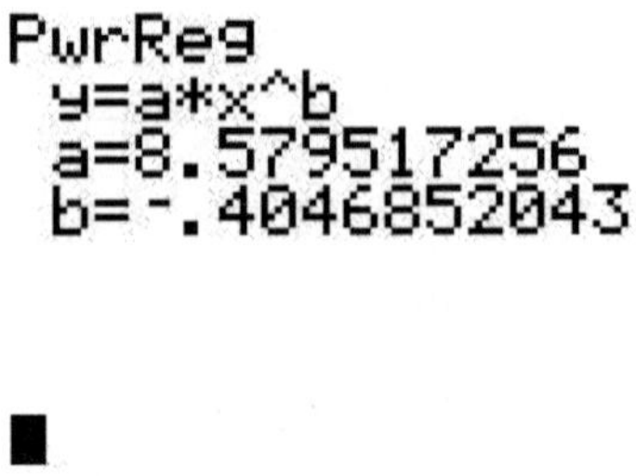

   ```
   PwrReg
   y=a*x^b
   a=8.579517256
   b=-.4046852043
   ```

   ■

(c) To compare the actual cost with the trend, we first put $W = 20{,}790$ grams into the model for $C$ from Part (b): $C = 8.58 \times 20{,}790^{-0.40} = 0.16$ milliliter of oxygen per gram per kilometer. This result is much lower than the actual cost of transport, which is $C = 0.43$ milliliter of oxygen per gram per kilometer. Thus, the penguin has a higher cost of transport than would be predicted given its weight. This does confirm the stereotype of penguins as awkward waddlers.

19. **Exponential growth rate and generation time**:

(a) The regression parameters for the power model are calculated in the figure on the left below. We find the power model $r = 0.8T^{-1.0}$, where we have rounded the parameters to one decimal place. The graph is shown on the right below using a horizontal span from 0 to 10 and a vertical span from 0 to 1.

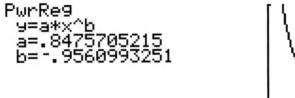

(b) From Example 5.9, the generation time in years of an organism is modeled by $1.1L^{0.8}$, if $L$ is the length in feet. In this exercise we measure generation time $T$ in days, so we multiply the model by 365:

$$T = 365 \times 1.1L^{0.8} = 401.5L^{0.8}.$$

This relationship holds over a wide range of organisms. From Part (a) we know that, for our group of lower organisms, $r = 0.8T^{-1}$. Since $T^{-1} = \dfrac{1}{T}$ , this can be written as $r = \dfrac{0.8}{T}$. We substitute into this formula for $r$ the model for $T$ in terms of $L$ to get

$$r = \frac{0.8}{T} = \frac{0.8}{401.5L^{0.8}}.$$

Since $\dfrac{0.8}{401.5}$ is about 0.002, we get

$$r = \frac{0.002}{L^{0.8}}.$$

This is the desired formula for $r$ in terms of $L$.

21. **Generation time as a function of length**: Below is a table that shows the natural logarithm of the data for $L$ and $T$. For example, $\ln 0.023 = -3.77$ (rounded to two decimal places).

| $L$ | 0.023 | 0.295 | 0.673 | 2.23 | 5.91 | 11.5 | 72.2 | 262 |
|-----|-------|-------|-------|------|------|------|------|-----|
| $T$ | 0.055 | 0.192 | 1 | 2.8 | 4 | 12.3 | 40 | 60 |
| $\ln L$ | $-3.77$ | $-1.22$ | $-0.40$ | 0.80 | 1.78 | 2.44 | 4.28 | 5.57 |
| $\ln T$ | $-2.90$ | $-1.65$ | 0 | 1.03 | 1.39 | 2.51 | 3.69 | 4.09 |

The figure below shows a plot of $\ln T$ (the fourth row of the able) versus $\ln L$ (the third row). The points nearly fall on a straight line, so a power model for $T$ as a function of $L$ may be appropriate.

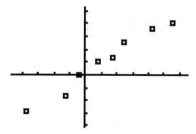

## Skill Building Exercises

S-1. **Broad jump record**: The school record increases by 2 inches over each 5-year period, and that represents a constant rate of change. This suggests a linear model.

S-3. **Further reflection regarding the rumor**: The rumor is spreading, but there is an upper bound to the number of people who will hear it, and this suggests a logistic model.

S-5. **An easy power formula**: When we inspect the data, the pattern emerges that the $y$ values are the cube of the $x$ values, so $y = x^3$.

S-7. **Modeling power data**: Power regression gives the formula $f = 3.60x^{1.30}$. A plot of the power model, together with the data, is shown below.

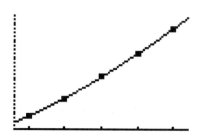

S-9. **Modeling almost power data**: Power regression gives the formula $f = 2.23x^{4.54}$. A plot of the power model, together with the data, is shown below.

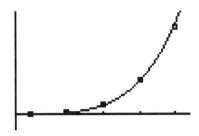

S-11. **Modeling almost power data**: Power regression gives the formula $f = 2.05x^{-0.90}$. A plot of the power model, together with the data, is shown below.

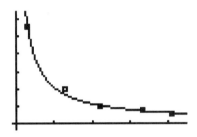

S-13. **Modeling almost power data**: Power regression gives the formula $f = 2.29x^{-1.59}$.

S-15. **Linear or power model**?

(a) Linear regression gives the model $y = 133.68x - 265.34$. The plot on the left below shows the data together with the linear model.

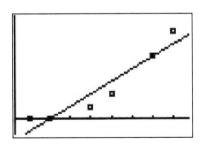 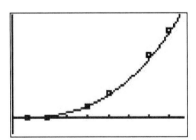

(b) Power regression gives the model $y = 2.65x^{2.81}$. The plot on the right above shows the data together with the power model.

(c) The power model is a better fit.

S-17. **Linear or power model?**

    (a) Linear regression gives the model $y = 0.01x + 1.89$. The plot on the left below shows the data together with the linear model.

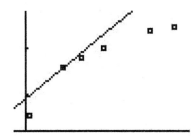 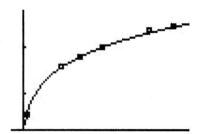

    (b) Power regression gives the model $y = 1.01x^{0.22}$. The plot on the right above shows the data together with the power model.

    (c) The power model is a better fit.

## 5.4 COMBINING AND DECOMPOSING FUNCTIONS

1. **Fixed and variable costs**: The fixed costs are present even if no items are produced (in which case the variable cost contributes nothing to the total cost). Thus, the fixed costs equal $C(0)$, which is 1400 dollars. The variable cost is the cost of producing a single item, so that is the slope of this linear function, namely 30 dollars per item produced. Thus, the fixed costs are $1400 per month, and the variable cost is $30 per item produced.

3. **River flow:**

    (a) We substitute the formula $C = wd$ into the formula $F = sC$ to find $F = s(wd)$. Therefore, a formula for $F$ in terms of $w$, $d$, and $s$ is $F = swd$.

    (b) We are given $w = 23$, $d = 2$, and $s = 24$, so by the formula from Part (a) we have $F = 24 \times 23 \times 2 = 1104$. Therefore, the flow rate for this river is 1104 cubic feet per minute.

5. **A skydiver:**

    (a) We want to find the initial value of $D$. Here $D = T - v$ and $T = 176$, so the initial value of $D$ is $176 -$ Initial value of $v$. Now $v$ starts at 0, so the initial value of $v$ is 0; thus the initial value of $D$ is $176 - 0 = 176$ feet per second.

(b) Let $t$ be time in seconds since the jump. Now $D$ is an exponential function of time, so it can be written as $D = Pa^t$. Here $P$ is the initial value of $D$, so from Part (a), $P = 176$. Thus $D = 176a^t$. On the other hand, $v(2) = 54.75$, so $D(2) = 176 - 54.75 = 121.25$, and thus $176a^2 = 121.25$. Solving for $a$ yields $a = \sqrt{\dfrac{121.25}{176}} = 0.83$, so an exponential formula for $D$ is $D = 176 \times 0.83^t$.

(c) To find a formula for $v$, note that we have $D = T - v$. Thus $D + v = T$ and so $v = T - D$. On the other hand, $T = 176$ and $D = 176 \times 0.83^t$, so a formula for $v$ is $v = 176 - 176 \times 0.83^t$.

(d) The velocity 4 seconds into the fall is expressed as $v(4)$ in functional notation. Its value is $v(4) = 176 - 176 \times 0.83^4 = 92.47$ feet per second.

7. **Immigration**: We have a population $N$ given by a formula

$$N = \frac{v}{r}(a^t - 1),$$

where $v$ is a positive number, $r$ is a number which is negative, and $a$ is a positive number less than 1. Now $a^t$ represents exponential decay (since $a$ is less than 1), so $a^t$ has a limiting value of 0. Thus the limiting value of the population $N$ is $\dfrac{v}{r}(0 - 1) = -\dfrac{v}{r}$.

9. **Biomass of haddock**:

(a) Now $D$ is the difference between the maximum length, which is 21 inches, and the length $L$, so $D(t) = 21 - L(t)$ for all $t$. The initial value of $L$ is 4 inches, so $L(0) = 4$, and therefore the initial value of $D$ is $D(0) = 21 - L(0) = 21 - 4 = 17$ inches.

(b) To find a formula for $L$, we start with $D = 21 - L$. This says that $D + L = 21$ and therefore $L = 21 - D$. To obtain a formula for $L$ in terms of $t$ we need a formula for $D$ in terms of $t$.

We know from Part (a) that $D$ is an exponential function with initial value 17, so $D = 17a^t$ for some $a$. Since a 6-year-old haddock is about 15.8 inches long, when $t = 6$, then $D = 21 - L = 21 - 15.8 = 5.2$. Using the formula for $D$, we have $5.2 = 17a^6$. Solving for $a$ gives $a = 0.82$. (For example, since $\dfrac{5.2}{17} = a^6$, $a = \left(\dfrac{5.2}{17}\right)^{1/6} = 0.82$). Thus $D = 17 \times 0.82^t$. Since $L = 21 - D$, then using the formula for $D$ above, we find that $L = 21 - 17 \times 0.82^t$.

(c) Weight $W$ is written as function of $L$ by the formula $W = 0.000293L^3$, and $L$ is written as a function of $t$ by the formula in Part (b) above, so $W$ can be written as a function of $t$ by function composition:

$$\begin{aligned} W &= 0.000293L^3 \\ W &= 0.000293(21 - 17 \times 0.82^t)^3. \end{aligned}$$

(d) The total biomass $B$ is the product of the number of fish $N$ and the weight per fish $W$. Each of $N$ and $W$ can be expressed as functions of the age $t$, so $B$ can be also:

$$B = N \times W$$
$$B = \left(1000e^{-0.2t}\right) \times \left(0.000293(21 - 17 \times 0.82^t)^3\right).$$

(e) The maximum value of the biomass $B$ can be determined by graphing $B$ as a function $t$ using the formula from Part (d). We are to consider ages up to 10 years, so the horizontal span should be from 0 to 10. Looking at a table of values (see the figure on the left below), we see that a vertical span of 0 to 400 should nicely display the graph. The figure on the right below shows the graph with the maximum calculated.

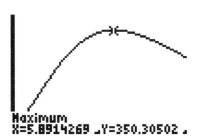

The maximum occurs at $t = 5.89$ with $B = 350.31$. So the biomass is at a maximum when the cohort is $5.89$ years old, or about $5.9$ years old.

11. **Waiting at a stop sign**: In this exercise $T = 5$, so the formula for the average delay $D$, in seconds, at a stop sign to enter a highway is given by

$$D = \frac{e^{5q} - 1 - 5q}{q},$$

where $q$ is the flow rate, in cars per second, of traffic on the highway.

(a) If the flow rate is 500 cars per hour, then $q = \dfrac{500}{60 \times 60} = 0.14$ car per second, and so the delay is

$$D = \frac{e^{5 \times 0.14} - 1 - 5 \times 0.14}{0.14} = 2.24 \text{ seconds.}$$

(b) The service rate $s$, in cars per second, is a function of $D$ given by the formula $s = D^{-1}$. To represent $s$ as a function of the flow rate $q$, we substitute the formula for $D$ in terms of $q$ into the formula for $s$ in terms of $D$:

$$s = D^{-1}$$
$$s = \left(\frac{e^{5q} - 1 - 5q}{q}\right)^{-1}$$
$$s = \frac{q}{e^{5q} - 1 - 5q}.$$

(c) If the service rate is 5 cars per minute, then $s = \dfrac{5}{60} = 0.083$ car per second. To solve the equation

$$s = \frac{q}{e^{5q} - 1 - 5q}$$
$$0.083 = \frac{q}{e^{5q} - 1 - 5q}$$

we can use the crossing-graphs method as shown below with a horizontal span of $q$ from 0 to 1 cars per second and a vertical span of $s$ from 0 to $0.15$ car per second.

This shows that when $s = 0.083$, $q = 0.42$ car per second, or 25.2 cars per minute.

13. **Probability of extinction**:

(a) We are given that the probability of extinction by time $t$ is

$$P(t) = \frac{d(e^{rt} - 1)}{be^{rt} - d}$$

and that $d = 0.24$, $b = 0.72$, and so $r = b - d = 0.48$. Substituting these values, we have

$$P(t) = \frac{0.24(e^{0.48t} - 1)}{0.72e^{0.48t} - 0.24}.$$

Scanning a table of values, for example the table below, we see quickly that the limiting value of $P$ is 0.33, so there is a 0.33 or 33% probability that the population will eventually become extinct.

| X | Y1 |
|---|---|
| 0 | 0 |
| 10 | .3315 |
| 20 | .33332 |
| 30 | .33333 |
| 40 | .33333 |
| 50 | .33333 |
| 60 | .33333 |

X=0

(b) We have a formula for $Q$ expressing $Q$ as a function of $P$, namely $Q = P^k$, and we already have a formula expressing $P$ in terms of $t$, $b$, $d$, and $r$ from Part (a). To find a formula for $Q$ in terms of these latter variables and $k$, we substitute the expression for $P$ from Part (a) into the formula for $Q$ in terms of $P$:

$$Q = P^k$$
$$Q = \left( \frac{d(e^{rt} - 1)}{be^{rt} - d} \right)^k.$$

(c) We are told that if $b > d$, then $Q$ can be expressed as

$$Q = \left( \frac{d(1 - a^t)}{b - da^t} \right)^k,$$

where $a < 1$. Since $a < 1$, the limiting value of $a^t$ is zero, so the limiting value of $Q$ is $\left( \dfrac{d}{b} \right)^k$. This is the probability that a population starting with $k$ individuals will eventually become extinct.

(d) We use the formula from Part (c). If $b$ is twice as large as $d$, then $\dfrac{d}{b} = \dfrac{1}{2} = 0.5$, and so the probability of eventual extinction is $\left( \dfrac{d}{b} \right)^k = 0.5^k$.

(e) As a function of $k$, the limiting value of $0.5^k$ is zero since $0.5 < 1$. In practical terms, this means that for a large initial population, the probability that the population will eventually become extinct is very small.

15. **Applications of the litter formula**:

(a) When $t = \dfrac{3}{k}$, then, replacing $t$ by $3/k$ in the formula for $A$, we find that

$$A = R - Re^{-kt} = R - Re^{-k \times 3/k} = R - Re^{-3} = R - 0.05R = 0.95R,$$

so the amount of litter present $A$ is 95% of $R$, the limiting value.

(b) By Part (a), we know that 95% of the limiting value is reached when $t = \dfrac{3}{k}$. In this case $k = 0.003$ per year, so 95% of the limiting value is reached in $t = \dfrac{3}{k} = \dfrac{3}{0.003} = 1000$ years.

(c) To reach 95% of the limiting value for forest litter in the Congo takes $t = \dfrac{3}{4} = 0.75$ of a year; in the Sierra Nevada Mountains it takes $t = \dfrac{3}{0.063} = 47.62$ years.

17. **Gray wolves in Michigan**:

(a) Since the gray wolves recolonized in the Upper Peninsula of Michigan it is reasonable to assume that this area is both suitable habitat and has far fewer wolves than it had historically. Under these conditions, we would expect exponential growth.

(b) We use regression to find an exponential model since the data is not exactly exponential. We find the model $W = 10.94 \times 1.39^t$, where $t$ is the number of years since 1990 and $W$ is the number of wolves.

(c) The figure below is a graph of the data, with $W$ as a function of $t$, together with the exponential model from Part (b).

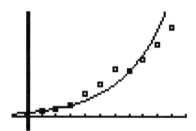

The data looks as if it follows two different exponential functions, neither of which is the exponential model we graphed. It would be better to use a piecewise-defined function.

(d) We find an exponential model for 1990 through 1996 by omitting later data. Using regression we get the exponential model $W = 7.95 \times 1.59^t$, where $t$ is time in years since 1990. This formula holds for $t$ from 0 through 6. We use the remaining data, from 1997 through 2000, to calculate the exponential model for $W$ as a function of $T$, where $T$ is time in years since 1997. Using regression we obtain $W = 112.22 \times 1.24^T$. This formula holds for $T$ from 0 through 3, which is the same as $t$ from 7 through 10. In fact, $T = t - 7$, and so the second exponential model can also be written as $W = 112.22 \times 1.24^{t-7}$.

(e) A single formula for $W$ as a piecewise-defined function, using the formulas from Part (d), is
$$W = \begin{cases} 7.95 \times 1.59^t & \text{for} \quad 0 \le t \le 6 \\ 112.22 \times 1.24^{t-7} & \text{for} \quad 7 \le t \le 10 \end{cases}$$

Graphing these two functions as a single piecewise-defined function shows an excellent fit with the data, as in the figure below.

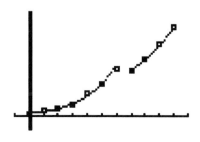

19. **Quarterly pine pulpwood prices**:

(a) The graph should start at the beginning of the year at $9 per ton, decrease steadily (so in a straight line) to $8 per ton at the end of the first quarter, remain at $8 per ton through the second and third quarters, and then rise steadily from there to $10 per ton by the end of the fourth quarter:

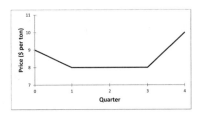

(b) From the beginning of the year to the end of the first quarter, the price $P$ decreases by $1 per ton from an initial value of $9 per ton. If $t = 0$ corresponds to the beginning of the year, $t = 1$ corresponds to the end of the first quarter, and so on, then the slope of $P$ as a linear function of $t$ over the interval from $t = 0$ to $t = 1$ is $-1$ dollar per ton per quarter. The value of $P$ when $t = 0$ is $9 per ton. Therefore, the formula is $P = -t + 9$ over the interval from $t = 0$ to $t = 1$.

(c) From the end of the first quarter to the end of the third quarter, the price is always $8 dollars per ton. Thus the formula for $P$ from $t = 1$ to $t = 3$ is $P = 8$.

(d) From the end of the third quarter to the end of the fourth quarter, the price $P$ increases by $2 per ton, from $8 per ton to $10 per ton. Therefore, the slope of $P$ as a function of $t$ over the interval from $t = 3$ to $t = 4$ is 2 dollars per ton per quarter. The value of $P$ when $t = 3$ is $8 per ton, and if we think of $P$ as a function of $t - 3$ this is the initial value. Therefore, the formula is $P = 2(t - 3) + 8$ over the interval from $t = 3$ to $t = 4$. This can also be written as $P = 2t + 2$.

(e) Using the formulas from Parts (b), (c), and (d), we can write a single formula for $P$ as a piecewise-defined function of $t$ as follows:

$$P = \begin{cases} -t + 9 & \text{for} \quad 0 \le t \le 1 \\ 8 & \text{for} \quad 1 \le t \le 3 \\ 2(t - 3) + 8 & \text{for} \quad 3 \le t \le 4 \end{cases}$$

## Skill Building Exercises

**S-1. Formulas for composed functions:** We have a formula for $w$ as a function of $s$ and a formula for $s$ as a function of $t$. To find a formula for $w$ as a function of $t$, we simply replace $s$ by its expression in terms of $t$. Since $s = t - 3$ and $w = s^2 + 1$, $w$ can also be written as $w = s^2 + 1 = (t - 3)^2 + 1$. This expresses $w$ as a function of $t$.

**S-3. Formulas for composed functions:** We have a formula for $w$ as a function of $s$ and a formula for $s$ as a function of $t$. To find a formula for $w$ as a function of $t$, we simply replace $s$ by its expression in terms of $t$. Since $s = e^t - 1$ and $w = \sqrt{2s + 3}$, $w$ can also be written as $w = \sqrt{2s + 3} = \sqrt{2(e^t - 1) + 3}$, which expresses $w$ as a function of $t$. This expression can be simplified to $w = \sqrt{2e^t + 1}$ since $2(e^t - 1) + 3 = 2e^t - 2 + 3 = 2e^t + 1$.

**S-5. Formulas for composed functions:** Here $f(x) = 3x + 1$ and $g(x) = \dfrac{2}{x}$, so $f(g(x)) = f\left(\dfrac{2}{x}\right) = 3\dfrac{2}{x} + 1 = \dfrac{6}{x} + 1$ and $g(f(x)) = g(3x + 1) = \dfrac{2}{3x + 1}$.

**S-7. Formulas for composed functions:** Here $f(x) = \dfrac{1}{x}$ and $g(x) = \dfrac{1}{x}$, so

$$f(g(x)) = f\left(\frac{1}{x}\right) = \frac{1}{\left(\frac{1}{x}\right)}.$$

This equals $x$, so $f(g(x)) = x$. Since $g = f$, $g(f(x)) = x$ also.

**S-9. Limiting values:** To find the limiting value of $7 + a \times 0.6^t$, we note that $a \times 0.6^t$ is a function which represents exponential decay, so the limiting value of $a \times 0.6^t$ is 0. Thus the limiting value of $7 + a \times 0.6^t$ is 7.

**S-11. Adding functions:** Since the function $f$ is the sum of two temperatures, its formula may be written as the sum of the formulas for the two temperatures. Thus $f = (t^2 + 3) + \dfrac{t}{t^2 + 1}$.

**S-13. Decomposing functions:** Now $f(x) = \sqrt{x^2 + 3}$ can be written as the composition of $g(x) = \sqrt{x}$ and $h(x) = x^2 + 3$; here $f(x) = g(h(x))$.

**S-15. Decomposing functions:** The cost is $C = 30 + 17n$ where $n$ is the number of books bought. The initial fee is the part of the formula which doesn't depend on $n$, so it is the initial value of $C$, which is $C(0) = 30$ dollars. The cost of each book is \$17 since each additional book adds \$17 to $C$.

**S-17. Combining functions:** Since $f(x) = x^2 - 1$ and $g(x) = 1 - x$, $f(g(x)) + g(f(x)) = f(1 - x) + g(x^2 - 1) = ((1 - x)^2 - 1) + (1 - (x^2 - 1))$. This can be simplified as $(1 - x)^2 - (x^2 - 1) = 1 - 2x + x^2 - x^2 + 1 = 2 - 2x$.

## 5.5  QUADRATIC FUNCTIONS

1. **Sales growth**:

   (a) To make a graph of $G$ versus $s$ we use a horizontal span of 0 to 4 since the model is valid up to a sales level of 4 thousand dollars. Using a table of values, we determine a vertical span of 0 to 1.3, and show the graph below.

   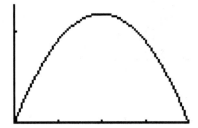

   (b) When the sales level is \$2260, $s = 2.26$ since $s$ is the sales level in thousands. The rate of growth in represented in functional notation by $G(2.26)$. Using the formula to evaluate this, we find that the growth is $G(2.26) = 1.2 \times 2.26 - 0.3(2.26)^2 = 1.18$ thousand dollars per year. This can also be found from a table of values or the graph.

   (c) The rate of growth $G$ is maximized at the maximum point of the graph. This is shown on the graph below:

   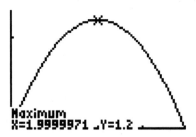

   The maximum growth rate is achieved when $s = 2.00$, that is, at a sales level of 2 thousand dollars.

3. **Cox's formula**:

(a) To graph the quadratic function $4v^2 + 5v - 2$ using a horizontal span from 0 to 10, a table of values suggests a vertical span from 0 to 400. The table of values is shown below on the left while the graph is shown on the right.

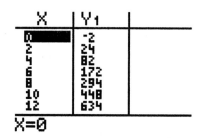
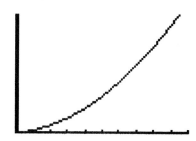

(b) If the velocity $v$ is 0.25 foot per second, then the value of $4v^2 + 5v - 2$ is $-0.5$. This would mean that the right side of Cox's formula is negative, whereas the left side of Cox's formula is $\dfrac{Hd}{L}$ and $H$, $d$, and $L$ are all positive numbers. Thus it is not possible for $v$ to be 0.25 foot per second.

(c) If $d = 4$, $L = 1000$, and $H = 50$, then Cox's formula implies that

$$\frac{50 \times 4}{1000} = \frac{4v^2 + 5v - 2}{1200}.$$

Evaluating the left-hand side and multiplying by 1200, we see that $4v^2 + 5v - 2 = 240$. Solving for $v$ shows that $v = 7.18$ feet per second.

(d) To find the largest pipe diameter $d$ for which $v$ does not exceed 10 feet per second, we calculate $d$ when $v = 10$, $L = 1000$, and $H = 50$:

$$\frac{50d}{1000} = \frac{4 \times 10^2 + 5 \times 10 - 2}{1200}.$$

Evaluating, we see that $d = 7.47$ inches is the largest such pipe diameter.

5. **Surveying vertical curves**:

(a) In this case, $g_1 = 1.35$, $g_2 = -1.75$, $L = 5$, and $E = 1040.63$, so the formula for the elevations of the vertical curve is

$$
\begin{aligned}
y &= \frac{-1.75 - 1.35}{2 \times 5}x^2 + 1.35x + 1040.63 - \frac{1.35 \times 5}{2} \\
&= -0.31x^2 + 1.35x + 1037.255.
\end{aligned}
$$

(b) The stations correspond to $x = 0$ through $x = 5$. The six stations are given in the table below, where we have used the formula from Part (a):

| Station number | 0 | 1 | 2 | 3 | 4 | 5 |
|---|---|---|---|---|---|---|
| Elevation | 1037.26 | 1038.30 | 1038.72 | 1038.52 | 1037.70 | 1036.26 |

Here the elevations are in feet.

(c) The highest point of the road on the vertical curve is at the maximum point on the graph of $y$. From the graph (see the figure below – we used a window with a horizontal span from 0 to 5 and a vertical span from 1035 to 1040) we find the maximum at station $x = 2.1774$, that is 217.74 feet along the vertical curve with an elevation of $E = 1038.72$ feet.

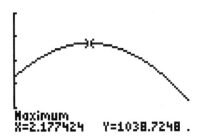

Maximum
X=2.177424    Y=1038.7248 .

7. **Traffic accidents**:

(a) Using quadratic regression as shown in the figure below, we see that the quadratic model is $R = 7.79s^2 - 514.36s + 8733.57$.

QuadReg
y=ax²+bx+c
a=7.785714286
b=-514.3571429
c=8733.571429
R²=.9879330701
∎

(b) Using the model from Part (a), we find that $R(50) = 7.79 \times 50^2 - 514.36 \times 50 + 8733.57 = 2490.57$, which means that commercial vehicles driving at night on urban streets at 50 miles per hour have traffic accidents at a rate of about 2491 per 100,000,000 vehicle-miles.

(c) The speed at which traffic accidents are minimized corresponds to the minimum point on the graph. This is calculated in the figure below:

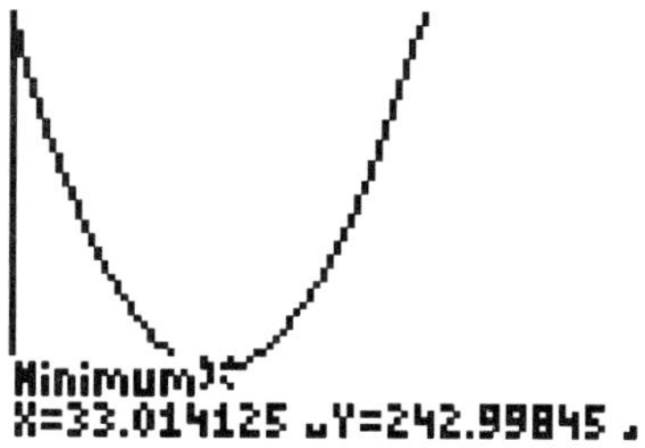

Minimum
X=33.014125 ˌY=242.99845 .

This shows that the minimum is when $s = 33.01$ or about 33 miles per hour.

9. **Women employed outside the home**:

   (a) Entering the data, and using $t$ as the number of years since 1942, we see the data shown in the figure below on the left and the calculation of the quadratic regression coefficients shown on the right.

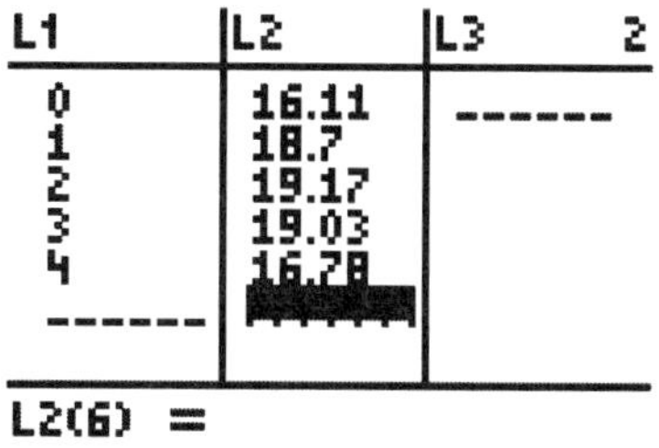

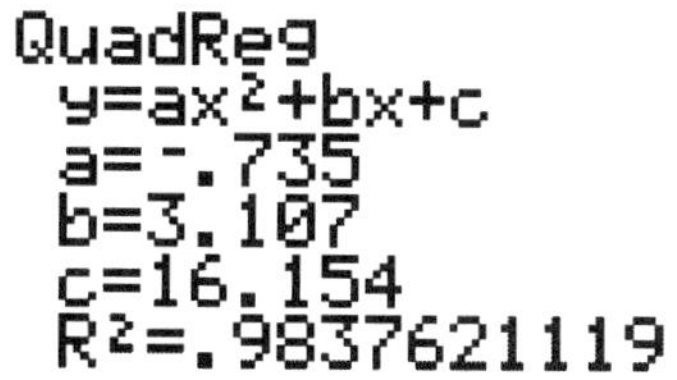

   Using $N$ as the number of women employed outside the home, in millions, the quadratic model is $N = -0.735t^2 + 3.107t + 16.154$.

   (b) In functional notation, the number of women working outside the home in 1947 is $N(5)$. The quadratic model from Part (a) gives that value as $N(5) = 13.31$ million.

   (c) The table below summarizes the quadratic prediction versus the actual numbers for 1947 and 1948:

| Year | $t$ | Quadratic prediction | Actual value |
|------|-----|---------------------|--------------|
| 1947 | 5   | 13.31               | 16.90        |
| 1948 | 6   | 8.34                | 17.58        |

   Here the figures are in millions. Clearly the predictions of the quadratic model are far from accurate, so a quadratic model is probably not appropriate for this time period. Another way to see this is to observe that the original data table shows a maximum in 1944, and the additional information in Part (c) shows that after 1946 the number started to increase. A quadratic model can't have both a maximum and a minimum.

11. **A falling rock:**

(a) We want to find $t$ so that $16t^2 + 3t = 400$. To solve this using the quadratic formula, we first subtract 400 from each side to get $16t^2 + 3t - 400 = 0$. This is of the form $at^2 + bt + c = 0$ with $a = 16$, $b = 3$, and $c = -400$. Plugging these values into the quadratic formula, we get:

$$t = \frac{-b + \sqrt{b^2 - 4ac}}{2a} = \frac{-3 + \sqrt{3^2 - 4 \times 16 \times (-400)}}{2 \times 16} = 4.91$$

$$t = \frac{-b - \sqrt{b^2 - 4ac}}{2a} = \frac{-3 - \sqrt{3^2 - 4 \times 16 \times (-400)}}{2 \times 16} = -5.09.$$

The second of these times has no physical meaning, but from the first we see that it takes the rock 4.91 seconds to fall 400 feet.

(b) We use the crossing graphs method to find $t$ so that $D(t) = 400$. In the figure below we have graphed $D = 16t^2 + 3t$ and the constant 400 using a horizontal span of 0 to 6 and a vertical span of 0 to 600. We see that the intersection occurs at $t = 4.91$. Thus, it takes the rock 4.91 seconds to fall 400 feet.

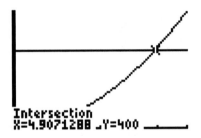

```
Intersection
X=4.9071288  Y=400
```

13. **Using the quadratic formula:** We want to find $x$ so that $-20\left(\dfrac{x}{250}\right)^2 + 0.5x = 0$. This equation is of the form $ax^2 + bx + c = 0$ with $a = -20/250^2$, $b = 0.5$, and $c = 0$. Plugging these values into the quadratic formula, we get:

$$x = \frac{-b + \sqrt{b^2 - 4ac}}{2a} = \frac{-0.5 + \sqrt{0.5^2 - 4 \times (-20/250^2) \times 0}}{2 \times (-20/250^2)} = 0$$

$$x = \frac{-b - \sqrt{b^2 - 4ac}}{2a} = \frac{-0.5 - \sqrt{0.5^2 - 4 \times (-20/250^2) \times 0}}{2 \times (-20/250^2)} = 1562.5.$$

The first of these distances is not relevant, but from the second we see that the cannonball travels 1562.5 feet.

15. **Profit:**

(a) Total cost $C$ is simply the fixed cost of \$2000 plus the variable cost of \$30 per widget times the number of widgets. Because $N$ is the number of widgets, $C = 30N + 2000$.

(b) The table of price $p$ as a function of number of widgets $N$ shows a constant decrease in price of 0.50 per 50 additional widgets sold, so $p$ is a linear function of $N$. The slope is $\dfrac{-0.50}{50} = -0.01$ dollar per widget, so $p = -0.01N + b$ for some initial value $b$. Now when $N = 200$, then $p = 41$, according to the table, so $41 = -0.01 \times 200 + b$, and thus $b = 43$. Thus the formula is $p = -0.01N + 43$.

(c) The total revenue $R$ is Price $\times$ Quantity sold, so $R = p \times N = (-0.01N + 43)N$ using the formula for $p$ from Part (b).

(d) The profit $P$ is Revenue $-$ Total cost, so $P = R - C = (-0.01N + 43)N - (30N + 2000)$. This is a quadratic function since, if multiplied out, there would be a term $-0.01N^2$.

(e) The manufacturer breaks even when $P = 0$. Solving the equation $(-0.01N + 43)N - (30N + 2000) = 0$, we find two solutions: $N$ is 178.30 or 1121.70 widgets.

## Skill Building Exercises

S-1. **Using the quadratic formula**: The quadratic formula gives the solutions of $ax^2 + bx + c = 0$ as
$$x = \frac{-b \pm \sqrt{b^2 - 4ac}}{2a}.$$
To solve $-2x^2 + 2x + 5 = 0$, we use $a = -2$, $b = 2$, and $c = 5$, so
$$x = \frac{-2 \pm \sqrt{2^2 - 4 \times (-2) \times 5}}{2 \times (-2)} = \frac{-2 \pm \sqrt{44}}{-4}$$
are the solutions. This expression can be simplified to $x = \dfrac{1 \pm \sqrt{11}}{2} = \dfrac{1}{2} \pm \dfrac{\sqrt{11}}{2}$ and gives the values $x = -1.16$ and $x = 2.16$.

S-3. **Using the quadratic formula**: The quadratic formula gives the solutions of $ax^2 + bx + c = 0$ as
$$x = \frac{-b \pm \sqrt{b^2 - 4ac}}{2a}.$$
To solve $x^2 + x - 3 = 0$, we use $a = 1$, $b = 1$, and $c = -3$, so
$$x = \frac{-1 \pm \sqrt{1^2 - 4 \times 1 \times (-3)}}{2 \times 1} = \frac{-1 \pm \sqrt{13}}{2}$$
are the solutions. This expression can be written as $x = -\dfrac{1}{2} \pm \dfrac{\sqrt{13}}{2}$ and gives the values $x = -2.30$ and $x = 1.30$.

S-5. **Using the quadratic formula**: The quadratic formula gives the solutions of $ax^2 + bx + c = 0$ as

$$x = \frac{-b \pm \sqrt{b^2 - 4ac}}{2a}.$$

To solve $2x^2 + 3x = 0$, we use $a = 2$, $b = 3$, and $c = 0$, so

$$x = \frac{-3 \pm \sqrt{3^2 - 4 \times 2 \times 0}}{2 \times 2} = \frac{-3 \pm \sqrt{9}}{4}$$

are the solutions. This expression can be simplified to $x = \dfrac{-3 \pm 3}{4}$ and gives the exact values $x = -3/2 = -1.5$ and $x = 0$.

S-7. **The single-graph method**: After consulting a table of values, we use a horizontal span of $-3$ to 3 and a vertical span of $-20$ to 6. Using the single-graph method, we find the solutions $x = -1.16$ and $x = 2.16$.

S-9. **The single-graph method**: After consulting a table of values, we use a horizontal span of $-3$ to 2 and a vertical span of $-4$ to 3. Using the single-graph method, we find the solutions $x = -2.30$ and $x = 1.30$.

S-11. **The single-graph method**: After consulting a table of values, we use a horizontal span of $-2$ to 0 and a vertical span of $-2$ to 2. Using the single-graph method, we find the exact solutions $x = -1.5$ and $x = 0$.

S-13. **Using quadratic regression**: The calculator gives the quadratic model $y = 0.85x^2 + 0.90x + 0.21$. The plot is below.

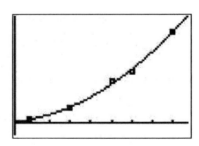

S-15. **Using quadratic regression**: The calculator gives the quadratic model $y = 2.47x^2 + 2.93x + 5.77$. The plot is below.

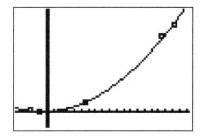

S-17. **Using quadratic regression**: The calculator gives the quadratic model $y = 2.70x^2 - 12.50x + 17.36$. The plot is below.

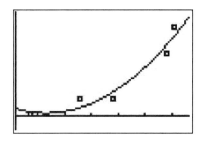

S-19. **Using quadratic regression**: The calculator gives the quadratic model $y = -3.88x^2 + 24.24x - 14.16$. The plot is below.

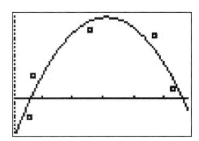

## 5.6  HIGHER-DEGREE POLYNOMIALS AND RATIONAL FUNCTIONS

1. **A dubious model of oil prices**:

   (a) Calculating the cubic regression coefficients, we find a cubic model for the data:
   $P = 0.0048t^3 + 0.1564t^2 - 1.5214t + 18.9911$. Here we rounded the parameters to
   four decimal places, as specified in the exercise.

   (b) The plot is below.

   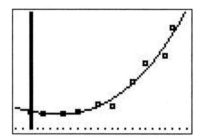

   (c) We use a horizontal span of 0 to 30 years. After consulting a table of values, we
   use a vertical span of 0 to 250 dollars per barrel. The graph is below.

   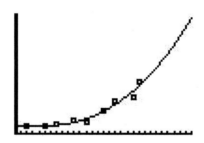

   (d) Because 2015 is 25 years after 1990, we take $t = 25$. Using the cubic model from
   part a, we calculate $P(25)$ to be 153.71. Thus, the estimate given by the model for
   oil prices in 2015 is $153.71 per barrel.

   (e) Care must always be taken when extrapolating beyond the limits of the data.

3. **Traffic accidents**:

   (a) Calculating the cubic regression coefficients, we find a cubic model for the data:
   $C = 0.000422s^3 - 0.0294s^2 + 0.562s - 1.65$. Here we rounded the parameters to
   two decimal places after the first non-zero digit, as specified in the exercise.

   (b) Using the cubic model from Part (a), we can calculate $C(42)$ as 1.36. This means
   that if commercial vehicles travel at 42 miles per hour at night on urban streets,
   the cost due to traffic accidents will be 1.36 cents per vehicle-mile.

(c) Graphing the cubic model, we find the minimum from the graph. Using a horizontal span of $s$ from 20 to 50 miles per hour and a corresponding vertical span of $C$ from $-2$ to 4, the graph is:

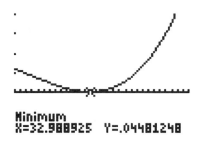

Minimum
X=32.988925  Y=.04481248

This shows the minimum to be when $s = 32.99$ or about 33 miles per hour.

5. **Population genetics**:

(a) The quadratic equation $\lambda^2 = \dfrac{1}{2}\lambda + \dfrac{1}{4}$ can be solved by any of a number of techniques. For example, using crossing graphs with a horizontal span of $\lambda$ from $-1$ to 1 and a vertical span from $-1$ to 1, we find the two solutions (see the figures below).

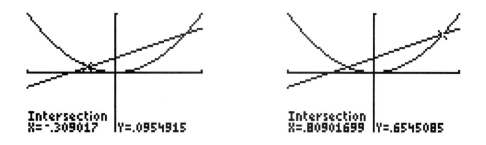

Intersection
X=-.309017  Y=.0954915

Intersection
X=.80901699  Y=.6545085

Here we see that the solutions are $\lambda = -0.31$ and $\lambda = 0.81$. Since $\lambda_1$ is the larger solution, $\lambda_1 = 0.81$.

(b) According to the problem, the proportion of the population which will be heterozygous (so not homozygous) after 5 generations is given by $\lambda_1^5 = 0.81^5 = 0.3487$ or 34.87%. Thus the homozygous proportion will be $100\% - 34.87\% = 65.13\%$.

(c) As in Part (b), the proportion of the population which will be heterozygous (so not homozygous) after 20 generations is given by $\lambda_1^{20} = 0.81^{20} = 0.0148$ or 1.48%. Thus the homozygous proportion will be $100\% - 1.48\% = 98.52\%$.

7. **Builder's Old Measurement**:

(a) In this exercise we are assuming that $L = 120$, so the formula for $T$ as a function of $W$ is
$$T = \frac{(120 - 0.6W)W^2}{188}.$$

We put $W = 25$ into this formula to obtain
$$T = \frac{(120 - 0.6 \times 25) \times 25^2}{188} = 349.07.$$

Therefore, for a ship of beam width 25 feet the Builder's Old Measurement gives a tonnage of 349.07.

(b) We want to solve the equation
$$\frac{(120 - 0.6W)W^2}{188} = 400.$$

This equation can be solved using the crossing-graphs method. The answer to Part (a) suggests that the solution here will be somewhat more than 25 feet. After consulting a table of values, we use a horizontal span of 0 to 30 feet and a vertical span of 0 to 500. In the figure below, we have graphed $T$ along with the constant function 400 (thick line). We see that they intersect when $W = 26.91$. Therefore, a beam width of 26.91 feet will give a tonnage of 400.

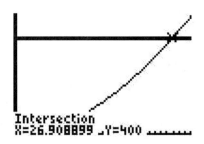

9. **An epidemic model**:

(a) The denominator in the formula for $A$ is 0 when $100 - q = 0$ or $q = 100$, and this is the value of $q$ that gives a pole.

(b) We saw in Part (a) that the pole occurs when $q = 100$, which corresponds to an immunization level at birth of 100%. The values of a rational function become larger and larger in size as we approach a pole. Therefore, when immunization levels at birth are near 100%, the average age of contraction is very large.

11. **Traffic signals**:

(a) The graph of

$$n = 1 + \frac{v}{30} + \frac{90}{v}$$

is shown below on the left, using a horizontal span of $v$ from 30 to 80 and a vertical span of $n$ from 4.4 to 5.1.

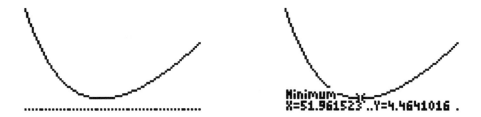

(b) In functional notation, the length of the yellow light when the approach speed is 45 feet per second is $n(45)$. The value of $n(45)$ is obtained using the formula given, a table of values, or the graph; we find that $n(45) = 4.5$ seconds.

(c) At $v$ tends towards 0, $n$ increases sharply. In practical terms, as the approach speed becomes very slow, the time needed to cross the intersection increases greatly. Therefore the yellow light needs to be much longer.

(d) Finding the minimum point on the graph, as shown in the figure above on the right, we see that the minimum occurs when $v = 51.96$ feet per second and $n = 4.46$ seconds. The minimum length of time for a yellow light is 4.46 seconds.

13. **Hill's law**:

(a) Using the given values for a fast twitch vertebrate muscle of $F_\ell = 300$, $a = 81$, and $b = 6.75$ in Hill's law, we obtain

$$S = \frac{6.75(300 - F)}{F + 81}.$$

(b) Graphing $S$ using a horizontal span of $F$ from 0 to 300 and a vertical of $S$ from 0 to 25, we obtain the figure below.

(c) As the force $F$ increases to 300, the muscle's contraction speed $S$ decreases towards 0.

(d) Here $S = 0$ when $F = 300$, so the muscle does not contract at all when the maximum force is applied to the muscle.

(e) The horizontal asymptote of $S$ is the limiting value. Using a table or graph, or formally calculating a limit, we see that the limiting value of $S$ is $-6.75$. Thus $S$ has a horizontal asymptote at $S = -6.75$. This makes no sense in terms of the muscle: as $F$ gets large, it becomes greater than the maximum force; moreover, a negative value of $S$ would not represent a muscle contraction.

(f) Here $S$ has a vertical asymptote at $F = -81$. A negative force is not the situation for which the equation applies.

15. **Inventory**:

(a) We use a horizontal span of 0 to 10 and a vertical span of 0 to 8500. The graph is shown below.

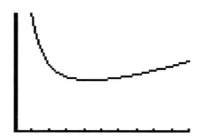

(b) From the graph we see that, near the pole, $E$ gets larger and larger as $Q$ gets smaller and smaller. This says that the yearly inventory expense gets larger and larger as the order size gets smaller and smaller. This makes sense: if the order sizes are very small then the dealer will have to place many orders, and so the total cost will be very high.

17. **Queues**:

(a) The rational function $L = \dfrac{a^2}{s(s-a)}$ has poles at $s = 0$ and $s - a$, that is $s = a$. Thus if the arrival rate $a$ is very close to the gap rate $s$, then the queue length $L$ becomes very long.

(b) We are given that $w = \dfrac{a}{s(s-a)}$ and that the gap length $s$ is 3, so $w = \dfrac{a}{3(3-a)}$. To determine $a$ such that $w = 2$, we need to solve $2 = \dfrac{a}{3(3-a)}$. Using any method, for example crossing graphs, we find that $a = 2.57$ cars per minute.

19. **Gliding pigeon**:

(a) The graph of $s = \dfrac{u^3}{2500} + \dfrac{25}{u}$ is shown below on the left using a horizontal span of $u$ from 0 to 15 and a vertical span of $s$ from 0 to 9.

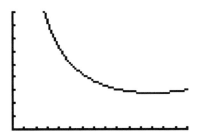 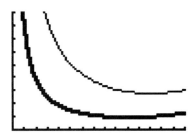

(b) Using the formula, a table of values, or a graph, we find that $s(10) = 2.9$ meters per second. In practical terms, a pigeon gliding with an airspeed of 10 meters per second will sink at a rate of 2.9 meters per second.

(c) The graph above on the right shows the performance diagrams for the falcon and the pigeon (the darker graph is that of the falcon). Since the falcon's graph lies below that of the pigeon, for the same airspeed, the falcon sinks much more slowly, so the falcon is the more efficient glider.

## Skill Building Exercises

S-1. **Cubic regression**: The calculator gives the cubic model $y = 0.073x^3 - 1.106x^2 + 4.920x - 2.946$. The plot is below.

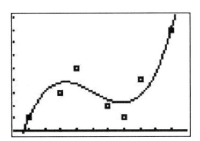

S-3. **Cubic regression**: The calculator gives the cubic model $y = -0.061x^3 + 0.589x^2 - 0.480x + 1.351$. The plot is below.

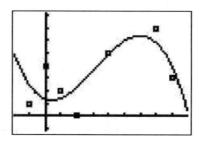

S-5. **Cubic regression**: The calculator gives the cubic model $y = 0.532x^3 + 4.684x^2 - 2.716x - 10.816$. The plot is below.

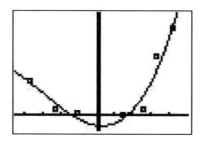

S-7. **Cubic regression**: The calculator gives the cubic model $y = -0.007x^3 + 0.247x^2 - 1.705x + 5.554$. The plot is below.

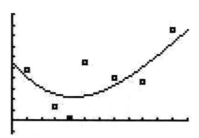

S-9. **Quartic regression**: The calculator gives the quartic model $y = -0.009x^4 + 0.224x^3 - 1.862x^2 + 6.718x - 3.697$. The plot is below.

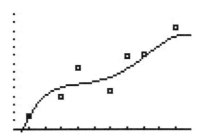

S-11. **Quartic regression**: The calculator gives the quartic model $y = -0.900x^4 + 4.988x^3 - 3.035x^2 - 5.636x + 2.515$. The plot is below.

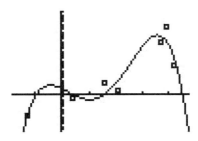

S-13. **Quartic regression**: The calculator gives the quartic model $y = -0.014x^4 + 0.403x^3 - 3.671x^2 + 12.437x - 8.399$. The plot is below.

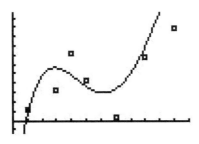

S-15. **Recognizing polynomials**: A polynomial is a function which can be written as a sum of power functions where each power is a nonnegative whole number. Clearly $f(x) = x^8 - 17x + 1$ is a polynomial of degree 8.

S-17. **Recognizing polynomials**: A polynomial is a function which can be written as a sum of power functions where each power is a nonnegative whole number. Clearly $f(x) = 9.7x - 53.1x^4$ is a polynomial of degree 4.

**S-19. Rational function?** The function $y = \dfrac{x - \sqrt{x}}{1 + x}$ is a ratio of $x - \sqrt{x}$ and $1 + x$, but $x - \sqrt{x}$ is not a polynomial function, so the ratio is not a rational function.

**S-21. Rational function?** The function $y = x + \dfrac{1}{x^2}$ can be written as $y = \dfrac{x^3 + 1}{x^2}$. This function is a ratio of $x^3 + 1$ and $x^2$, each of which is a polynomial. Thus, the ratio is a rational function.

**S-23. Finding poles:** The poles of $y = \dfrac{x + 1}{x^2 + 7x}$ are the values of $x$ for which the denominator $x^2 + 7x$ is zero but the numerator $x + 1$ is not zero. In this case, $x^2 + 7x = 0$ when $x = 0$ and $x = -7$, so these are the poles.

**S-25. Horizontal asymptotes:** The horizontal asymptotes of $y = \dfrac{6x^4 - 5}{2x^4 + x}$ are the limiting values of this expression. Using a table of values, for example, it can be determined that the limit is 3, so the horizontal asymptote is the line $y = 3$.

## Chapter 5 Review Exercises

1. **Logistic model:** The formula for a logistic model is $N = \dfrac{K}{1 + be^{-rt}}$. We are given that $N(0) = 100$ and $K = 2500$, so $b = \dfrac{K}{N(0)} - 1 = \dfrac{2500}{100} - 1 = 24$. Since $r = 0.02$, our logistic model is $N = \dfrac{2500}{1 + 24e^{-0.02t}}$.

2. **Logistic formula:**

   (a) The equation given in the exercise matches the form of the logistic equation, so we read the $r$ value from the formula: The $r$ value for this population is 1.4 per year.

   (b) The numerator is 25, so the carrying capacity for this population is 25 thousand.

   (c) The optimum yield level is half of the carrying capacity, so it is 12.5 thousand.

3. **Inflection point:** For logistic growth the inflection point corresponds to the maximum growth rate, and this occurs at half the carrying capacity. Thus the carrying capacity is $2 \times 320 = 640$.

4. **Homogeneity:** In this case $f$ is a power function with power $k = 3.2$. By the homogeneity property, if $x$ is increased by a factor of $t$, then $f$ is increased by a factor of $t^{3.2}$. In this case, $x$ is doubled, so $t = 2$. Therefore $f$ is increased by a factor of $2^{3.2}$, that is, by a factor of $9.19$.

5. **Power**: From 1 to 10, 1 is increased by a factor of $t = 10$. Thus $f$ is increased by a factor of $t^k = 10^k$. Since $f(10)$ is twice the size of $f(1)$, $2 = t^k = 10^k$. Solving for $k$ yields $k = 0.30$.

6. **Flow rate**: The radius $R$ is a power function with power 0.25, so if the flow rate $F$ changes by a factor of $t$, then $R$ changes by a factor of $t^{0.25}$.

   (a) If the flow rate is doubled, then $t = 2$, so the radius changes by a factor of $2^{0.25} = 1.19$.

   (b) If the radius is tripled, then $3 = t^{0.25}$. Solving for $t$ yields $t = 81$, so the flow rate is multiplied by 81.

7. **Falling object**:

   (a) For Earth the formula is $T = 0.25s^{0.5}$. We put in $s = 16$ and find $T = 0.25 \times 16^{0.5} = 1$. Thus it takes 1 second for an object to fall 16 feet on Earth.

   (b) We know that $T = 5$ when $s = 20$, so $5 = c \times 20^{0.5}$. Thus $c = \dfrac{5}{20^{0.5}} = 1.12$.

   (c) The time $T$ is a power function with power 0.5. By the homogeneity property, if $s$ is increased by a factor of $t$, then $T$ is increased by a factor of $t^{0.5}$. In this case, $s$ is multiplied by 4 in going from 10 feet to 40 feet, so $t = 4$. Therefore $T$ is increased by a factor of $4^{0.5}$, that is, by a factor of 2. Hence it takes an object $2 \times 1 = 2$ seconds to fall 40 feet. This can also be found by first solving for $c$ using the method of Part (b) and then plugging $s = 40$ into the resulting formula.

8. **Modeling almost power data**: Power regression gives the power model $f = 3.50x^{-1.20}$.

9. **Gas cost**:

   (a) Power regression gives the power model $D = 22.02g^{-1.00}$.

   (b) The distance $D$ is a power function with power $-1$. Now an increase of 50% in the price of gas corresponds to multiplying $g$ by 1.5 or 3/2. By the homogeneity property, if $g$ is multiplied by 1.5, then $D$ is multiplied by $1.5^{-1} = (3/2)^{-1} = 2/3$. Thus the distance is multiplied by 2/3, or reduced by 33.33%.

   (c) If a gallon of gas costs \$1 then the distance can you drive on \$1 worth of gas is the distance you can drive on 1 gallon of gas, and that is your gas mileage. Thus we want to find the value of $D$ when $g = 1$. Now if we put $g = 1$ into the formula from Part (a) we find $D = 22.02 \times 1^{-1} = 22.02$. Thus your gas mileage is 22.02, or about 22, miles per gallon. Another way to do this is to note that the gas mileage is the product of the cost $g$ with the distance $D$ (check the units!), and by the formula $D = 22.02g^{-1}$ from Part (a) we have $g \times D = g \times 22.02g^{-1} = 22.02$.

10. **Falling rocks**:

   (a) Power regression gives the power model $T = 0.20s^{0.50}$.

   (b) We put $s = 70$ into the formula from Part (a) and find $T = 0.2 \times 70^{0.5} = 1.67$. Thus it takes 1.67 seconds for a rock to fall 70 feet.

   (c) We want to find the value of $s$ for which $T = 2$, so (by the formula from Part (a)) we want to solve the equation $0.2s^{0.5} = 2$ for $s$. Using a table of values or a graph (or solving by hand), we find $s = 100$. Thus a rock falls 100 feet in 2 seconds.

   (d) The time $T$ is a power function with power 0.5. By the homogeneity property, if $s$ is multiplied by 2, then $T$ is multiplied by $2^{0.5} = 1.41$. Thus the time is multiplied by 1.41.

11. **Composing functions**: Because $x(t) = t - 1$ and $y(x) = 3x^2 + 5x$, we have $y(x(t)) = 3(t - 1)^2 + 5(t - 1)$. Thus the formula is $y = 3(t - 1)^2 + 5(t - 1)$.

12. **Decomposing functions**: The balance is $B = 120 \times 1.02^t$, and this is an exponential function. The initial value is 120, so the balance at the start of 2012 was $120. The growth factor is 1.02, so the yearly percentage growth rate is 2%, and this is the annual interest rate.

13. **Population growth**:

   (a) We show the graph below with a horizontal span of 0 to 200 and a vertical span of 0 to 0.015.

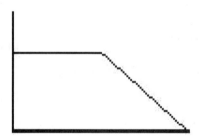

(b) Because the population level $N$ is between 100 and 200, we use the formula

$$R = 0.02 \left( 1 - \frac{N}{200} \right).$$

Putting in $N = 150$ gives

$$R = 0.02 \left( 1 - \frac{150}{200} \right) = 0.005.$$

Thus the per capita growth rate is 0.005 per year.

14. **Volume**:

(a) We put $1.5 + 0.1t$ in place of $V$ in the formula for $R$ and find

$$R = \left( \frac{3(1.5 + 0.1t)}{4\pi} \right)^{1/3}.$$

(b) We put $t = 2$ into the formula from Part (a) and find

$$R = \left( \frac{3(1.5 + 0.1 \times 2)}{4\pi} \right)^{1/3} = 0.74.$$

Thus the radius is 0.74 inch.

15. **Quadratic formula**: The quadratic formula gives the solutions of $ax^2 + bx + c = 0$ as

$$x = \frac{-b \pm \sqrt{b^2 - 4ac}}{2a}.$$

To solve $2x^2 - x - 1 = 0$, we use $a = 2$, $b = -1$, and $c = -1$. The solutions are

$$x = \frac{-(-1) \pm \sqrt{(-1)^2 - 4 \times 2 \times (-1)}}{2 \times 2} = \frac{1 \pm \sqrt{9}}{4} = \frac{1 \pm 3}{4}.$$

This expression gives the values $x = 1$ and $x = -1/2$.

16. **Quadratic regression**: To find a quadratic model for this data set using regression, we enter the data in the first two columns and use the calculator to find directly the quadratic regression coefficients. This gives the quadratic model $f(x) = -3x^2 + 5x - 2$.

17. **Chemical reaction**:

(a) Quadratic regression gives the model $R = 0.005x^2 - 0.75x + 25$.

(b) Using the model from Part (a), we find that $R(24) = 0.005 \times 24^2 - 0.75 \times 24 + 25 = 9.88$ moles per cubic meter per second. This means that the reaction rate at a concentration of 24 moles per cubic meter is 9.88 moles per cubic meter per second.

(c) We want to find the value of $x$ for which $R = 6$. We can do this using the crossing-graphs method or the quadratic formula. The result is $s = 32.28$ moles per cubic meter. Thus the reaction rate is 6 moles per cubic meter per second at a concentration of 32.28 moles per cubic meter.

18. **Cubic regression**: To use cubic regression, we enter the data in the first two columns and then use the calculator to determine the cubic regression coefficients. This gives the cubic model $y = 0.1x^3 - 0.2x^2 + 3$.

19. **Finding poles**: The poles of $\dfrac{2x - 5}{x^2 + 4x + 3}$ are the values of $x$ for which the denominator $x^2 + 4x + 3$ is zero but the numerator $2x - 5$ is not zero. In this case, $x^2 + 4x + 3 = 0$ when $x = -1$ and $x = -3$, so these are the poles.

20. **Expanding balloon**:

    (a) Calculating the cubic regression coefficients, we find a cubic model for the data:
    $$V = 0.03t^3 + 0.50t^2 + 2.51t + 4.19.$$

    (b) The volume after 5 seconds is expressed in functional notation as $V(5)$. Plugging $t = 5$ into the model from Part (a) gives the value 32.99 cubic inches.

21. **Travel time**:

    (a) The graph is shown below. We used a horizontal span of 0 to 70 and a vertical span of 0 to 50.

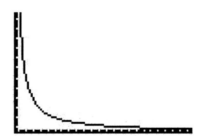

    (b) The value of $T(25)$ can be found by putting $s = 25$ into the formula given or by using a table of values (or the graph). The result is $T(25) = 4$ hours. This means that 4 hours is the time required to drive 100 miles if the average speed is 25 miles per hour.

    (c) At $s$ tends towards 0, $T$ increases sharply. In practical terms, as the average speed becomes very slow, the time needed to drive 100 miles increases greatly.

## A FURTHER LOOK: FITTING LOGISTIC DATA USING RATES OF CHANGE

1. **Using logistic regression**: Entering the data and using linear regression as shown in figure below, we find that the regression line is $G = -0.002N + 0.21$.

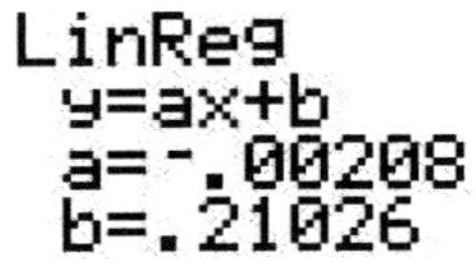

```
LinReg
 y=ax+b
 a=-.00208
 b=.21026
```

Since the vertical intercept equals $r$, we have $r = 0.21$ per year. Since the slope equals $-\dfrac{r}{K}$, we have $-0.002 = -\dfrac{r}{K} = -\dfrac{0.21}{K}$. Thus $-0.002K = -0.21$, and so $K = 0.21/0.002 = 105$.

2. **Using logistic regression**: Entering the data and using linear regression as shown in figure below, we find that the regression line is $G = -0.002N + 0.41$.

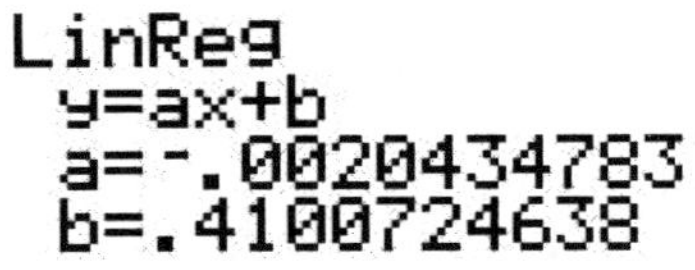

```
LinReg
 y=ax+b
 a=-.0020434783
 b=.4100724638
```

Since the vertical intercept equals $r$, we have $r = 0.41$ per year. Since the slope equals $-\dfrac{r}{K}$, we have $-0.002 = -\dfrac{r}{K} = -\dfrac{0.41}{K}$. Thus $-0.002K = -0.41$, and so $K = 0.41/0.002 = 205$.

3. **Negative growth rates**:

   (a) The regression line computed in Example 5.18 is $G = -0.012N + 0.24$. Using it to estimate $G$ for $N = 24$ and $N = 32$, we get $G(24) = -0.012 \times 24 + 0.24 = -0.05$ per day and $G(32) = -0.012 \times 32 + 0.24 = -0.14$.

   (b) Since both values for $N$ used in Part (a) are greater than the carrying capacity of 20, the population cannot be sustained at $N = 24$ or $N = 32$, so the population decreases, whence the negative growth rates.

4. **Logistic rate of change**:

   (a) The $r$ value is the initial value for $G$ as a linear function of $N$, that is, the value when $N = 0$, which is given as 0.13 per year.

   (b) To find the carrying capacity we use the fact that $G = 0$ when $N$ equals the carrying capacity, so the carrying capacity $K$ is 378.

   (c) The maximum growth rate occurs at half of the carrying capacity, so that is when $N = 378/2 = 189$.

5. **World population from two points**:

   (a) We know that $G = 0.0219$ when $N = 3.21$ and that $G = 0.0114$ when $N = 6.08$. The slope of $G$ as a function of $N$ is $m = \dfrac{0.0114 - 0.0219}{6.08 - 3.21} = -0.0037$, so $G = -0.0037N + b$. To find $b$ we plug in the value for 1963: $0.0219 = -0.0037 \times 3.21 + r$, so $r = 0.0219 + 0.0037 \times 3.21 = 0.034$, so the linear function is $G = -0.0037N + 0.034$.

   (b) The $r$ value is the initial value of $G$ as a function of $N$, so $r = 0.034$ per year.

   (c) The slope is $-0.0037$ and also equals $-\dfrac{r}{K}$, so $-0.0037 = -\dfrac{r}{K} = -\dfrac{0.034}{K}$. Solving for $K$, we have $0.0037K = 0.034$ and so $K = \dfrac{0.034}{0.0037} = 9.19$ billion.

6. **Maximum growth rate for the logistic model**:

   (a) We have two equations to combine: $G = -\dfrac{r}{K}N + r$ and $G = \dfrac{\text{Growth rate}}{N}$, since $G$ is the percentage growth rate. Using the second equation and then the first, we have

   $$\text{Growth rate} = G \times N = \left(-\frac{r}{K}N + r\right) \times N = -\frac{r}{K}N^2 + rN.$$

   (b) The formula from Part (a) factors as $N\left(-\dfrac{r}{K}N + r\right)$. Thus the growth rate is zero when $N = 0$ and also when

   $$-\frac{r}{K}N + r = 0$$
   $$-\frac{r}{K}N = -r$$
   $$N = r \times \frac{K}{r} = K.$$

   (c) The vertex of the parabola Growth rate $= aN^2 + bN + c$ represents the maximum growth rate and occurs at

   $$N = \frac{-b}{2a} = \frac{-r}{2\left(-\dfrac{r}{K}\right)} = \frac{rK}{2r} = \frac{K}{2}.$$

7. **Maximum growth rate for tuna**:

   (a) We have the values for $r$ and $K$, and so

   $$\text{Growth rate} = -\frac{r}{K}N^2 + rN = -\frac{2.61}{148}N^2 + 2.61N = -0.01764N^2 + 2.61N.$$

   (b) Using a horizontal span of 0 to 150 and a vertical span of 0 to 160, the graph is shown in the left-hand figure below.

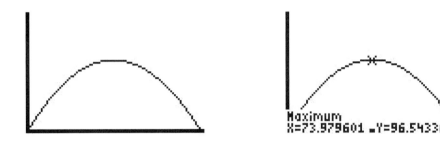

   Maximum
   X=73.979601  Y=96.543367

   (c) The right-hand figure above shows the maximum occurs when $N = 73.98$ thousand tons.

   (d) The maximum occurs on the graph at $N = 73.98$, which is about the same as $K/2 = 148/2 = 74$ thousand tons.

8. **Maximum growth rate**:

   (a) We have Growth rate $= -\dfrac{r}{K}N^2 + rN$ and we want to evaluate this when $N = \dfrac{K}{2}$, so

   $$\text{Maximum growth rate} = -\frac{r}{K}\left(\frac{K}{2}\right)^2 + r\frac{K}{2} = -\frac{r}{K}\frac{K^2}{4} + \frac{rK}{2} = -\frac{rK}{4} + \frac{rK}{2} = \frac{rK}{4}.$$

   (b) For the Pacific sardine population from Example 5.1, we have $r = 0.338$ and $K = 2.4$, so the maximum growth rate is $\dfrac{rK}{4} = \dfrac{0.338 \times 2.4}{4} = 0.20$ million tons per year.

# A FURTHER LOOK: FACTORING POLYNOMIALS, BEHAVIOR AT INFINITY

1. **Finding the degree of a polynomial**: Since the polynomial has no complex zeros and no repeated zeros, the degree is exactly the number of real zeros, that is, the number of times the graph crosses the horizontal axis. Thus the degree of this polynomial is 5.

2. **Choosing a model**: If the data set shows three maxima and two minima, then the smallest-degree polynomial that could possibly fit the data exactly would be degree 6.

3. **Finding polynomials:** The polynomial has zeros 2, 3, and 4, so by the factor theorem, the polynomial is divisible by $(x-2)(x-3)(x-4)$, which also has degree 3 and leading coefficient 1. Thus the polynomial must be $(x-2)(x-3)(x-4)$. This can be written out as $x^3 - 9x^2 + 26x - 24$.

4. **Finding polynomials:** The polynomial has zeros 2, $-2$, and 0, so by the factor theorem, the polynomial is divisible by $(x-2)(x+2)(x-0)$, which has degree 3 and leading coefficient 1. To obtain a leading coefficient of 3, we must multiply by 3, and so the polynomial must be $3(x-2)(x+2)(x-0)$. This can be written out as $3x(x-2)(x+2) = 3x^3 - 12x$.

5. **Finding polynomials:** The polynomial has only a zero at $x = 1$ and has degree 5, so by the factor theorem, the polynomial is divisible by $x - 1$, has degree 5 and has leading coefficient 1, so a polynomial with these properties is $(x-1)^5$. This can be written out as $x^5 - 5x^4 + 10x^3 - 10x^2 + 5x - 1$.

6. **Finding polynomials:** The polynomial has only a zero at $x = 0$ and degree 10, so by the factor theorem, the polynomial is divisible by $x$, has degree 10 and has leading coefficient 5, so a polynomial with these properties is $5x^{10}$.

7. **Calculating limits:** To calculate the limiting values of these rational functions, we use the fact that we need consider only the leading terms.
$$\lim_{x \to \infty} \frac{5x^4 + 4x^3 + 7}{2x^4 - x^2 + 4} = \lim_{x \to \infty} \frac{5x^4}{2x^4} = \lim_{x \to \infty} \frac{5}{2} = \frac{5}{2} = 2.5.$$

8. **Calculating limits:** To calculate the limiting values of these rational functions, we use the fact that we need consider only the leading terms.
$$\lim_{x \to \infty} \frac{6x^3 + 4x^2 + 5}{3x^3 - 5x + 14} = \lim_{x \to \infty} \frac{6x^3}{3x^3} = \lim_{x \to \infty} \frac{6}{3} = \frac{6}{3} = 2.$$

9. **Calculating limits:** To calculate the limiting values of these rational functions, we use the fact that we need consider only the leading terms.
$$\lim_{x \to \infty} \frac{3x^4 + 4x^2 - 9}{2x^5 + 4x^3 - 8} = \lim_{x \to \infty} \frac{3x^4}{2x^5} = \lim_{x \to \infty} \frac{3}{2x} = 0.$$

10. **Calculating limits:** To calculate the limiting values of these rational functions, we use the fact that we need consider only the leading terms.
$$\lim_{x \to \infty} \frac{ax^5 + bx^2 + c}{dx^5 - ex^2 + f} = \lim_{x \to \infty} \frac{ax^5}{dx^5} = \lim_{x \to \infty} \frac{a}{d} = \frac{a}{d}.$$

11. **Getting a polynomial from points**: We are looking for a quadratic $ax^2 + bx + c$ to pass through the three given points. Substituting the values for $x$ into the quadratic gives three linear equations in terms of $a$, $b$ and $c$ which we solve by elimination. For $y = ax^2 + bx + c$ to pass through $(1, 5)$, $(-1, 3)$, and $(2, 9)$, the following equations need to hold:

$$
\begin{aligned}
a + b + c &= 5 \\
a - b + c &= 3 \\
4a + 2b + c &= 9
\end{aligned}
$$

Subtracting the first equation from the second, and subtracting 4 times the first equation from the third, yield:

$$
\begin{aligned}
a + b + c &= 5 \\
-2b &= -2 \\
-2b - 3c &= -11
\end{aligned}
$$

The second equation shows that $b = 1$. Substituting into the third equation, we find $-2 - 3c = -11$ and so $c = 3$. Substituting into the first equation, we get $a + 1 + 3 = 5$ and so $a = 1$. Thus the desired quadratic is $x^2 + x + 3$.

12. **Getting a polynomial from points**: We are looking for a quadratic $ax^2 + bx + c$ to pass through the three given points. Substituting the values for $x$ into the quadratic gives three linear equations in terms of $a$, $b$ and $c$ which we solve by elimination. For $y = ax^2 + bx + c$ to pass through $(1, 5)$, $(2, 4)$, and $(3, 19)$, the following equations need to hold:

$$
\begin{aligned}
a + b + c &= 5 \\
4a + 2b + c &= 4 \\
9a + 3b + c &= 19
\end{aligned}
$$

Subtracting 4 times the first equation from the second, and subtracting 9 times the first equation from the third, yield:

$$
\begin{aligned}
a + b + c &= 5 \\
-2b - 3c &= -16 \\
-6b - 8c &= -26
\end{aligned}
$$

Subtracting 3 times the second equation from the third yields:

$$a + b + c = 5$$
$$-2b - 3c = -16$$
$$c = 22$$

Thus $c = 22$. Substituting into the second equation, we find $-2b - 3 \times 22 = -16$ and so $b = -25$. Substituting into the first equation, we get $a - 25 + 22 = 5$ and so $a = 8$. Thus the desired quadratic is $8x^2 - 25x + 22$.

13. **Getting a polynomial from points**: We are looking for a quadratic $ax^2 + bx + c$ to pass through the three given points. Substituting the values for $x$ into the quadratic gives three linear equations in terms of $a$, $b$ and $c$ which we solve by elimination. For $y = ax^2 + bx + c$ to pass through $(1, -2)$, $(0, -5)$, and $(-1, -2)$, the following equations need to hold:

$$a + b + c = -2$$
$$c = -5$$
$$a - b + c = -2$$

Thus $c = -5$. Substituting into the first and third equations yields:

$$a + b = 3$$
$$a - b = 3$$

Adding the equations, we find $2a = 6$, so $a = 3$. Substituting in either equation shows that $b = 0$. Thus the desired quadratic is $3x^2 - 5$.

14. **Getting a polynomial from points**: We are looking for a quadratic $ax^2 + bx + c$ to pass through the three given points. Substituting the values for $x$ into the quadratic gives three linear equations in terms of $a$, $b$ and $c$ which we solve by elimination. For $y = ax^2 + bx + c$ to pass through $(-1, 1)$, $(0, 0)$, and $(1, 1)$, the following equations need to hold:

$$a - b + c = 1$$
$$c = 0$$
$$a + b + c = 1$$

Thus $c = 0$. Substituting into the first and third equations yields:

$$a - b = 1$$
$$a + b = 1$$

Adding the equations, we find $2a = 2$, so $a = 1$. Substituting in either equation shows that $b = 0$. Thus the desired quadratic is $x^2$.

15. **Lagrangian polynomials**:

(a) Since the Lagrangian polynomials $P_i$ are similarly constructed, we consider $P_1$ and compute $P_1(x_2)$:

$$P_1(x_2) = \frac{(x_2 - x_2)(x_2 - x_3)}{(x_1 - x_2)(x_1 - x_3)} = 0.$$

In like fashion $P_1(x_3)$ is also zero due to the $(x - x_3)$ term in the numerator. In general $P_i(x_j) = 0$ whenever $i \neq j$.

(b) We compute $P_1(x_1)$:

$$P_1(x_1) = \frac{(x_1 - x_2)(x_1 - x_3)}{(x_1 - x_2)(x_1 - x_3)} = 1.$$

Similarly $P_i(x_i) = 1$ for all $i$.

(c) Let $P = y_1 P_1 + y_2 P_2 + y_3 P_3$ and consider $P(x_1)$:

$$P(x_1) = y_1 P_1(x_1) + y_2 P_2(x_1) + y_3 P_3(x_1) = y_1 \times 1 + y_2 \times 0 + y_3 \times 0 = y_1.$$

In similar fashion $P(x_2) = y_2$ and $P(x_3) = y_3$.

(d) Using $x_1 = 1$, $x_2 = 2$, and $x_3 = 3$, we have

$$
\begin{aligned}
P_1(x) &= \frac{(x-2)(x-3)}{(1-2)(1-3)} = \frac{(x-2)(x-3)}{2} = \frac{1}{2}x^2 - \frac{5}{2}x + 3 \\
P_2(x) &= \frac{(x-1)(x-3)}{(2-1)(2-3)} = \frac{(x-1)(x-3)}{-1} = -x^2 + 4x - 3 \\
P_3(x) &= \frac{(x-1)(x-2)}{(3-1)(3-2)} = \frac{(x-1)(x-2)}{2} = \frac{1}{2}x^2 - \frac{3}{2}x + 1.
\end{aligned}
$$

Using $y_1 = 2$, $y_2 = 5$, and $y_3 = 6$, then using Part (c), we find that the quadratic passing through the points is

$$
\begin{aligned}
P &= y_1 P_1 + y_2 P_2 + y_3 P_3 \\
&= 2\left(\frac{1}{2}x^2 - \frac{5}{2}x + 3\right) + 5\left(-x^2 + 4x - 3\right) + 6\left(\frac{1}{2}x^2 - \frac{3}{2}x + 1\right) \\
&= x^2 - 5x + 6 - 5x^2 + 20x - 15 + 3x^2 - 9x + 6 \\
&= -x^2 + 6x - 3.
\end{aligned}
$$

This can also be written as

$$(x-2)(x-3) - 5(x-1)(x-3) + 3(x-1)(x-2).$$

It is easy to check that $P(1) = 2$, $P(2) = 5$, and $P(3) = 6$.

# Solution Guide for Chapter 6: Rates of Change

## 6.1 VELOCITY

1. **From New York to Miami again**: In the left-hand picture below, we have marked the extreme locations and the places where the distance north is zero. Note that while the plane is on the ground at Miami its distance north from Richmond is negative. Hence it is marked below the horizontal axis. In the right-hand picture we have completed the graph.

    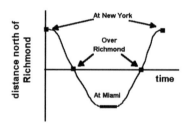

Next we make the graph of the velocity of the plane relative to Richmond. The velocity is zero when the plane is at rest at New York and Miami. The velocity is negative when the plane is headed south toward Miami and positive when the plant is headed north toward New York. We have marked these regions in the left-hand picture below. In the right-hand figure, we have completed the graph of velocity. Note that the graph is below the horizontal axis where the graph of distance is decreasing and above the horizontal axis where the graph of distance is increasing. Note also that the graph of velocity is the same as the one shown in Figure 6.14 in the text.

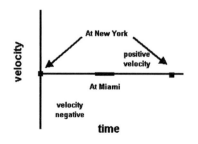    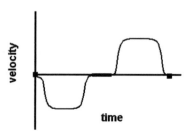

3. **The rock with a formula**:

(a) To select a graphing window, we look at the table of values below. We see that the rock is back on the ground before 2 seconds and that it reaches its peak somewhere around 14 feet. Allowing a little extra room, we set our window using a horizontal span of 0 to 2 and a vertical span of 0 to 20. The graph below shows time on the horizontal axis and height on the vertical axis.

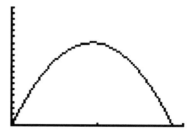

(b) The rock reaches it maximum height at the peak of the graph. From the left-hand figure below, we see that the rock reaches a maximum height of 14.06 feet at the time 0.94 seconds after the toss.

(c) The rock strikes the ground where the graph crosses the horizontal axis. That is, we need to find the root (or zero) of the function. From the right-hand figure below, we see that the rock strikes the ground at 1.875 seconds, or about 1.88 seconds, after the toss.

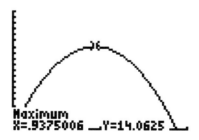
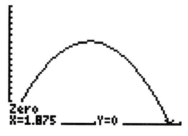

(d) When the rock is moving up from the ground, the velocity will be positive. The velocity will be zero at the point of greatest height, and it will be negative when the rock is dropping back toward the ground. In the left-hand figure below, we have marked the corresponding regions. The graph is completed in the right-hand figure below. (As in Exercise 2, the graph is a straight line. Students may have no way of knowing this, and other graphs could be acceptable.)

5. **Walking and running**: We measure location as west from your home. Since you live east of campus, your initial location is a positive number, so the graph starts above the horizontal axis. You are walking home at a constant rate, so the graph of location decreases in a straight line until it reaches the horizontal axis (home), where it stays for 5 minutes. After that, you are going back west, running now, so the graph rises quickly in a straight line until it reaches the level corresponding to your destination, where it remains, as a horizontal line, for 10 minutes. The graph of location is below on the left.

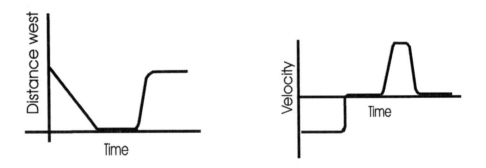

Velocity is the rate of change of location. When the graph of location is a straight line, then velocity is constant, so it forms a horizontal line. As you walk home, the velocity is negative, since the graph of location is decreasing, so initially the graph of velocity is a horizontal line below the axis. When you reach home, velocity is 0, so the graph lies on the horizontal axis. When you run back west, velocity is constant, but positive, so the graph of velocity is a horizontal line above the axis. It should be higher above the axis than the line below the axis was below it, since you run faster than you walk. Finally, when you rest for 10 minutes, location is not changing, so velocity is 0 and the graph is on the horizontal axis. The graph of velocity is above on the right.

7. **Gravity on Earth and on Mars**:

Since Mars is much smaller than Earth, the acceleration due to gravity will be greater on Earth than on Mars. So the rock on Earth will hit the ground much faster than the rock on Mars. Since we are measuring the distance from the top of the cliff, the beginning position is zero, and distance increases until the rock hits the ground. Our graph is shown in the left-hand figure below.

The velocity of both rocks will start at zero and then be positive since the rocks are moving away from the top of the cliff. Since the acceleration on Earth is greater than that on Mars, the velocity on Earth will be greater. Our graph is shown in the right-hand figure below.

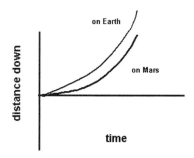 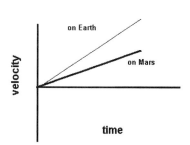

9. **Making up a story about a car trip**:

(a) A graph of velocity that matches the information in the exercise is shown in the left-hand figure below.

(b) The distance from home increases slightly at first and then levels off when the velocity is zero. Then the distance increases at a greater rate as the velocity increases. When the velocity is constant, the distance is still increasing, but now as a linear function. When the velocity is negative, the distance begins to decrease. Our graph of distance is shown in the right-hand figure below.

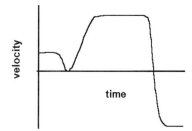 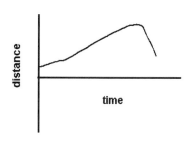

(c) Answers to this part will vary greatly. Here is one example of an answer: I started out of my driveway very slowly and then came to a stop at a traffic light at the end of the street. I pulled out on to the freeway and rapidly increased my speed to match the other cars, and the speed became constant when I set my cruise control at the posted speed limit. Just then I remembered that I had forgotten my briefcase, so I slowed down, looking for an exit. When I found one I turned around and drove back home fast.

11. **A car moving in an unusual way**:

(a) The graph of the velocity should be a horizontal line above the horizontal axis where the graph of position is going up. It jumps immediately to a horizontal line below the horizontal axis when the graph starts down.

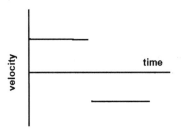

(b) The car's velocity is a positive constant over the first segment and then instantly negative after the peak in the distance graph. Thus it appears that the car has changed instantly from traveling forward to traveling backward without slowing down for the change. It is not possible to drive a car in that manner.

## Skill Building Exercises

S-1. **Velocity**: The rate of change in directed distance is velocity.

S-3. **Sign of velocity**: When the graph of directed distance is decreasing, velocity is negative, so the graph of velocity is below the horizontal axis.

S-5. **Constant velocity**: When the graph of directed distance is a straight line, then the rate of change is constant, so the velocity is constant. Thus the graph of velocity is the graph of a constant function, that is, it is a horizontal line.

S-7. **A car**: We know that directed distance is a linear function. The slope of that linear function is its rate of change, which is its velocity. Thus the slope is 60 miles per hour.

S-9. **A rock:**

(a) The rock is going upwards 1 second after the toss since it has not yet reached its peak. Directed distance is increasing, so the velocity is positive.

(b) The rock has reached its peak 2 seconds after the toss. Directed distance is momentarily not changing, so the velocity is 0.

(c) The rock is going downwards 3 seconds after the toss, since it has reached its peak and is on its way down. Directed distance is decreasing, so the velocity is negative.

S-11. **Change direction:** When the graph of directed distance is increasing velocity is positive, and when the graph of directed distance is decreasing velocity is negative. Thus if the graph of directed distance switches from increasing to decreasing then velocity switches from positive to negative (passing through 0).

## 6.2  RATES OF CHANGE FOR OTHER FUNCTIONS

1. **Marginal profit from advertising:**

   (a) The expression $\dfrac{dP}{dt}$ is the additional profit expected if 1 more dollar is invested in advertising.

   (b) Because $\dfrac{dP}{dt}$ is positive, the function $P$ is increasing. Thus, more money should be invested in advertising (because that will cause profits to rise).

   (c) Because $\dfrac{dP}{dt}$ is negative, the function $P$ is decreasing. Thus, less money should be invested in advertising (because that will cause profits to rise).

3. **Estimating rates of change:** The graph of $f(x) = x^3 - 5x$ is shown at the left below with a horizontal span of $-3$ to $3$ and a vertical span of $-10$ to $10$.

   (a) The figure at the left below shows that the function is increasing at $x = 2$. Thus $\dfrac{df}{dx}$ is positive there.

   (b) Now $\dfrac{df}{dx}$ is negative wherever the function is decreasing. The figure at the right below shows that the function begins to decrease at $x = -1.29$, and it decreases until $x = 1.29$. Thus any point on the graph for which $x$ lies between $-1.29$ and $1.29$ is correct.

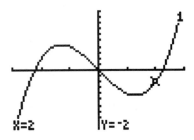

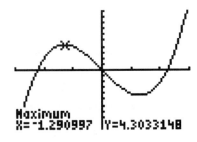

5. **Mileage for an old car**:

   (a) The expression $\dfrac{dM}{dt}$ gives us the change in the car's gas mileage we expect over a year's time.

   (b) Since we expect that the gas mileage of the car will decrease as the car gets older, then we expect $\dfrac{dM}{dt}$ to be negative.

7. **Hiking**:

   (a) The expression $\dfrac{dE}{dt}$ is the change in elevation we expect over one unit of time.

   (b) When $\dfrac{dE}{dt}$ is a large positive number, then elevation is increasing quickly, and so we might be climbing up a hill.

   (c) When $\dfrac{dE}{dt}$ is briefly zero we might have reached the top of a hill, or we might have stopped to catch our breath.

   (d) When $\dfrac{dE}{dt}$ is a large negative number, then elevation is decreasing quickly, and so we might be descending from the top of a hill.

9. **Health plan**:

   (a) We want to make the graph of $10e^{0.1t} - 12$. Since we are interested in the first 5 years of the plan, we use a horizontal span of 0 to 5. The table below leads us to choose a vertical span of $-3$ to 5. The horizontal axis of the graph is years since the beginning of the plan, and the vertical axis is the rate of change in account balance.

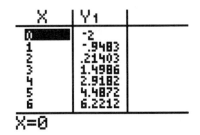

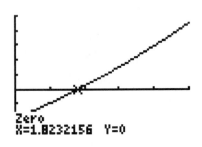

(b) The account balance $B$ is decreasing whenever $\dfrac{dB}{dt}$ is negative. In the graph above, we see that the graph of $\dfrac{dB}{dt}$ is zero when $t = 1.82$. Since $\dfrac{dB}{dt}$ is negative before this time, the account balance is decreasing during the first 1.82 years.

(c) The account balance $B$ is at its minimum when $\dfrac{dB}{dt}$ switches from negative to positive. As we noted above, this occurs at 1.82 years.

11. **The acceleration due to gravity**: Acceleration is the rate of change in velocity with respect to time. In symbols this means that $\dfrac{dV}{dt}$ is acceleration. Since acceleration $\dfrac{dV}{dt}$ has a constant value of 32 feet per second per second, velocity $V$ is a linear function of $t$ with slope 32.

13. **A population of bighorn sheep**:

(a) The expression $\dfrac{dN}{dt}$ is the change in the sheep population we expect over one unit of time. In more familiar terms, it is the rate of population growth.

(b) We expect $\dfrac{dN}{dt}$ to be positive when the conditions are favorable for the sheep because we would expect the population to grow.

(c) The population would grow less rapidly, making $\dfrac{dN}{dt}$ smaller than before. If the problem is acute, the population might stop increasing, making $\dfrac{dN}{dt}$ zero, or it might decline, making $\dfrac{dN}{dt}$ negative.

(d) The population will change slowly if at all. Thus $\dfrac{dN}{dt}$ will be near zero.

15. **Visiting a friend**: Assume that the car travels at the speed limit whenever this is possible. The distance graph should be a straight line with slope 35 until the stop sign is reached. The graph then changes to a straight line with slope 65. The trip back should be a reflection of the trip to the friend's home. This is shown in the left-hand graph below.

The velocity is constant at 35 up to the stop sign, constant at 65 from the stop sign to the friend's home, constant at $-65$ from the friend's home back to the stop sign, and finally constant at $-35$ on the final part of the trip. The graph of velocity is shown in the right-hand figure below.

 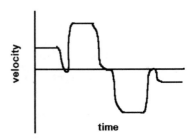

Acceleration is zero where the velocity is constant; that is for most of the graph. Filling in the spaces between is simply a matter of remembering that acceleration is positive when velocity is increasing and negative where velocity is decreasing. Our graph is shown below.

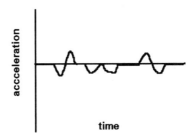

17. **Velocity of an airplane**: The distance from Atlanta increases until the plane nears Dallas. While the airplane is circling, its distance from Atlanta increases and then decreases with each circle. For much of the trip, we expect velocity to be constant, and so distance should be linear. This is reflected in the left-hand figure below.

The velocity is constant over most of the trip. Near the airport, we need to remember that velocity is positive when distance is increasing and negative when distance is decreasing. Velocity is shown in the right-hand figure below.

 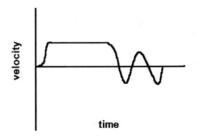

## Skill Building Exercises

**S-1. Marginal cost**: The marginal cost $\dfrac{dC}{dn}$ is the additional cost expected if 1 additional item is produced.

**S-3. Buying for the short term**: To make a profit, we want the function $P$ to be increasing. The function will be increasing if its rate of change is positive, so we should buy when $\dfrac{dP}{dt}$ is positive.

**S-5. Meaning of rate of change**:

(a) The rate of change of directed distance as a function of time is called velocity.

(b) The rate of change of velocity as a function of time is called acceleration.

(c) The rate of change of tax due as a function of income is called the marginal tax rate.

(d) The rate of change of profit as a function of dollars invested is called the marginal profit.

**S-7. Sign of the derivative**:

(a) If the function $f$ is increasing, then its rate of change is positive, so $\dfrac{df}{dx}$ is positive.

(b) If the graph of $f$ has reached a peak, then its rate of change is momentarily 0, so $\dfrac{df}{dx}$ is 0.

(c) If the function $f$ is decreasing, then its rate of change is negative, so $\dfrac{df}{dx}$ is negative.

(d) If the graph of $f$ is a horizontal line, then the value of $f$ is not changing, so $\dfrac{df}{dx}$ is zero.

**S-9. A value for the rate of change**: The slope of $f$ is precisely its rate of change, so since $\dfrac{df}{dx}$ is 10, the slope is also 10.

**S-11. Marginal tax rate**: If you earn some extra income working at $20 per hour, you will get to keep $20 less the taxes on $20. If your marginal tax rate is 34%, then you get to keep $100\% - 34\% = 66\%$ of $20, that is, $0.66 \times 20 = 13.20$ dollars of the hourly wage.

**S-13. Graph of $f$**: If the graph of $\dfrac{df}{dx}$ is below the horizontal axis, then $\dfrac{df}{dx}$ is negative, so the graph of $f$ is decreasing.

**S-15. Gasoline prices**: When the price reaches a minimum the rate of change is momentarily 0, so $\frac{dP}{dt}$ is 0 at that time.

## 6.3 ESTIMATING RATES OF CHANGE

1. **Minimum wage**: The average rate of change from 2008 to 2009 is

$$\frac{\text{Change in wage}}{\text{Elapsed time}} = \frac{7.25 - 6.55}{1} = 0.70 \text{ dollar per hour per year.}$$

Using this to approximate the rate of change in 2009, we estimate that $\frac{dW}{dt}$ in 2009 was about \$0.70 per hour per year. (Note that the average rate of change from 2008 to 2009 is the same as the average rate of change from 2009 to 2008.)

3. **Population growth**:

   (a) The average rate of change from 1955 to 1960 is

   $$\frac{\text{Change in population}}{\text{Elapsed time}} = \frac{2793 - 678}{5} = 423 \text{ reindeer per year.}$$

   Using this to approximate the rate of change at 1955, we get $\frac{dN}{dt} = 423$ reindeer per year. This means that we can expect the population to grow by about 423 reindeer during the year 1955.

   (b) For each year past 1955 we expect the population to increase by 423 reindeer, so in 1957 we expect the population to be around $678 + 2 \times 423 = 1524$ reindeer.

   (c) If the reindeer population is more closely exponential, then our answer is too large. The reason is that the rate of change after 1955 is larger than it is in 1955 if the data is exponential, since then the function is increasing at an increasing rate. This means that the average rate of change from 1955 to 1960 is larger than the actual rate of change in 1955.

5. **Deaths from heart disease**:

   (a) The average rate of change for $H_m$ from 2004 to 2007 is $\frac{288.8 - 312.8}{2007 - 2004} = -8.00$ deaths per 100,000 per year, and this is the approximate value of $\frac{dH_m}{dt}$ in 2004.

   (b) The number in Part (a) means that the number of deaths per 100,000 males caused by heart disease is decreasing (because of the negative sign) in 2004 by about 8 per year.

   (c) Since the number of deaths is decreasing by about 8 per year then in 2006 we would expect there to be about $312.8 - 2 \times 8 = 296.8$ deaths per 100,000 in 2006.

   (d) The average rate of change in $H_f$ from 2004 to 2007 is $\frac{117.9 - 131.5}{2007 - 2004} = -4.53$ deaths per 100,000 per year, and this is the approximate value of $\frac{dH_f}{dt}$ in 2004.

(e) In 2004, the proportion of males dying from heart disease was much greater than the proportion of females dying from heart disease; however, for males that number was decreasing much more rapidly than the number for females.

7. **Falling with a parachute**:

(a) Based on the table of values at the left below, we use a horizontal span of 0 to 10 and a vertical span of 0 to 200. In the figure at the right below we have made the graph.

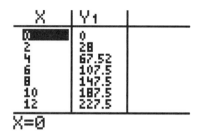 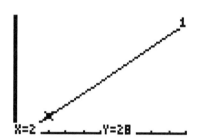

(b) On the graph above we have evaluated $S$ at $t = 2$, and we see that the parachutist falls 28 feet in 2 seconds.

(c) In the graph below we have calculated $\dfrac{dS}{dt}$ at $t = 2$, and the result is 19.20. This tells us that, after he falls 2 seconds, we expect the parachutist to fall 19.20 feet over the next second. In other words, after falling 2 seconds his velocity is 19.20 feet per second.

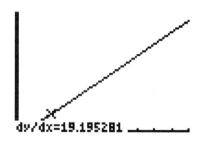

9. **A yam baking in the oven**:

   (a) We use a horizontal span of 0 to 45 as indicated. Based on the table of values at the left below, we use a vertical span of 0 to 300. In the figure at the right below we have made the graph.

   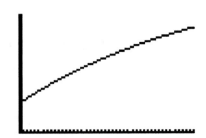

   (b) In the graph below at the left we have calculated $\dfrac{dY}{dt}$ at $t = 10$, and the result is 5.32 degrees per minute.

   (c) In the graph below at the right we have calculated $\dfrac{dY}{dt}$ at $t = 30$, and the result is 3.57 degrees per minute.

   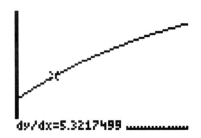 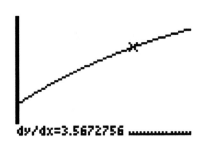

   (d) A comparison of the answers in Parts (b) and (c) shows that the temperature of the yam increases more slowly the longer it has been in the oven. This is also indicated by the fact that the graph is concave down.

11. **A pond**:

   (a) The expression $\dfrac{dG}{dt}$ tells us the change (measured in gallons per minute) in the amount of water in the pond we expect over one minute's time.

   (b) Since the amount of water in the pond is decreasing, we expect $\dfrac{dG}{dt}$ to be negative.

   (c) Since the amount of water in the pond decreases by 8000 gallons each minute, we know that $\dfrac{dG}{dt}$ at $t = 30$ is $-8000$ gallons per minute.

   (d) At $t = 30$ there are 2,000,000 gallons in the pond. The pond is losing 8000 gallons each minute, and we want to estimate the amount of water at $t = 35$, or 5 minutes after $t = 30$. Thus we expect $G(35)$ to be $2{,}000{,}000 - 5 \times 8000 = 1{,}960{,}000$ gallons.

## Skill Building Exercises

**S-1. Rate of change for a linear function**: If $f$ is the linear function $f = 7x - 3$, then its slope is 7. The rate of change of $f$ is therefore 7, so the value of $\dfrac{df}{dx}$ is 7.

**S-3. Rate of change from data**: If $f(4) = f(4.01) = 7$, then we can estimate the value of $\dfrac{df}{dx}$ at $x = 4$ using the average rate of change. The average rate of change is

$$\frac{f(4.01) - f(4)}{4.01 - 4} = \frac{7 - 7}{0.01} = 0,$$

so we estimate the value of $\dfrac{df}{dx}$ at $x = 4$ as 0.

**S-5. Rate of change from data**: If $f(3) = 8$ and $f(3.005) = 7.972$, then we can estimate the value of $\dfrac{df}{dx}$ at $x = 3$ using the average rate of change. The average rate of change is

$$\frac{f(3.005) - f(3)}{3.005 - 3} = \frac{7.972 - 8}{0.005} = -5.6,$$

so we estimate the value of $\dfrac{df}{dx}$ at $x = 3$ as $-5.6$.

**S-7. Estimating rates of change**: To estimate the value of $\dfrac{df}{dx}$ for $f(x) = \dfrac{1}{x^2}$ at $x = 4$, we calculate the average rate of change in $f$ from $x = 4$ to $x = 4.0001$ using the formula for $f$:

$$\frac{f(4.0001) - f(4)}{4.0001 - 4} = \frac{(1/4.0001^2) - (1/4^2)}{0.0001} = \frac{0.0624968751 - 0.0625}{0.0001} = -0.0312488282.$$

Based on this, we estimate the value of $\dfrac{df}{dx}$ at $x = 3$ to be about $-0.03$.

**S-9. Estimating rates of change with the calculator**: Using a horizontal span for $x$ from 1 to 4 and a vertical span for $y = x + \dfrac{1}{x}$ from 0 to 5, we can see from the figure below that the calculated rate of change at $x = 3$ is 0.89.

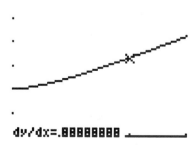

dy/dx=.88888888

S-11. **Estimating rates of change with the calculator**: Using a horizontal span for $x$ from 0 to 4 and a vertical span for $y = 3^{-x}$ from $-1$ to 1, we can see from the figure below that the calculated rate of change at $x = 3$ is $-0.04$.

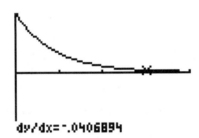

dy/dx=-.0406894

S-13. **Limiting value of a rate of change**: The graph from Exercise S-9 looks like a straight line for large values of $x$, so we expect the rate of change to be nearly constant there. We enlarge the horizontal span to go from 0 to 100 and the vertical span to go from 0 to 110, and then we calculate $\dfrac{df}{dx}$ at $x = 100$. As shown in the figure below, the value we find for the rate of change is 0.9999, which is about 1. Thus $\dfrac{df}{dx}$ is near 1 for large values of $x$.

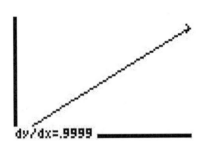

dy/dx=.9999

## 6.4 EQUATIONS OF CHANGE: LINEAR AND EXPONENTIAL FUNCTIONS

1. **Target**: Because $R$ satisfies the equation of change $\dfrac{dR}{dt} = 1647.7$, $R$ is a linear function with slope 1647.7 million dollars per year. Because the initial revenue is 66,726.4 million dollars, we have $R = 1647.7t + 66{,}726.4$.

3. **Looking up**:

   (a) The exercise tells us that $g = 32$, so $\dfrac{dV}{dt} = -32$. Thus $V$ is a linear function with slope $-32$. Since the initial velocity is 40, we have $V = -32t + 40$.

(b) The rock will reach the peak of its flight when the velocity is zero. We must solve the equation $-32t + 40 = 0$ for $t$. If we do this, we find that the rock will reach its peak when $t = \dfrac{40}{32} = 1.25$ seconds.

(c) The rock will reach its peak halfway through its flight, and so the rock will strike the ground at $t = 2 \times 1.25 = 2.5$ seconds.

5. **A better investment**:

(a) The account has continuous compounding with $r = 0.0575$ (the APR in decimal form) and an initial balance of \$250. Let the balance $B$ be measured in dollars. It is an exponential function of time $t$ in years since the initial investment. In standard form, $B = Pa^t$ where $P$ is the initial value and $a = e^r$. Thus $P = 250$ and $a = e^{0.0575} = 1.0592$, and so $B = 250 \times 1.0592^t$. In alternative form, $B = Pe^{rt}$, which is simply $B = 250e^{0.0575t}$.

(b) After 5 years, the balance is expected to be $B = 250 \times 1.0592^5 = 333.30$ dollars using the standard form, and $B = 250e^{0.0575 \times 5} = 333.27$ dollars using the alternative form. The alternative form gives a more accurate answer because it was not rounded until the very end, whereas for the standard form there was early rounding in reporting $e^{0.0575}$ as 1.0592.

7. **Borrowing money**:

(a) We know that $B$ is an exponential function because its equation of change has the form $\dfrac{df}{dx} = rf$ where $r$ is a constant. This is described in Key Idea 6.6.

(b) From the equation of change we have an exponential growth rate of $r = 0.07$ per year. The initial value for $B$ is 10,000 dollars. Thus the formula for $B$ in the alternative form is $B = 10,000e^{0.07t}$.

(c) The yearly growth factor here is $e^{0.07} = 1.073$, and the initial value is 10,000, so the formula in standard form is $B = 10,000 \times 1.073^t$.

(d) The account balance will double when it reaches \$20,000. Thus we need to solve the equation $10,000e^{0.07t} = 20,000$. We do this using the crossing graphs method. Based on the table of values below for $B$ and the constant function 20,000, we use a horizontal span of 0 to 15 years and a vertical span of 0 to 30,000 dollars. In the figure at the right below we have graphed $B$ and the target balance of 20,000 (thick line). We see that the intersection occurs at a time of $t = 9.90$ years. Thus the account will double after 9.90 years.

| X | Y₁ | Y₂ |
|---|---|---|
| 0 | 10000 | 20000 |
| 5 | 14191 | 20000 |
| 10 | 20138 | 20000 |
| 15 | 28577 | 20000 |
| 20 | 40552 | 20000 |
| 25 | 57546 | 20000 |
| 30 | 81662 | 20000 |

X=0

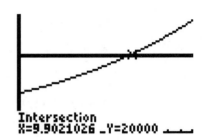

Intersection
X=9.9021026 _Y=20000

9. **A population of bighorn sheep:**

(a) As an equation of change, the sentence becomes $\dfrac{dN}{dt} = 0.04N$.

(b) From the equation of change we have that $N$ is an exponential function with an exponential growth rate of $r = 0.04$ per year. The initial value is 30, so the formula in the alternative form is $N = 30e^{0.04t}$. (For the standard form, note that the yearly growth factor is $e^{0.04} = 1.04$, while the initial value is 30, so the formula is $N = 30 \times 1.04^t$.)

(c) We want to find when $N$ is 50, so we want to solve the equation $30e^{0.04t} = 50$. We do this using the crossing graphs method. Based on the table of values below for $N$ and the constant function 50, we use a horizontal span of 0 to 15 years and a vertical span of 0 to 60. In the figure at the right below we have graphed $N$ and the target population of 50. We see that the intersection occurs at a time of $t = 12.77$ years, and this is when the population will have grown to a level of 50. (If we use the standard form for $N$, we get a time of $t = 13.02$ years.)

| X | Y₁ | Y₂ |
|---|---|---|
| 0 | 30 | 50 |
| 5 | 36.642 | 50 |
| 10 | 44.755 | 50 |
| 15 | 54.664 | 50 |
| 20 | 66.766 | 50 |
| 25 | 81.548 | 50 |
| 30 | 99.604 | 50 |

X=0

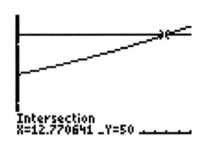

Intersection
X=12.770641 _Y=50

11. **Growing child:**

(a) Since the girl grew steadily at a rate of 5.5 pounds per year, the weight $W$ has a constant rate of change of 5.5 pounds per year, and thus is a linear function with slope 5.5.

(b) Since the rate of change is the constant 5.5, the equation of change is $\dfrac{dW}{dt} = 5.5$.

(c) From Part (a), $W$ is a linear function of $t$ with slope 5.5, and here we are told that $W = 30$ when $t = 3$. We need to find the initial value $b$ of $W$. We have $W = 5.5t + b$, and using the fact that $W = 30$ when $t = 3$ gives $30 = 5.5 \times 3 + b$, or $30 = 16.5 + b$. Thus $b = 30 - 16.5 = 13.5$. Hence the formula is $W = 5.5t + 13.5$.

13. **Radioactive decay**:

   (a) The equation of change for $A$ has the form $\dfrac{df}{dx} = rf$ where $r$ is a constant, and thus $A$ is an exponential function with exponential growth rate $r = -0.05$ per day.

   (b) The initial value is 3 grams, so in the alternative form the formula is $A = 3e^{-0.05t}$. (For the standard form, note that the daily decay factor is $e^{-0.05} = 0.95$, while the initial value is 3, so the formula is $B = 3 \times 0.95^t$.)

   (c) Half the initial amount is 1.5 grams, so we want to find when $A = 1.5$. That is, we want to solve the equation $3e^{-0.05t} = 1.5$. We do this using the crossing graphs method. Based on the table of values below for $A$ and the constant function 1.5, we use a horizontal span of 0 to 15 days and a vertical span of 0 to 3 grams. In the figure at the right below we have graphed $A$ and the target amount of 1.5. We see that the intersection occurs at a time of $t = 13.86$ days, and this is the half-life. (If we use the standard form for $A$, we get a time of $t = 13.51$ days.)

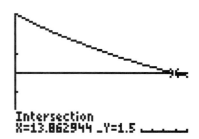

| X | Y₁ | Y₂ |
|---|----|----|
| 0 | 3 | 1.5 |
| 5 | 2.3364 | 1.5 |
| 10 | 1.8196 | 1.5 |
| 15 | 1.4171 | 1.5 |
| 20 | 1.1036 | 1.5 |
| 25 | .85951 | 1.5 |
| 30 | .66939 | 1.5 |

X=0

Intersection
X=13.862944 _Y=1.5

## Skill Building Exercises

S-1. **Writing an equation of change**: Because $P$ is a linear function with slope 2300 dollars per year, its equation of change is $\dfrac{dP}{dt} = 2300$.

S-3. **New equation of change?** The equation of change for $I$ shows that $I$ is a linear function. If only the initial income were altered, the equation of change would not be altered because the slope would be unaltered. Therefore, the equation of change would still be $\dfrac{dI}{dt} = 2300$.

S-5. **Technical terms**: An equation of change is known more formally by the common mathematical term "differential equation."

S-7. **Slope**: If $f$ satisfies the equation of change $\dfrac{df}{dx} = 5$, then the rate of change of $f$ is 5, so $f$ is a linear function of slope 5.

S-9. **Solving an equation of change**: If $f$ satisfies the equation of change $\dfrac{df}{dx} = 8f$, then $f$ is an exponential function and hence can be written as $f = Ae^{ct}$. The value of $c$ is 8.

S-11. **A leaky balloon**: If the balloon leaks air at a rate of one third the volume per minute, the rate of change is $-\dfrac{1}{3}V$ (the minus sign because the balloon is leaking and so the volume is decreasing). Thus an equation of change for $V$ is $\dfrac{dV}{dt} = -\dfrac{1}{3}V$.

S-13. **Solving an equation of change**: The equation of change $\dfrac{df}{dx} = 3$ shows that $f$ has a constant rate of change of 3, so $f$ is a linear function with slope 3. Since the initial value of $f$ is 7, we have $f = 3x + 7$.

S-15. **Filling a tank**: The rate of change in the height $H$ of the water level with respect to time $t$ is always 4 feet per minute because the level rises 4 feet every minute. Thus an equation of change is $\dfrac{dH}{dt} = 4$.

## 6.5 EQUATIONS OF CHANGE: GRAPHICAL SOLUTIONS

1. **Small business loan**:

   (a) The equilibrium solution satisfies the equation $0.06B - 5000 = 0$. This is a linear equation in $B$, and solving gives $B = \dfrac{5000}{0.06} = 83{,}333.33$. Thus, the equilibrium solution is $B = 83{,}333.33$ dollars.

   (b) At the equilibrium solution the balance does not change, which means that payments exactly match accrued interest.

   (c) If the small business expects to pay off the loan, the rate of change $\dfrac{dB}{dt}$ should be negative. This means that $0.06B - 5000$ should be negative, so the account balance $B$ should be smaller than the equilibrium solution.

3. **Equation of change for logistic growth**:

(a) Since $Y_1$ corresponds to $\dfrac{dN}{dt}$ and $X$ corresponds to $N$, the function we want to graph is $Y_1 = 0.8X(1 - X/177)$. Based on the table of values at the left below, we use a horizontal span of 0 to 200 and a vertical span of $-20$ to 40. In the figure at the right below we have made the graph.

Equilibrium solutions are those for which $\dfrac{dN}{dt} = 0$, or $Y_1 = 0$ in the above correspondence. As shown in the figure at the right below, $Y_1 = 0$ occurs when $X$ is 0 or 177. Thus the equilibrium solutions are $N = 0$ and $N = 177$. The physical significance of the equilibrium solution $N = 0$ is that there will never be any deer if we start with none. The solution $N = 177$ is the environmental carrying capacity for the deer in this reserve. If the number of deer ever reaches 177, then environmental limitations match growth tendencies, and the population stays the same.

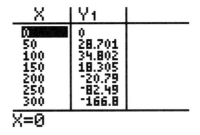

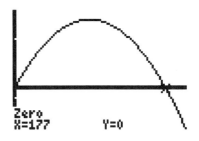

(b) The relevant information we get from the graph in Part (a) is summarized in the table below.

| Range for $N$ | from 0 to 177 | greater than 177 |
|---|---|---|
| Sign of $\dfrac{dN}{dt}$ | positive | negative |
| Effect on $N$ | increasing | decreasing |

If $N(0) = 10$ then the population will increase toward the equilibrium solution of $N = 177$. If $N(0) = 225$ then the population will decrease toward the equilibrium solution of $N = 177$. The graphs are shown below.

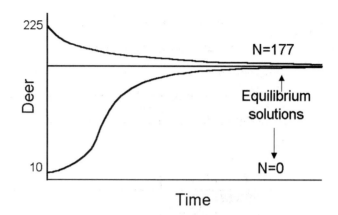

(c) The starting value for $N$ corresponding to the given solution is the initial value of that solution:

$$N(0) = \frac{6.21}{0.035 + 0.45^0} = 6.$$

5. **Experimental determination of the drag coefficient:**

(a) Terminal velocity occurs when $\dfrac{dV}{dt} = 0$, so $32 - rV$ must equal 0 when $V$ equals terminal velocity. Since terminal velocity $V$ is 176 feet per second, we have

$$32 - r \times 176 = 0.$$

Solving for $r$ yields $r = \dfrac{32}{176} = 0.1818.$

(b) Reasoning as in Part (a), we put in $V = 4$ for the terminal velocity, and we get $32 - r \times 4 = 0$. Thus the drag coefficient is $r = \dfrac{32}{4} = 8.$

7. **Fishing for sardines:**

(a) We need to find the value of $F$ such that 1.8 million tons is an equilibrium solution for the equation of change. This means that $\dfrac{dN}{dt} = 0$ when $N = 1.8$. Using these values in the equation of change, we get

$$0.338 \times 1.8 \left(1 - \frac{1.8}{2.4}\right) - F = 0.$$

We can easily solve for $F$:

$$F = 0.338 \times 1.8 \left(1 - \frac{1.8}{2.4}\right) = 0.1521.$$

Thus the fishing level for Pacific sardines should be set at about 0.15 million tons per year.

(b)    i. First we graph $\dfrac{dN}{dt}$ versus $N$. Since $Y_1$ corresponds to $\dfrac{dN}{dt}$ and $X$ corresponds to $N$, the function we want to graph is $Y_1 = 0.338X(1 - X/2.4) - 0.1$. (Remember that $F = 0.1$.) Based on the table of values at the left below, we use a horizontal span of 0 to 3 and a vertical span of $-0.15$ to $0.15$. In the figure at the right below we have made the graph.

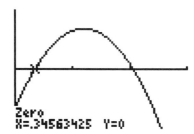

The equilibrium solutions occur where $\dfrac{dN}{dt} = 0$, so in the graph above we want to find where $Y_1 = 0$. There are two such points, and we find them using the single graph method. The graph above shows that one zero occurs when $N = 0.35$, and finding the other gives $N = 2.05$. Thus the two equilibrium solutions are $N = 0.35$ million tons and $N = 2.05$ million tons.

ii. The relevant information from Part i above is summarized in the following table.

| Range for $N$ | less than 0.35 | from 0.35 to 2.05 | greater than 2.05 |
|---|---|---|---|
| Sign of $\dfrac{dN}{dt}$ | negative | positive | negative |
| Effect on $N$ | decreasing | increasing | decreasing |

From the table we see that the population increases when $N$ is between 0.35 and 2.05 million tons. The population decreases when $N$ is less 0.35 million tons and when $N$ is greater than 2.05 million tons.

iii. If $N(0) = 0.3$ then the population will decrease to 0. Note that 0 is not an equilibrium solution, and in fact $\dfrac{dN}{dt}$ is negative even at $N = 0$, so with $N(0) = 0.3$ the population will reach 0 quickly. If $N(0) = 1.0$ then the population will increase to the equilibrium solution at 2.05. If $N(0) = 2.3$ then the population will decrease to the equilibrium solution at 2.05.

The graphs are shown below.

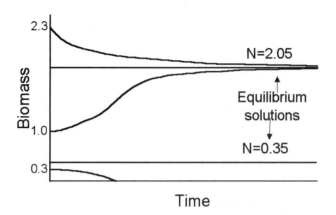

iv. In practical terms, if the initial population is 0.3 million tons, then the population will die out quickly. If the initial population is 1 million tons, then it will increase and slowly approach 2.05 million tons. If the initial population is 2.3 million tons, then it will decrease and slowly approach 2.05 million tons. If the initial population is 0.35 or 2.05 million tons, then the population will remain at that level.

9. **Logistic growth with a threshold**:

(a) Since $r = 0.338$, $K = 2.4$, and $S = 0.8$, the equation of change is

$$\frac{dN}{dt} = -0.338N \left(1 - \frac{N}{0.8}\right)\left(1 - \frac{N}{2.4}\right).$$

(b) Since $Y_1$ corresponds to $\frac{dN}{dt}$ and $X$ corresponds to $N$, we want to graph the function $Y_1 = -0.338X(1 - X/0.8)(1 - X/2.4)$. Based on the table of values at the left below, we use a horizontal span of 0 to 3 and a vertical span of $-0.25$ to $0.25$. In the figure at the right below we have made the graph.

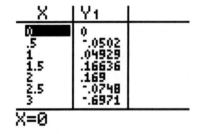

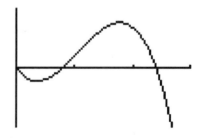

The equilibrium solutions are at the zeros of the function graphed above. Clearly one such zero is at 0, and the graph shows that another is at 0.8. In a similar way, we find that the third zero is at 2.4. Thus the equilibrium solutions are $N = 0$, $N = 0.8$ and $N = 2.4$ (all measured in millions of tons). The solution $N = 0.8$

corresponds to $S$, the survival threshold, while the solution $N = 2.4$ corresponds to $K$, the carrying capacity.

(c) The relevant information from Part (b) above is summarized in the following table.

| Range for $N$ | from 0 to 0.8 | from 0.8 to 2.4 | greater than 2.4 |
|---|---|---|---|
| Sign of $\dfrac{dN}{dt}$ | negative | positive | negative |
| Effect on $N$ | decreasing | increasing | decreasing |

From the table we see that the population increases when $N$ is between 0.8 and 2.4 million tons. The population decreases when $N$ is between 0 and 0.8 million tons and when $N$ is greater than 2.4 million tons.

(d) If the initial population is 0.7 million tons, then, since $N(0)$ is less than 0.8, the population will decrease to 0. This makes sense, because the initial population is less than the survival threshold, and so the population can be expected to dwindle away to nothing.

(e) The population is growing the fastest when $\dfrac{dN}{dt}$ is the greatest. This occurs at the maximum point on the graph in Part (b) above, and the figure below shows that this corresponds to $N = 1.77$ million tons.

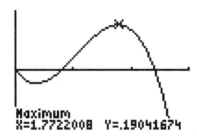

Maximum
X=1.7722008  Y=.19041674

11. **Competition between bacteria**:

(a) Since $a = 2.3$ and $b = 1.7$, we have $a - b = 2.3 - 1.7 = 0.6$, so the equation of change is $\dfrac{dP}{dt} = 0.6P(1 - P)$. This fits the form of a logistic equation of change with $r = 0.6$ and $K = 1$.

(b) We first find the equilibrium solutions, and we do so by graphing $\dfrac{dP}{dt}$ versus $P$. Since $Y_1$ corresponds to $\dfrac{dP}{dt}$ and $X$ corresponds to $P$, we want to graph the function $Y_1 = 0.6X(1 - X)$. Based on the table of values at the left below, we use a horizontal span of 0 to 1 and a vertical span of $-0.1$ to 0.2. In the figure at the right below we have made the graph.

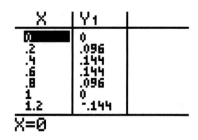

From the table or the graph we see that the equilibrium solutions are $P = 0$ and $P = 1$. (This is what we expect from a logistic equation with carrying capacity $K = 1$.)

If initially $P$ is 0 then, since $P = 0$ is an equilibrium solution, $P$ will stay at 0. This makes sense since $P(0) = 0$ says that there are only type $B$ bacteria and no type $A$ bacteria, so there never will be any type $A$ bacteria.

(c) If $P(0) = 1$ then, since $P = 1$ is an equilibrium solution, $P$ will stay at 1. Also, from the graph in Part (b) above, we see that $\dfrac{dP}{dt}$ is always positive for $P$ strictly between 0 and 1. Thus if $P(0)$ is positive but less than 1 then $P$ will increase to 1. In this case it does not matter what the exact value of $P(0)$ is. This means that as long as some amount of the type $A$ bacteria (the type with the larger growth rate) is initially present, then it will dominate, and eventually there will be only type $A$ bacteria.

13. **Grazing sheep**:

(a) In practical terms, $\dfrac{dC}{dV}$ tells how much more food (in pounds) we can expect a merino sheep to consume in a day if we increase the amount of vegetation available by one pound per acre.

(b) Since $Y_1$ corresponds to $\dfrac{dC}{dV}$ and $X$ corresponds to $C$, we want to graph the function $Y_1 = 0.01(2.8 - X)$. We are told to use a horizontal span of 0 to 3, and, based on the table of values at the left below, we use a vertical span of $-0.01$ to $0.05$. In the figure at the right below we have made the graph.

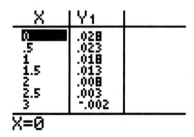

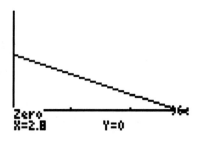

(c) We first find the equilibrium solution. In the graph above we see that it corresponds to $C = 2.8$. (Since the graph is linear, this can also be found by hand.) From that graph we also see that $\dfrac{dC}{dV}$ is positive for $C$ less than 2.8 pounds. Thus the most we would expect the merino sheep to consume in a day is 2.8 pounds.

## Skill Building Exercises

S-1. **Equilibrium solutions**: An equilibrium solution of an equation of change is a solution which does not change, that is, it remains constant.

S-3. **Finding equilibrium solutions**: The equilibrium solutions of an equation of change occur when $\dfrac{df}{dx} = 0$. To find the equilibrium solutions of $\dfrac{df}{dx} = 2f - 6$ is the same as solving $0 = 2f - 6$, so the equilibrium solution is $f = 3$.

S-5. **Finding equilibrium solutions**: The equilibrium solutions of an equation of change occur when $\dfrac{df}{dx} = 0$. Finding the equilibrium solutions of $\dfrac{df}{dx} = (f - 1)(f - 2)(f - 3)$ is the same as solving $(f - 1)(f - 2)(f - 3) = 0$. The latter equation says $f = 1$, or $f = 2$, or $f = 3$. Thus, the equilibrium solutions are $f = 1$, $f = 2$, and $f = 3$.

S-7. **Water**: If the process of water flowing in and some of it draining out continues for a long time, then we expect the water volume to reach some type of equilibrium. The equilibrium solution to $\dfrac{dv}{dt} = 5 - \dfrac{v}{3}$ occurs when $\dfrac{dv}{dt} = 0$, that is, $0 = 5 - \dfrac{v}{3}$. Solving, we see that $v = 15$ is the equilibrium solution, so there will be 15 cubic feet of water in the tank.

S-9. **Water**: If there are 8 cubic feet of water in the tank, so $v = 8$, then the equation of change gives $\dfrac{dv}{dt} = 5 - \dfrac{v}{3} = 5 - \dfrac{8}{3} = 2.33$. Since $\dfrac{dv}{dt}$ is positive, the volume $v$ is increasing.

S-11. **Population**: If the population is 4238, so $N = 4238$, then the equation of change gives $\dfrac{dN}{dt} = 0.03N\left(1 - \dfrac{N}{6300}\right) = 0.03 \times 4238\left(1 - \dfrac{4238}{6300}\right) = 41.61$. Since $\dfrac{dN}{dt}$ is positive, the population $N$ is increasing.

S-13. **Equation of change**: When $f = 1$ then $5f - 7 = 5 \times 1 - 7 = -2$, so $\frac{df}{dx}$ is negative. Hence $f$ is decreasing when $f = 1$.

## Chapter 6 Review Exercises

1. **Maximum distance**: At a maximum of directed distance the velocity is 0.

2. **Constant slope**: When the graph of directed distance is a straight line, then the rate of change is constant, so the velocity is constant. The value of that constant is the slope $-3$. Hence the velocity is a constant $-3$.

3. **A casual walk**: We measure location as distance east from your home. Your initial location is at home, so the graph starts at the origin. For the first 10 minutes you are walking east at a constant speed, so the graph of location increases in a straight line over that time period. While you rest at the park your location is unchanging, so the graph is a horizontal line segment over that 5-minute period. After that you resume going east, so the graph begins to rise again. The graph of location is below on the left.

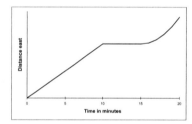 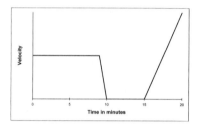

Velocity is the rate of change of location. During the first 10 minutes, the graph of location is a straight line, so velocity is constant and its graph is a horizontal line. Also, during this period the velocity is positive, since the graph of location is increasing. Thus initially the graph of velocity is a horizontal line above the axis. While you are at the park, the graph of location is a horizontal line. Hence over that time period the velocity is 0, so the graph of velocity lies on the horizontal axis. When you resume going east, location is increasing. Then the velocity is positive, so the graph of velocity is above the axis. The graph of velocity is above on the right.

4. **Velocity given**:

   (a) We use a horizontal span of 0 to 3 and a vertical span of $-12$ to 12. The graph of $V$ is on the left below.

   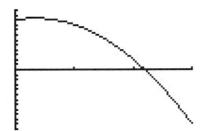
   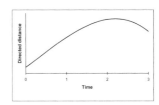

   (b) The velocity $V$ is positive until it reaches a point that is just greater than 2. Using the single-graph method shows that this point is $t = 2.19$. After this point the velocity is negative. Hence the directed distance is increasing from $t = 0$ to $t = 2.19$ and decreasing after that. One possible graph of directed distance is shown on the right above.

5. **Marginal profit**: Because the rate of change $\dfrac{dP}{dn}$ is negative, the profit $P$ is a decreasing function of $n$. Hence increasing the number of widgets made decreases the profit.

6. **Water depth**: After time $t = 5$ the depth of water $H$ does not change, so the rate of change $\dfrac{dH}{dt}$ is 0.

7. **Helicopter**:

   (a) The expression $\dfrac{dA}{dt}$ represents the change in altitude we expect in 1 unit of time.

   (b) If the rate of change $\dfrac{dA}{dt}$ is 0 for a period of time then the altitude $A$ is not changing over that time period.

   (c) If the helicopter is coming down for a landing then the altitude $A$ is decreasing, so the rate of change $\dfrac{dA}{dt}$ is negative.

8. **Account balance:**

   (a) We want to make a graph of $10 + 5t - 3t^2$ versus $t$. We use a horizontal span of 0 to 3 and a vertical span of $-5$ to 15. The graph is below.

   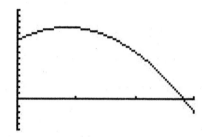

   (b) The balance $B$ is increasing when the rate of change $\dfrac{dB}{dt}$ is positive. From the graph above the rate of change is positive until it reaches a point that is just less than 3. Using the single-graph method shows that this point is $t = 2.84$. Thus the balance $B$ is increasing from 0 to 2.84 years.

   (c) The balance $B$ is at its maximum when it changes from increasing to decreasing, and that is when the rate of change $\dfrac{dB}{dt}$ changes from positive to negative. From the graph in Part (a) and the calculation in Part (b) we find that this occurs at $t = 2.84$. Thus the balance $B$ is at its maximum at 2.84 years.

9. **Rate of change from data**: If $f(1) = 3$ and $f(1.002) = 3.012$, then we can estimate the value of $\dfrac{df}{dx}$ at $x = 1$ using the average rate of change. The average rate of change is

$$\frac{f(1.002) - f(1)}{1.002 - 1} = \frac{3.012 - 3}{0.002} = 6,$$

so we estimate the value of $\dfrac{df}{dx}$ at $x = 1$ as 6.

10. **Estimating rates of change with the calculator**: Using the suggested horizontal and vertical spans, we can see from the figure below that the calculated rate of change at $x = 4$ is 8.31.

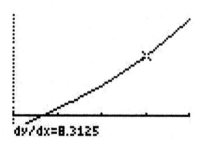

dy/dx=8.3125

11. **Ebola**:

    (a) The average rate of change for $N$ from $t = 1$ to $t = 8$ is $\dfrac{25{,}515 - 25{,}178}{8 - 1} = 48.14$ cases per day, and this is the approximate value of $\dfrac{dN}{dt}$ at $t = 1$.

    (b) From the 1st to the 2nd day we expect the cumulative number of cases of Ebola to increase by about 48.

    (c) The average rate of change for $N$ from $t = 18$ to $t = 29$ is $\dfrac{26{,}298 - 25{,}863}{29 - 18} = 39.55$ cases per day, and this is the approximate value of $\dfrac{dN}{dt}$ at $t = 18$. Comparing this number with the result in Part (a), we see that the rate of growth in the cumulative number of cases of Ebola has decreased from day $t = 1$ to day $t = 18$.

12. **Account balance**:

    (a) We use a horizontal span of 0 to 15 and a vertical span of 0 to 1500. In the figure on the left below we have made the graph.

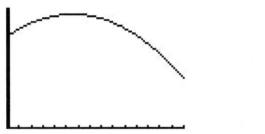
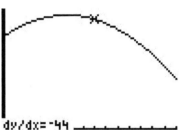

    (b) In the graph on the right above we have calculated $\dfrac{dB}{dt}$ at $t = 8$, and the result is $-44$ dollars per month.

    (c) We expect the balance to decrease \$44 from month 8 to month 9.

13. **Finding an equation of change**: The rate of change of depth $H$ as a function of time $t$ is a constant $-0.5$ foot per minute. Hence an equation of change satisfied by $H$ is $\dfrac{dH}{dt} = -0.5$.

14. **Solving an equation of change**: From the equation of change $\dfrac{df}{dx} = 6f$ we know that $f$ is an exponential function of the form $f = Pe^{rx}$. Here $r = 6$ and $P$ is the initial value 3. The solution is $f = 3e^{6x}$.

15. **Traffic signals**:

  (a) The expression $\dfrac{dn}{dw}$ tells us the change we expect in the number of seconds for the yellow light when the width of the crossing street is increased by 1 foot.

  (b) From the equation of change

  $$\frac{dn}{dw} = 0.02$$

  we know that $n$ is a linear function of $w$ with slope 0.02. We need to find the initial value $b$ of $n$. We have $n = 0.02w + b$, and using the fact that $n = 4.7$ when $w = 70$ gives $4.7 = 0.02 \times 70 + b$. Thus $b = 4.7 - 0.02 \times 70 = 3.3$. Hence the formula is $n = 0.02w + 3.3$.

16. **Atmospheric pressure**:

  (a) The expression $\dfrac{dP}{dh}$ tells us the change we expect in the atmospheric pressure when the altitude increases by 1 kilometer.

  (b) From the equation of change

  $$\frac{dP}{dh} = -0.12P$$

  we know that $P$ is an exponential function.

  (c) From the equation of change we have that $P$ is an exponential function with an exponential growth rate of $r = -0.12$. The initial value is 1035 because that is the atmospheric pressure when $h = 0$. Thus the formula in the alternative form is $P = 1035e^{-0.12h}$. (For the standard form, note that the decay factor is $e^{-0.12} = 0.89$, while the initial value is 1035, so the formula is $P = 1035 \times 0.89^h$.)

17. **Equilibrium solution**: The equilibrium solutions of such an equation of change occur when $\dfrac{df}{dx} = 0$. Thus to find an equilibrium solution of $\dfrac{df}{dx} = 7f - 14$ we solve $7f - 14 = 0$. We obtain the equilibrium solution $f = 2$.

18. **Equation of change**: When $f = 1$ we have $\dfrac{df}{dx} = 7 \times 1 - 14 = -7$. Thus when $f = 1$ the rate of change is negative, and so $f$ is decreasing.

19. **Sales growth**:

(a) First we make a graph of $\dfrac{ds}{dt}$ versus $s$. Since $Y_1$ corresponds to $\dfrac{ds}{dt}$ and $X$ corresponds to $s$, we want to graph the function $Y_1 = 0.3X(4-X)$. We use a horizontal span of 0 to 5 and, based on a table of values, a vertical span of $-1.5$ to 1.5. In the figure on the left below we have made the graph.

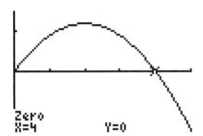 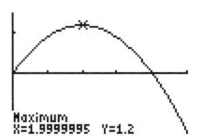

The equilibrium solutions are at the zeros of the function graphed above. Clearly one such zero is at 0, and the graph shows that another is at 4. (This can also be seen just by looking at the equation of change.) The other relevant information from the graph is summarized in the following table.

| Range for $s$ | from 0 to 4 | greater than 4 |
|---|---|---|
| Sign of $\dfrac{ds}{dt}$ | positive | negative |
| Effect on $s$ | increasing | decreasing |

From the table we see that sales increase between 0 and 4 thousand dollars and decrease above a level of 4 thousand dollars. Thus a level of 4 thousand dollars will be attained in the long run. Another way to do this is to note that the equation of change can be written in the form

$$\frac{ds}{dt} = 0.3 \times 4 \times s \left(1 - \frac{s}{4}\right),$$

which shows that this equation represents logistic growth with a carrying capacity of 4 thousand dollars.

(b) The largest rate of growth in sales is the maximum value of $\dfrac{ds}{dt}$. In the figure on the right above we have located the maximum point on the graph, and we see that the largest rate of growth in sales is 1.2 thousand dollars per year.

20. **Critical threshold**:

(a) We substitute $r = 0.5$ and $T = 10$ into the equation and obtain

$$\frac{dN}{dt} = -0.5N\left(1 - \frac{N}{10}\right).$$

(b) Since $Y_1$ corresponds to $\dfrac{dN}{dt}$ and $X$ corresponds to $N$, we want to graph the function $Y_1 = -0.5X(1 - X/10)$. We use a horizontal span of 0 to 12 and, based on a table of values, a vertical span of $-1.5$ to $1.5$. In the figure below we have made the graph.

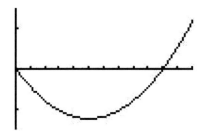

The equilibrium solutions are at the zeros of the function graphed above. Clearly one such zero is at 0, and using the single-graph method (or just looking at the original equation) shows that another is at 10. Note that 10 is the threshold level $T$. Thus the equilibrium solutions are $N = 0$ and the threshold $N = T = 10$.

(c) The relevant information from the graph is summarized in the following table.

| Range for $N$ | from 0 to 10 | greater than 10 |
| --- | --- | --- |
| Sign of $\dfrac{dN}{dt}$ | negative | positive |
| Effect on $N$ | decreasing | increasing |

From the table we see that $N$ increases for $N$ greater than 10 and that $N$ decreases for $N$ between 0 and 10.

(d) If initially $N$ is smaller than 10 then $N$ will decrease to 0. If initially $N$ is larger than 10 then $N$ will increase.